AF389400

Renewable Energy and Energy Saving: Worldwide Research Trends

Renewable Energy and Energy Saving: Worldwide Research Trends

Editor

Alberto-Jesus Perea-Moreno

MDPI • Basel • Beijing • Wuhan • Barcelona • Belgrade • Manchester • Tokyo • Cluj • Tianjin

Editor
Alberto-Jesus Perea-Moreno
University of Cordoba
Spain

Editorial Office
MDPI
St. Alban-Anlage 66
4052 Basel, Switzerland

This is a reprint of articles from the Special Issue published online in the open access journal *Sustainability* (ISSN 2071-1050) (available at: https://www.mdpi.com/journal/sustainability/special_issues/saving).

For citation purposes, cite each article independently as indicated on the article page online and as indicated below:

LastName, A.A.; LastName, B.B.; LastName, C.C. Article Title. *Journal Name* **Year**, *Volume Number*, Page Range.

ISBN 978-3-0365-4665-0 (Hbk)
ISBN 978-3-0365-4666-7 (PDF)

Contents

About the Editor . . . . **vii**

Preface to "Renewable Energy and Energy Saving: Worldwide Research Trends" . . . . **ix**

Alberto-Jesus Perea-Moreno
Renewable Energy and Energy Saving: Worldwide Research Trends
Reprinted from: *Sustainability* **2021**, *13*, 13261, doi:10.3390/su132313261 . . . . **1**

Abdelhakim Mesloub, Ghazy Abdullah Albaqawy and Mohd Zin Kandar
The Optimum Performance of Building Integrated Photovoltaic (BIPV) Windows Under a
Semi-Arid Climate in Algerian Office Buildings
Reprinted from: *Sustainability* **2020**, *12*, 1654, doi:10.3390/su12041654 . . . . **5**

**Oscar Danilo Montoya, Walter Gil-González, Luis Grisales-Noreña,
Quetzalcoatl Hernandez-Escobedo and Alberto-Jesus Perea-Moreno**
Optimal Placement and Sizing of Wind Generators in AC Grids Considering Reactive Power
Capability and Wind Speed Curves
Reprinted from: *Sustainability* **2020**, *12*, 2983, doi:10.3390/su12072983 . . . . **43**

Van-Hai Bui, Akhtar Hussain, Thai-Thanh Nguyen and Hak-Man Kim
Multi-Objective Stochastic Optimization for Determining Set-Point of Wind Farm System
Reprinted from: *Sustainability* **2021**, *13*, 624, doi:10.3390/su13020624 . . . . **63**

**Erika Winquist, Michiel Van Galen, Simon Zielonka, Pasi Rikkonen, Diti Oudendag,
Lijun Zhou and Auke Greijdanus**
Expert Views on the Future Development of Biogas Business Branch in Germany, The
Netherlands, and Finland until 2030
Reprinted from: *Sustainability* **2021**, *13*, 1148, doi:10.3390/su13031148 . . . . **79**

Wai Fang Wong, AbdulLateef Olanrewaju and Poh Im Lim
Value-Based Building Maintenance Practices for Public Hospitals in Malaysia
Reprinted from: *Sustainability* **2021**, *13*, 6200, doi:10.3390/su13116200 . . . . **99**

**Andrés Alfonso Rosales Muñoz, Luis Fernando Grisales-Noreña, Jhon Montano,
Oscar Danilo Montoya and Alberto-Jesus Perea-Moreno**
Application of the Multiverse Optimization Method to Solve the Optimal Power Flow Problem
in Direct Current Electrical Networks
Reprinted from: *Sustainability* **2021**, *13*, 8703, doi:10.3390/su13168703 . . . . **115**

Artem Korzhenevych and Charles Kofi Owusu
Renewable Minigrid Electrification in Off-Grid Rural Ghana: Exploring Households
Willingness to Pay
Reprinted from: *Sustainability* **2021**, *13*, 11711, doi:10.3390/su132111711 . . . . **143**

Wai Fang Wong, AbdulLateef Olanrewaju and Poh Im Lim
Importance and Performance of Value-Based Maintenance Practices in Hospital Buildings
Reprinted from: *Sustainability* **2021**, *13*, 11908, doi:10.3390/su132111908 . . . . **161**

**Vadim A. Golubev, Viktoria A. Verbnikova, Ilia A. Lopyrev, Daria D. Voznesenskaya,
Rashid N. Alimov, Olga V. Novikova and Evgenii A. Konnikov**
Energy Evolution: Forecasting the Development of Non-Conventional Renewable Energy
Sources and Their Impact on the Conventional Electricity System
Reprinted from: *Sustainability* **2021**, *13*, 12919, doi:10.3390/su132212919 . . . . **175**

About the Editor

Alberto-Jesus Perea-Moreno

Dr. Alberto-Jesus Perea-Moreno (Ph.D.), associate professor at the Department of Applied Physics, Radiology and Physical Medicine at the University of Cordoba (Spain), received his M.S. in agricultural engineering and Ph.D. in geomatics at the University of Cordoba (Spain). He has published over 60 papers in JCR journals (http://orcid.org/0000-0002-3196-7033), and has an H-index of 16. His main interests are renewable energy, energy saving, biomass, sustainability, and remote sensing. Awards: 2019 winner of the Sustainability Best Paper Awards.

Preface to "Renewable Energy and Energy Saving: Worldwide Research Trends"

Today, there is an increasing concern for the environment, especially regarding the already-evident rise in the planet's temperature. This circumstance has led to technological progress in the use of natural resources for energy and their availability to all productive sectors. Faced with the current situation, it is vital to propose actions to put aside production models based on fossil fuels and opt for more sustainable models based on a circular economy, energy saving, and renewable energy. Energy saving and renewable energy allow us to save our scarce economic resources, postpone the depletion of our scarce fossil resources (on which our energy supply depends for the most part), and finally seem to be some of the best alternatives for reducing CO_2 emissions. This book aims to advance the contribution of the use of renewable energies and energy saving in order to achieve a more sustainable world.

Alberto-Jesus Perea-Moreno
Editor

sustainability

MDPI

Editorial

Renewable Energy and Energy Saving: Worldwide Research Trends

Alberto-Jesus Perea-Moreno

Departamento de Física Aplicada, Radiología y Medicina Física, Universidad de Córdoba, Campus de Rabanales, 14071 Córdoba, Spain; g12pemoa@uco.es

Citation: Perea-Moreno, A.-J. Renewable Energy and Energy Saving: Worldwide Research Trends. *Sustainability* **2021**, *13*, 13261. https://doi.org/10.3390/su132313261

Received: 25 November 2021
Accepted: 28 November 2021
Published: 30 November 2021

Energy is a very important resource for the development of the residential and industrial sectors, and it should be used with high efficiency, low environmental impact, and at the lowest possible cost [1]. Historically, economic development has been closely correlated with increased energy consumption and greenhouse gas emissions, leading to significant environmental impacts and heavy dependence on fossil energy sources [2].

Today, there is an increasing concern for the environment, especially regarding the already evident rise in the planet's temperature. This circumstance has led to technological progress in the use of natural resources for energy and their availability to all productive sectors [3].

Energy saving, responsible energy consumption, and efficient use of energy sources are essential at all levels. The importance of energy saving and efficiency measures is manifested in the need to reduce the energy bill, restrict energy dependence on the outside world, and reduce the emission of Greenhouse Gases (GHG) and the purchase of emission rights in order to meet the commitments acquired with the ratification of the Kyoto Protocol. Energy efficiency, including residential, industrial, and municipal energy savings, is critical to meeting national energy and climate change targets set by countries around the world [4]. Energy efficiency remains the least-cost option for meeting national climate change commitments.

After several years of negotiations, UN member states agreed on a roadmap that would help leaders and governments meet the EU's 2016–2030 goals, which include promoting sustainable development and environmental protection for all countries. The 2030 agenda for sustainable development is made up of 17 sustainable development goals and 169 priority action targets for the period 2016–2030, highlighting in the energy sector goal 7: Ensure access to affordable, secure, sustainable, and modern energy for all people.

The climate alert raised by the scientific community has led to CO_2 emissions becoming the main vector for the transformation of the energy sector by 2020–2050. The systematic and deep decarbonization of the energy system is the priority political horizon to be achieved in the 21st century. The landmark Paris Climate Change Agreement (2015) aims, at a minimum, to keep the global average temperature increase "well below 2 °C" this century compared to pre-industrial levels. Renewables, coupled with a rapid increase in energy efficiency, are the cornerstone of a viable climate solution [5,6]. There are a number of global policies to achieve these goals. As an example, the European Union is committed to achieving climate neutrality by 2050. Achieving this goal will require a transformation of European society and the European economy, which will have to be cost-effective, fair, and socially balanced [7]. In China, there has been strong growth in the renewable energy sector compared to the fossil fuel and nuclear energy sectors. China aims to achieve carbon neutrality by 2060 and peak emissions by 2030 [8,9].

Faced with the current situation, it is vital to propose actions to put aside production models based on fossil fuels and opt for more sustainable models based on the circular economy, energy saving, and renewable energy [10,11].

Energy saving and renewable energy allow us to save our scarce economic resources, postpone the depletion of our scarce fossil resources (on which our energy supply depends for the most part), and finally seem to be some of the best alternatives for reducing CO_2 emissions [4,12].

This Special Issue aims to advance the contribution of use of renewable energies and energy saving in order to achieve a more sustainable world. Leading authors have published important publications in the field of renewable energies and energy saving:

Abdelhakim Mesloub, Ghazy Abdullah Albaqawy, and Mohd Zin Kandar developed a study on the use of Building Integrated Photovoltaic windows under semi-arid climate in Algerian office buildings. This study was realized in terms of energy production, heating and cooling load, and artificial lighting. The results demonstrate the great advantages of the use of such windows in terms of energy savings compared to energy demand.

Walter Gil-González, Oscar Danilo Montoya, Luis Fernando Grisales-Noreña, Alberto-Jesús Perea-Moreno, and Quetzalcóatl Hernández-Escobedo developed an optimization model for the optimal placement and sizing of wind turbines, considering wind speed and demand curves and their reactive power capacity. The authors employed the General Algebraic Modeling System to develop the optimization model.

Van-Hai Bui, Akhtar Hussain, Thai-Thanh Nguyen, and Hak-Man Kim proposed a multi-objective stochastic optimization model to determine the set-point for a wind farm (WF) system.

Erika Winquist, Michiel Van Galen, Simon Zielonka, Pasi Rikkonen, Diti Oudendag, Lijun Zhou, and Auke Greijdanus analyzed how the biogas business will develop until 2030 in three countries (Germany, the Netherlands and Finland). This study is based on expert opinions.

Wai Fang Wong, AbdulLateef Olanrewaju, and Poh Im Lim evaluated the causal relationships between value factors and value outcomes of building maintenance in public hospitals in Malaysia. They concluded that value-adding practices and value co-creation have a positive influence on value outcomes in hospitals.

Andrés Alfonso Rosales-Muñoz, Luis Fernando Grisales-Noreña, Jhon Montano, Oscar Danilo Montoya, and Alberto-Jesus Perea-Moreno developed a methodology to optimize the power flow problem in direct current (DC) employing and optimization algorithm based on the multiverse (master stage) and the numerical method (slave stage) theories.

Artem Korzhenevych and Charles Kofi Owusu estimated the willingness to pay (WTP) values for renewable-generated electricity in five off-grid communities with a pilot renewable minigrid. The authors used data obtained from surveys of the inhabitants of these five communities.

Wai Fang Wong, AbdulLateef Olanrewaju and Poh Im Lim studied the critical success factors for improving the level of maintenance service delivery of hospitals. They conducted a total of 66 surveys among maintenance staff of public hospitals in Malaysia.

Vadim A. Golubev, Viktoria A. Verbnikova, Ilia A. Lopyrev, Daria D. Voznesenskaya, Rashid N. Alimov, Olga V. Novikova, and Evgenii A. Konnikov developed a model to study the evolution of non-conventional renewable energies over time, in order to guarantee their continuity and reliable and uninterrupted supply, taking into account the existing electricity system.

List of Contributions:

1. Mesloub, A.; Albaqawy, G.A.; Kandar, M.Z. The Optimum Performance of Building Integrated Photovoltaic (BIPV) Windows Under a Semi-Arid Climate in Algerian Office Buildings.
2. Gil-González, W.; Montoya, O.D.; Grisales-Noreña, L.F.; Perea-Moreno, A.-J.; Hernandez-Escobedo, Q. Optimal Placement and Sizing of Wind Generators in AC Grids Considering Reactive Power Capability and Wind Speed Curves.
3. Bui, V.-H.; Hussain, A.; Nguyen, T.-T.; Kim, H.-M. Multi-Objective Stochastic Optimization for Determining Set-Point of Wind Farm System.
4. Winquist, E.; Van Galen, M.; Zielonka, S.; Rikkonen, P.; Oudendag, D.; Zhou, L.; Greijdanus, A. Expert Views on the Future Development of Biogas Business Branch in Germany, The Netherlands, and Finland until 2030.

5.   Wong, W.F.; Olanrewaju, A.; Lim, P.I. Value-Based Building Maintenance Practices for Public Hospitals in Malaysia.
6.   Rosales-Muñoz, A.A.; Grisales-Noreña, L.F.; Montano, J.; Montoya, O.D.; Perea-Moreno, A.-J. Application of the Multiverse Optimization Method to Solve the Optimal Power Flow Problem in Direct Current Electrical Networks.
7.   Korzhenevych, A.; Owusu, C.K. Renewable Minigrid Electrification in Off-Grid Rural Ghana: Exploring Households Willingness to Pay.
8.   Wong, W.F.; Olanrewaju, A.; Lim, P.I. Importance and Performance of Value-Based Maintenance Practices in Hospital Buildings.
9.   Golubev, V.A.; Verbnikova, V.A.; Lopyrev, I.A.; Voznesenskaya, D.D.; Alimov, R.N.; Novikova, O.V.; Konnikov, E.A. Energy Evolution: Forecasting the Development of Non-Conventional Renewable Energy Sources and Their Impact on the Conventional Electricity System.

Funding: This research received no external funding.

Institutional Review Board Statement: Not applicable.

Informed Consent Statement: Not applicable.

Data Availability Statement: Not applicable.

Conflicts of Interest: The author declares no conflict of interest.

References

1.   Abeykoon, C.; McMillan, A.; Nguyen, B.K. Energy efficiency in extrusion-related polymer processing: A review of state of the art and potential efficiency improvements. *Renew. Sustain. Energy Rev.* **2021**, *147*, 111219. [CrossRef]
2.   Qudrat-Ullah, H.; Kayal, A.; Mugumya, A. Cost-effective energy billing mechanisms for small and medium-scale industrial customers in uganda. *Energy* **2021**, *227*, 120488. [CrossRef]
3.   Renna, P.; Materi, S. A Literature Review of Energy Efficiency and Sustainability in Manufacturing Systems. *Appl. Sci.* **2021**, *11*, 7366. [CrossRef]
4.   De la Cruz-Lovera, C.; Perea-Moreno, A.-J.; De la Cruz-Fernández, J.-L.; Alvarez-Bermejo, J.A.; Manzano-Agugliaro, F. Worldwide Research on Energy Efficiency and Sustainability in Public Buildings. *Sustainability* **2017**, *9*, 1294. [CrossRef]
5.   DeConto, R.M.; Pollard, D.; Alley, R.B.; Velicogna, I.; Gasson, E.; Gomez, N.; Sadai, S.; Condron, A.; Gilford, D.M.; Ashe, E.L.; et al. The Paris Climate Agreement and future sea-level rise from Antarctica. *Nature* **2021**, *593*, 83–89. [CrossRef] [PubMed]
6.   Potrč, S.; Čuček, L.; Martin, M.; Kravanja, Z. Sustainable renewable energy supply networks optimization—The gradual transition to a renewable energy system within the european union by 2050. *Renew. Sustain. Energy Rev.* **2021**, *146*, 111186. [CrossRef]
7.   Ortega-Cabezas, P.-M.; Colmenar-Santos, A.; Borge-Diez, D.; Blanes-Peiró, J.-J. Can eco-routing, eco-driving and eco-charging contribute to the european green deal? Case study: The city of alcalá de henares (madrid, spain). *Energy* **2021**, *228*, 120532. [CrossRef]
8.   Qiu, S.; Lei, T.; Wu, J.; Bi, S. Energy demand and supply planning of china through 2060. *Energy* **2021**, *234*, 121193. [CrossRef]
9.   Xie, Y.; Liu, X.; Chen, Q.; Zhang, S. An integrated assessment for achieving the 2 °C target pathway in china by 2030. *J. Clean. Prod.* **2020**, *268*, 122238. [CrossRef]
10. Klemeš, J.J.; Varbanov, P.S.; Walmsley, T.G.; Foley, A. Process integration and circular economy for renewable and sustainable energy systems. *Renew. Sustain. Energy Rev.* **2019**, *116*, 109435. [CrossRef]
11. Klemeš, J.J.; Varbanov, P.S.; Ocłoń, P.; Chin, H.H. Towards Efficient and Clean Process Integration: Utilisation of Renewable Resources and Energy-Saving Technologies. *Energies* **2019**, *12*, 4092. [CrossRef]
12. Sadeghian, O.; Moradzadeh, A.; Mohammadi-Ivatloo, B.; Abapour, M.; Anvari-Moghaddam, A.; Shiun Lim, J.; Garcia Marquez, F.P. A comprehensive review on energy saving options and saving potential in low voltage electricity distribution networks: Building and public lighting. *Sustain. Cities Soc.* **2021**, *72*, 103064. [CrossRef]

Article

The Optimum Performance of Building Integrated Photovoltaic (BIPV) Windows Under a Semi-Arid Climate in Algerian Office Buildings

Abdelhakim Mesloub [1],*, Ghazy Abdullah Albaqawy [1] and Mohd Zin Kandar [2]

[1]   Department of Architectural Engineering, Ha'il University, Ha'il 2440, Saudi Arabia; g.albaqawy@uoh.edu.sa
[2]   Department of Architecture, Universiti Sains Islam Malaysia, Nilai, Negeri Sembilan 71800, Malaysia; mzin@usim.edu.my
*   Correspondence: a.maslub@uoh.edu.sa

Received: 5 January 2020; Accepted: 19 February 2020; Published: 22 February 2020

Abstract: Recently, Building Integrated Photovoltaic (BIPV) windows have become an alternative energy solution to achieve a zero-energy building (ZEB) and provide visual comfort. In Algeria, some problems arise due to the high energy consumption levels of the building sector. Large amounts of this energy are lost through the external envelope façade, because of the poorness of the window's design. Therefore, this research aimed to investigate the optimum BIPV window performance for overall energy consumption (OEC) in terms of energy output, heating and cooling load, and artificial lighting to ensure visual comfort and energy savings in typical office buildings under a semi-arid climate. Field measurements of the tested office were carried out during a critical period. The data have been validated and used to develop a model for an OEC simulation. Extensive simulations using graphical optimization methods are applied to the base-model, as well as nine commercially-available BIPV modules with different Window Wall Ratios (WWRs), cardinal orientations, and tilt angles. The results of the investigation from the site measurements show a significant amount of energy output compared to the energy demand. This study revealed that the optimum BIPV window design includes double-glazing PV modules (A) with medium WWR and 20% VLT in the southern façade and 30% VLT toward the east–west axis. The maximum energy savings that can be achieved are 60% toward the south orientation by double-glazing PV module (D). On the other hand, the PV modules significantly minimize the glare index compared to the base-model. The data extracted from the simulation established that the energy output percentages in a 3D model can be used by architects and designers in early stages. In the end, the adoption of optimum BIPV windows shows a significant enough improvement in their overall energy savings and visual comfort to consider them essential under a semi-arid climate.

Keywords: BIPV window; WWR; overall energy; tilt angle; visual comfort; energy saving; semi-arid

---

1. Introduction

Buildings need to be energy efficient and fully utilize renewable energy to cover their energy demands. Global environmental awareness and expanding energy demands are increasing alongside the stable progress in renewable energy technologies seeking to create new prospects for renewable energy resources [1]. Many studies show that this new sector plays a vital role and many countries, such as Algeria, have taken measures to ensure sustainability in the utilization of global alternative energy resources [2,3]. The vital portion of energy consumption in the building sector is marked by a solid annual growth rate of 6.28%, which needs to be considered in terms of energy savings [4]. In contrast, the sector of energy consumption among Algerian buildings alone consumes around 42% of the overall amount used by all sectors. From another perspective, a study showed that there are no building regulations or any recommendations for daylighting and window-to-wall ratios (WWR) for public buildings. Thus, poor window designs and the absence of regulations in Algeria are leading to higher energy consumption [5]. However, electric lighting now comprises 25% of the total energy consumption, making it one of the main consumers of electricity in buildings [6]. The Algerian government has acknowledged advancing sustainable solutions, such as solar photovoltaic energy and greenhouse gas emissions, to fight climate change and facilitate the reduction of fossil fuels [7].

To preserve energy, windows can be used. This can be fulfilled by using Photovoltaic (PV) cells embedded into the windows. With the increase in the usage of glass and windows in the facades of the buildings, it has become a trend to produce electricity from windows and glass [8]. This can be achieved by using PV panels embedded in the windows. The design considerations for Building Integrated Photovoltaic (BIPV) windows in an office require examining the climate and solar conditions that are affected by the location and building type. Therefore, several design variables can strongly influence the impact of BIPV windows on energy performance and visual comfort. Various architectural variables, such as orientation, size of the window (WWR), and BIPV window types and daylight control are used to carry out simulations [9–11].

A pleasant and visual indoor environment is offered by daylighting as a natural lighting source for people in office rooms. The light passing through a building façade helps to achieve daylighting (any semi-transparent material or window glazing). There have been a few studies carried out on the performance of BIPV window daylighting and lighting energy. Certain authors have proposed methods to evaluate and optimize the daylighting and visual comfort for first generation BIPV window (STPV) applications. While some studies used daylight autonomy metrics [12,13], across the Diva and Dysim software tools under a tropical climate [14,15]. Other studies conducted a luminous test under real conditions to achieve a visual comfort level in accordance with European standards. By comparing the two BIPV window modules with 20% transparency, the results indicated the energy consumption of lighting for the mono-crystalline (m-Si) module to be slightly lower than that of the Copper Indium Selenide (CIS) [16]. Miyazaki et al. used a continuous dimming control for artificial lighting metric and proposed 10%–80% energy savings for different transmittance solar cells by considering the center of the office as the reference point. This result revealed the optimum solar cell transmittances to be 80% and 30% WWR in the Japanese context, although smaller cell transmittance values contributed less to electricity consumption [17]. Another perspective concerning the application of BIPV windows (poly-crystalline modules) is the skylight, as this study was conducted on a residential building roof with a south orientation and a 30-degree tilt angle. Since the direct daylight illuminance calculation model was used for the evaluation, semi-transparent PV top light systems were found to have contributed significantly towards lighting energy savings compared to the conventional glass used in Japan [18].

Modern architecture rarely employs Building Integrated Photovoltaic (BIPV) windows. However, only a few studies have investigated the overall energy performance of BIPV windows instead of conventional windows. These studies considered the three main aspects of the overall energy of BIPV windows—the energy output (electrical), daylighting (optical), and heat gain/loss (thermal)—through the use of modeling and experimental approaches.

The optimum PV inclination and orientation level, only in terms of energy output, also depend on the local climate, load consumption temporal profile, and latitude [19]; for example, the south orientation was found to be the ideal orientation for building façades facing the northern hemisphere near the equator [20]. Nevertheless, skylights facing the equator with an inclined angle against the building altitude will maximise their generation of electricity [21]. The performance of PV arrays at different orientations and tilt angles for Guangzhou city (latitude 27°N) was investigated by Chen Wei et al. According to their report, the monthly average energy output of the PV arrays at different angle-settings has nearly the same trend in the spectrum according to their monthly average solar radiation incidence. This finding proves that the amount of solar radiation on a PV array is the major factor that determines its system's efficiency. For the energy output plots of PV arrays, it was also concluded that the optimum yearly energy output value can be achieved from a PV array facing south with a tilt angle of 19° [22]. Yang et al. investigated the optimal tilt angle and azimuth angle for a wide range of locations in China by means of a specifically developed mathematical equation based on an anisotropic model. The results showed the optimal tilt angle for the maximum yearly solar radiation to be usually smaller than the local latitude, except for areas where the beam radiation occupies a great portion of the total solar radiation [23].

A previous study focused on the estimation of energy savings in a Japanese office building using different transmittance values for the semi-transparent solar cell by modeling a standard floor of an office building based on the Architectural Institute (AIJ) in Japan and applying an amorphous silicon solar cell under Japan's climate. Consequently, compared to the standard model, the total reduction was 55% [17]. Ng, Mithraratne, and Kua evaluated the overall energy performance of six commercially available semi-transparent PV modules under a tropical climate, based on their Net Electricity Benefit (NEB), and compare them with conventional windows used in Singapore in terms of total electricity consumption. Their findings revealed that even in orientations that do not receive direct solar gains, BIPV can be adopted; moreover, PV efficiencies and good thermal properties are essential to achieve a better NEB performance [24]. Lu and Law (2013) estimated the overall energy performance corresponding to the five orientations of a semi-transparent BIPV window system installed in a typical office in Hong Kong by integrating the simulation results of thermal, power, and visual behaviours. The main finding of the work was that the system would lead to an annual electrical benefit of about 1300 kWh [10]. In the same context, a comparison of the overall energy performance was carried out in Hong Kong using a semi-transparent BIPV window and double-glazed window. The results revealed that a semi-transparent BIPV window can save up to 16% total electricity per year, and the best orientation is south–west [11]. Another comparison study was performed in five cities in China between double skin façades and an insulating glass unit through experimentation and simulations. The results show that the performance of the Photovoltaic Insulated Glass Unit (PV-IGU) offered 2% better performance than the ventilated PV-Double skin façade. In contrast, the PV-DSF had a better reduction in solar heat gain compared to PV-IGU, while the PV-IGU was better than the PV-DSF in terms of thermal insulation [25]. An et al. focused only on the cooling and heating performance of an amorphous silicon PV module in Korea using a comparison study between different types of PV glass (single, double glazing). Their results revealed a reduction of 18% in cooling and heating loads [26]. Georgios Martinopoulos et al. assessed a nine story office building in terms of its energy performance and thermal comfort for a number of various building integrated retrofitting measures; they found that the shading scenario can reduce total energy consumption by 33% [27].

Recently, European countries have given the utmost importance to this technology to help achieve a Zero Energy Building (ZEB). In this context, an assessment of four BIPV configurations, namely: vertical, tilted PV façade, semi-transparent BIPV and photovoltaic shading device, in terms of energy demand and visual comfort under semi-continental climate. The results indicated that semi-transparent BIPV eliminated the disturbing glare and resulted in a decrease of 20% on the cooling demand compared to the reference case. On the other hand, the annual heating and lighting loads were covered by the annual PV energy output and also a reduction in the cooling loads, approaching the zero-energy standard in most of the cases [28]. As a consequence, to date, a balanced solution for the overall energy performance of semi-transparent BIPV windows has been limited in terms of the WWR, transparency, and efficiency of the available product markets seeking to maximize the energy savings of the BIPV module in buildings and ensure visual comfort at the same time, particularly under the semi-arid climate in Algeria.

2. Methods

2.1. Field Measurement

As shown in Figure 1, a typical office room in Tebessa city facing the east–south was selected to study the energy output, heat gain, and heat loss, as well as for daylighting measurements. The selected room is on the 3rd floor and free from shading and any obstacles to ensure it receives the maximum amount of solar radiance. The experiment was conducted during the critical period of summer, with the field-measured data recorded systematically using proper interval times. The floor area of the room was representative of the office room size for all government staff (12 m^2), according to the DLEP and DUC Algerian standard. An off-grid photovoltaic window system was designed and used to measure, evaluate, and validate the maximum power direct current (DC) from the PV window module. This system consists of four main components: a photovoltaic window with visible light transmittance (VLT) 20%, (2) solar charge controllers (maximum power point tracker (MPPT) with a built-in data logger (DC energy output), a battery, a load element (fluorescent lamp at 12 volts). These elements were covered with a box during the experiment, so they had an effect on the daylighting measurements. Furthermore, four illuminance meters (HOBO Pendant Temperature/Light Data Logger) were arranged inside to measure the indoor temperature and internal WPI, and one illuminance light meter was placed outside the office to measure the external illuminance and outdoor temperature; a heat flux meter (fluxDaq+) was used to measure the heat gain and loss of the PV module. The details of the electrical and thermo–optical characteristics of the PV module are stated in Table 1.

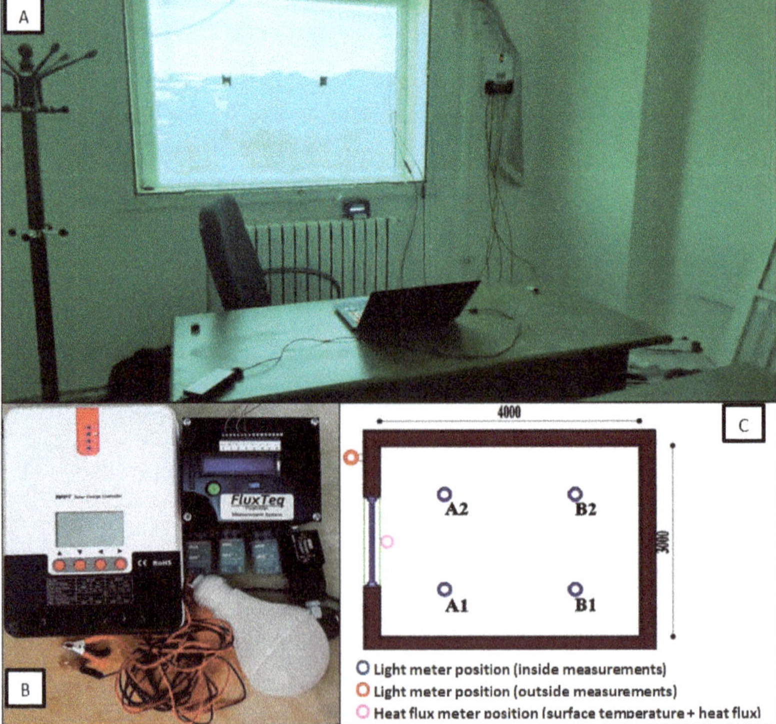

Figure 1. (**A**) The tested BIPV window placed in the office building chosen for experiment; (**B**) the instruments used during the experiment; (**C**) the arrangement of the instruments for measurements.

Table 1. Electrical and thermo–optical characteristics of the Building Integrated Photovoltaic (BIPV) windows used in this experiment.

Amorphous silicon (ASG090)	
Dimension (Length, Width, Thickness)	1400 x 1100 x 6.8 mm
Electrical Properties (STC)	
Efficiency of Module (η)	4.50%
Max power (Pmax)	90 Watt
Max power Voltage (Vpm)	78 V
Max power Current (Ipm)	1.15 A
Open circuit voltage	100 V
Short circuit current	1.43 A
Temperature Coefficient (β)	$-0.0033/^\circ$C
Temperature Coefficient (α)	$-0.0009/^\circ$C
Temperature Coefficient (γ)	$-0.002/^\circ$C
Optical properties	
Transmittance (VLT)	20%
Thermal Properties	
U-value	5.11 at Summer Daytime 5.65 at Winter Night-time
Solar Heat Gain Coefficient (SHGC)	0.34

To produce correct results and predict the actual consumption, validation is a major approach that has to be done using building simulation tools. Therefore, the accuracy of the comprehensive performance of the energy factors (energy output, heat gain/loss, indoor and outdoor temperature, and cell temperature) was validated through Energy Plus, whereas the indoor and outdoor illuminance was validated through IES-VE by comparing the experimental data to the simulated results. This validation is based on the mean bias error (MBE) and the coefficient of variation of the root mean square error (CV RMSE) indicators, which are strongly recommended for energy models by the ASHRAE14 Guidelines.

$$MBE(\%) = \frac{\sum_{i=1}^{Np}(mi - si)}{\sum_{i=1}^{Np}(mi)} \tag{1}$$

$$CV\ RMSE = \frac{\sqrt{\sum_{i=1}^{Np}(mi - si)^2 / Np}}{m} \tag{2}$$

where *mi* is the measured value, *si* is the simulated value, and *n* is the number of measured data points. The results of the accepted model should be less than 10% for MBE and 30% for CV RMSE [29].

2.2. Computer Simulation with Energy Plus and IES-VE

To adequately assess the characteristics of the current overall energy design practice of the BIPV window, a series of simulations were performed for the basic model (Geometric A) that represents the common construction practices of offices in Algeria [30] with the same PV module used in the experiment. Moreover, an extensive simulation was carried out by modeling 65 BIPV windows at different tilt angles and orientations at once in order to obtain a general picture of the design implications for various BIPV window systems (Geometric B) and to estimate their application impacts on building energy production and energy savings as shown in Figure 2. Algeria–Tebessa TMY data were used Meteonorm data-base, to determine the office buildings' energy use for artificial lighting, cooling, and heating electricity usage, as well as the photovoltaic electricity generated [31]. Several parameters evaluated the overall energy performance of the thin-film PV windows (amorphous silicon, micromorph). These parameters were the size of the window, the main orientations, the tilt angle, and the different levels of visible effective transmittance (VLT), Solar Heat Gain Coefficient (SHGC), U-value, and energy efficiency (n), as seen in Table 2.

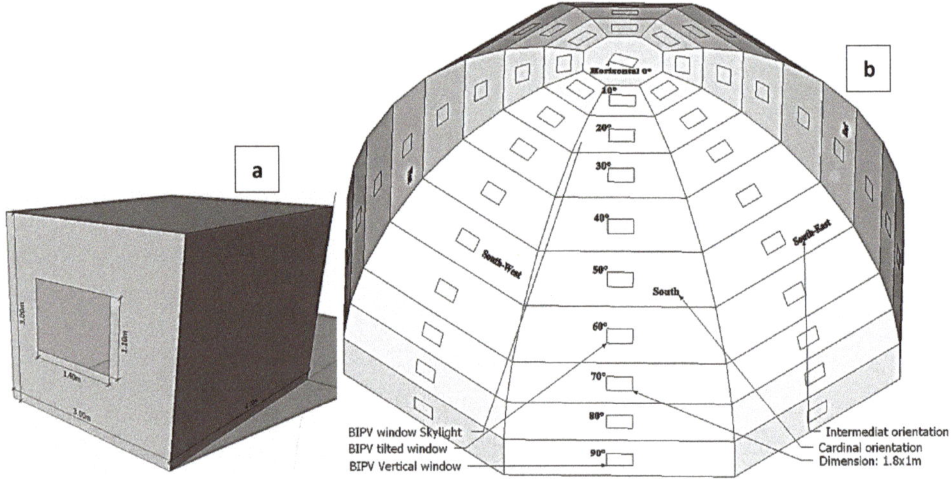

Figure 2. (a) Base-model office building used for the daylighting validation sketch-up version 2015 (Geometric A); **(b)** the proposed model for evaluation of the energy output of BIPV windows at different tilt angles and cardinal/intermediate orientations (Geometric B).

Table 2. The optical and thermal properties of the thin-film photovoltaics used in the experimental simulations, S.G: single glazing, D.G: Double glazing.

Technology	Configuration Module	VLT (%)	U-value (W/m²K)	SHGC	Efficiency (%)
	Module (A1) S.G	10	5.70	0.29	4
	Module (A2) S.G	20	5.70	0.34	3.4
	Module (A3) S.G	30	5.70	0.41	2.8
(Thin-film)	Module (B1) D.G	10	2.70	0.11	4
a-silicon	Module (B2) D.G	20	2.70	0.14	3.4
and	Module (B3) D.G	30	2.70	0.19	2.8
Micro-morph	Module (C) D.G	6.91	1.674	0.154	4.75
	Module (D) S.G	9.17	5.076	0.289	8.02
	Module (E) S.G	5.19	4.795	0.413	5.90

In this study, the nine commercially-available BIPV modules selected were the Auria Solar (Micromorph) Red, Schott Solar double-glazed amorphous silicon (Voltarlux ASI-ISO-E1.2), Hanwa Makmax single-glazed silicon (KN-42), and six modules of Onyx Solar single laminated and double-glazed silicon, with transparency values ranging from 10% to 30% (See Figure A7). All of the modules were made from thin-film solar technologies that can accommodate the studied climate (semi-arid) and had shown better performance at high temperature levels, as well as better shade tolerance, than alternative modules [32]. These modules included both single- and double-glazed units, which consisted of different constructions and technologies as shown in Table 2.

The position and WWR of the BIPV windows used in geometric model C were based on daylighting and view designs, rather than energy output and heat gain–loss factors. The daylight zone, which is associated with window size, is defined as an area with a depth that is two times the window's height (measured from the ground) in a direction parallel to the window; the daylight area spreads horizontally, equal to the window width, plus one metre on either side of the aperture. Consequently, all of the BIPV windows were positioned in the middle part of the wall. To achieve the minimum line of sight, the distance between the floor and the bottom part of the BIPV window (10%–70% WWR) was set at 0.9 m, which is slightly higher than the working plane of 0.85 m. As shown in Figure 3, a distance of 2.1 m between the floor and the top portion of the BIPV window (10%–40% WWR) was found to be the most optimal height for the full penetration of daylight into the room.

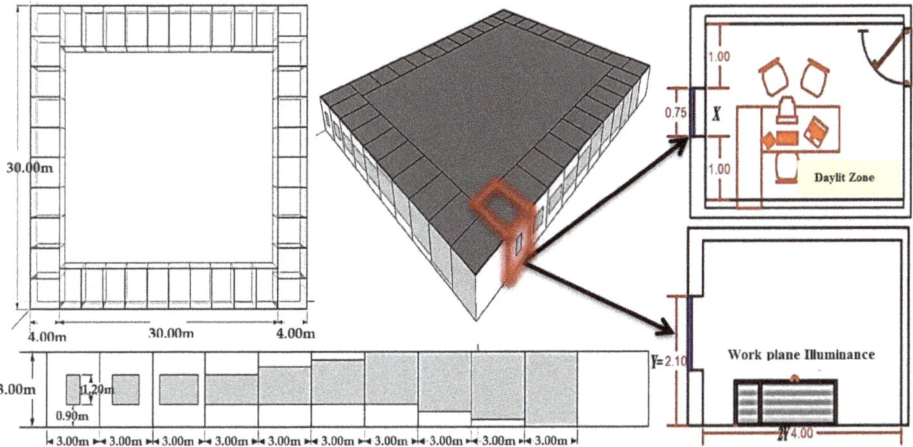

Figure 3. Proposed 3D model for the optimization simulation stage (Geometric C).

The assessment of energy savings and the optimisation procedure consisted of three stages:

- The first stage involved estimating the current situation's overall energy consumption for the base-model;
- The second stage involved using different BIPV window modules instead of conventional windows to estimate the energy output based on the cardinal and intermediate orientation, with different tilted angles from horizontal to vertical, with an interval of 10 degrees (geometric b; Figure 2). This stage also involved evaluating conversion efficiency (n) and PV cell temperature, the heating and cooling load, as well as artificial lighting and a constant internal load (computer load);
- The last stage included optimizing the design of the BIPV window by balancing the WWR and physical factors through the use of the geometric model (C). The potential total electricity savings and their visual comfort criteria when installed in an office building were then estimated and compared with those of the base model.

The following sections provide the simulation of the overall energy models, as shown in Figure 4.

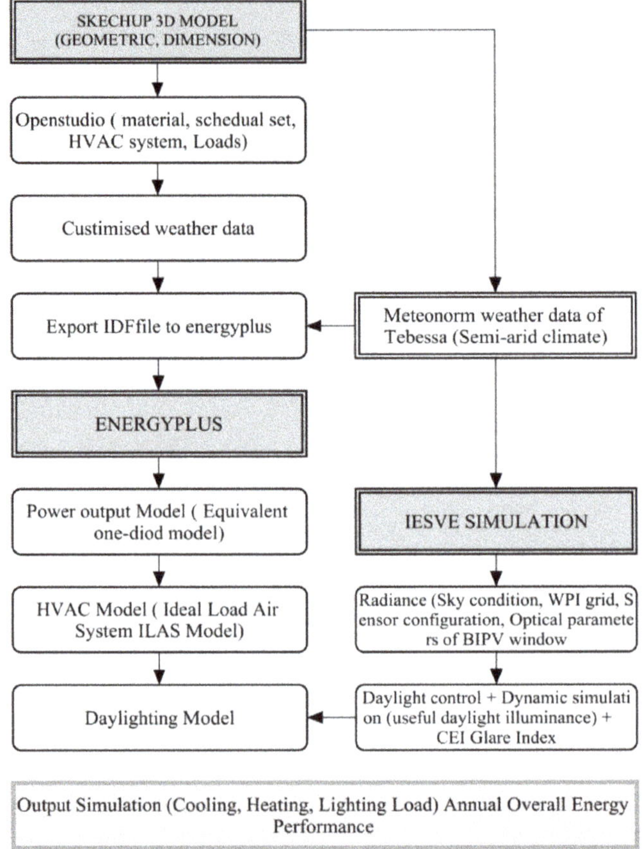

Figure 4. Schematic diagram of the simulation method.

2.2.1. Energy Output Model Simulation

This study employed *the equivalent one-diode model*. This model uses an empirical relationship to predict the operating performance of the PV, based on conditions such as the PV cell temperature and an estimation of the conversion efficiency for each time-step. This model consists of a diode, a DC current source, and a series of resistors [33]. The following equation presents the equivalent one diode module (3)

$$I = IL - I_0 \left[\exp\left(\frac{q}{\gamma ktc} (V + IRs) \right) - 1 \right] \tag{3}$$

where I is current [A], V is voltage [V], R_s is the resistance of the module series [Ω], T_c is the module temperature [K], I_o is the diode that reverses the saturation current, γ is the empirical PV curve-fitting parameter, I is the current, I_L is the module's photocurrent, q is the electron charge constant. However, the four parameters in this model, I_L, I_o, γ, and R_s, are empirical values that cannot be determined directly through physical measurements. The EnergyPlus model calculates these values from manufactures' catalog data. Meanwhile, the *Integrated Surface Outside Face* option is applied to estimate the cell temperature [11]. The energy exported from the surface as electricity becomes a sink in the internal source modeling for the heat transfer surface [34].

2.2.2. Thermal Simulation

The thermal transmittance coefficient (U-value) and solar heat gain coefficient (SHGC) are the two parameters typically used for thermal characterization, comparisons between BIPV window systems, and as the input for building energy performance simulations [35]. The SHGC is estimated through energy balance equations integrated into EnergyPlus models, which can be employed to express conductive, convective, and radiative heat transfer phenomena [36]. The heating and cooling load is a necessary critical factor required in the overall energy consumption assessment. For this study, the Ideal Loads Air Systems (ILAS) component that was built in Energy Plus was used to represent an ideal HVAC system. This component is assumed to supply cooling or heating air to the related zone, to meet the zone load up to the specified limit required by the user. As shown in Table 3, with a coefficient of performance (COP) of 1, the ILAS model is connected to the outdoor air and supplies the necessary quantity of cooling and heating energy required to meet the temperature set points of the building's indoor air temperature. However, the heating and cooling systems were only turned on during the working hours, as stipulated by the Algerian regulation policy schedule. The thermal properties of external building element characteristics are summarized in Table 4.

Table 3. Simulation parameters and operation conditions of the thermostat set point.

Occupancy Density	One Person
Heating set-point	21 °C (08:00–17:00) and the rest of the day Off
Cooling set-point	24 °C (08:00–17:00) and the rest of the day Off

Table 4. Thermal properties of the external boundary of a typical office building in Algeria.

	Thickness (mm)	Conductivity (W/mK)	Density (kg/m3)	Specific Heat (J/kg K)
Roof				
Roof membrane	1	0.16	1121	1460
reinforced concrete slab	40	1.4	2400	300
Hollow block	160	1.2	2400	946
Cement mortar	5	1.5	1900	1080
Coating of plaster	10	0.5	1900	1080
Interior wall				
Plaster coating	10	0.5	1900	1080
Cement mortar	20	1.15	1900	1080
brick	100	0.44	1100	940
Cement mortar	20	1.15	1900	1080
Coating of plaster	10	0.5	1900	1080
Exterior wall				
Coating of plaster	10	0.5	1900	1080
Brick	150	0.44	1100	940
Wall air space	10	0.6	800	1000
Brick	100	0.44	1100	940
Coating of plaster	10	0.5	1900	1080
Ceiling/Floor				
Herission	30	1	1100	828
Floating slab	100	1.75	2400	946
Cement mortar	5	1.5	1900	1080
Surface finish	20	1.2	2000	800

2.2.3. Daylighting Simulation

In this study, a combination of the daylighting control method and dynamic daylight climate-based metrics (Useful Daylight illuminance) were used to evaluate daylighting performance and energy savings [37,38]. The daylighting performance of the BIPV windows demonstrated the amount of illuminance that is accepted and the capability of the lighting on the systems under both cloudy and clear sky conditions to perform a transitional shift and be displayed similarly to regular glass material [39]. The selected performance indicator of daylighting quantity is based on the International Standard (ISO), as shown in Table 5. The approach of IES-VE was used to model the potential for the daylighting of BIPV windows. This method was used considering the finding that visible light transmittance (VLT) is viewed as the most fitting element for estimating work plane illuminance (WPI) in the tested offices at the reference points [39]. The finishing materials of the offices were selected from among materials that are generally used in Algerian office buildings:

1. Floor: stone coverings, with a reflection coefficient of 25%;
2. Walls: cream paint, with a reflection coefficient of 70%;
3. Ceiling: white paint, with a reflection coefficient of 90%;
4. Ground coverings: concrete, with a reflection coefficient of 30%.

Table 5. Summary of Performance Indicator Criteria for Daylight Simulations (ISO).

Analysis	Criteria	Performance Indicator	Work Plane Height
Quantitate **+** **Qualitative**	*Useful daylight illuminance (UDI)*	300 lux < Dark area (needs artificial light) 300 lux–750 lux: comfortable at least 50% of the time > 750 lux: too bright with thermal discomfort	0.85 m
	WPI	WPI Recommended 300–750 lux	
	Mean CGI	19 for sedentary status situations are acceptable 22 for transient situations are acceptable	

It is assumed that there is no building in the vicinity that could obstruct direct light from entering the windows.

2.3. Criteria for Optimum BIPV Window Design

The yearly evaluation of energy performance employed a graphical optimization method that combines (1) the visual comfort criteria shown by UDI, where the shaded areas (in the graph) cover less than 50% of the occupancy hours, with a value of 300–750 lux for the UDI curve. Beyond this area are the given visual criteria to be met. Next, in agreement with the CEI Glare Index (CGI), only the best PV modules achieved the requirements of UDI to assess the quantity of glare in the four design days alongside the base-model. (2) By considering the WWR in a cardinal orientation, the overall energy consumption (OEC) is then calculated, where a lower value indicates a more energy efficient building, as shown in Figure 5. The following relation is used to evaluate the overall energy consumption (OEC)

$$
\begin{aligned}
\textit{Overall Energy } \quad &\textit{Consumption } (OEC) \\
&= \textit{ Cooling energy consumption} + \textit{Heating energy consumption} \\
&+ \textit{Lighting energy consumption} + \textit{Electrical equipement energy consumption} \\
&- \textit{PV energy output} \left[\tfrac{kWh}{m^2} \right]
\end{aligned}
\tag{4}
$$

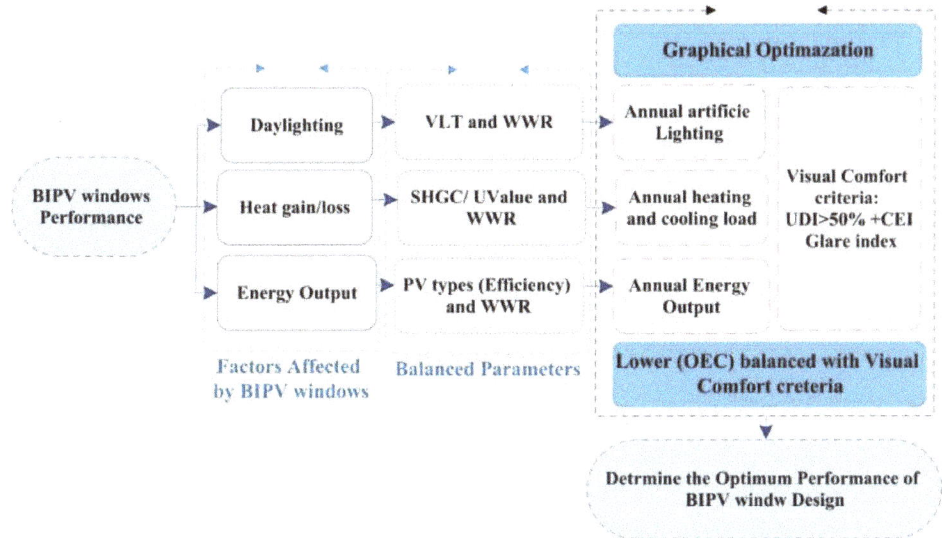

Figure 5. A flowchart for determining the optimum BIPV window design.

2.4. Climatic Conditions (Vertical Solar Radiance)

The consideration of climatic conditions has been demonstrated to be an important factor in investigating the energy output of BIPV window applications, especially solar radiance [40]. Global solar radiation and sunshine duration values are available on a mean daily or monthly basis. However, the diffuse, direct, and cloud cover data are rarely recorded on an hourly basis. Some present databases have been created based on the interpolation and extrapolation of the available data for estimating solar radiation at each point in the world. These databases include the Meteonorm database, which is used in this research [31].

Figure 6 identifies the annual horizontal and tilted 90° (vertical) global radiation information together with its diffused component. As is clearly shown, the amount of global horizontal irradiance (GHI) which reaches 1929 kWh annually, is more than the vertical radiance on a different azimuth. The highest amount of radiation is received on the south azimuth, with 72% compared to GHI, and the lowest amount is found on the north azimuth (24%), primarily due to the absence of direct solar radiance. However, the east and west azimuth reveal a difference of 60% which is approximately the same amount as the symmetric incidence of solar radiance, as shown in the sun's path. The diffuse radiance on the vertical façade of the azimuth represents a significant percentage difference compared to GHI, which can reach up to 90% for the north azimuth and between 40% and 46% for S, E, and W.

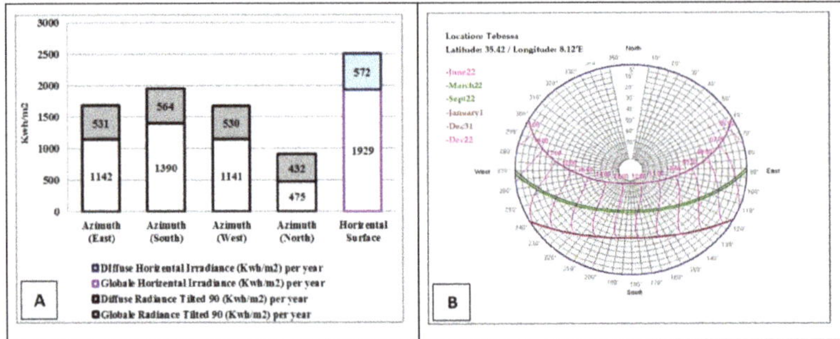

Figure 6. (A) Yearly global horizontal and diffuse radiance in different orientations (E, N, W, and S); (B) the sun path of the city of Tebessa by IES-VE 2017.3. Result and Discussion.

3. Result and Discussion

3.1. Empirical Validation of the Overall Energy Performance BIPV Window

Figure 7 shows that the simulated WPI in reference points A1 and B1 is similar to the measured result. The highest illuminance level was 370 lux during the morning period for reference point A1, while the B reference point does not exceed 200 lux due to the low transparency and its distance from the tested PV module. On the other hand, the indoor air temperature was between 25 and 28 °C, which is slightly higher than the comfortable air temperature inside the office. This result demonstrates that the measured and simulated air temperatures were consistent (less than 2.3%). This outcome provides a good prediction for the cooling and heating loads during the next part of the simulation. The outcomes from the EnergyPlus dynamic reproduction include hourly heat gains and losses, and achievements via BIPV window indicate great dependability with thw experiment model, where the Mean bias error of heat gain is 2.61%, and 8.64% for heat loss. However, the heat loss happens around evening time of the summer season, which is not within the investigation period from 8a.m. to 17p.m.

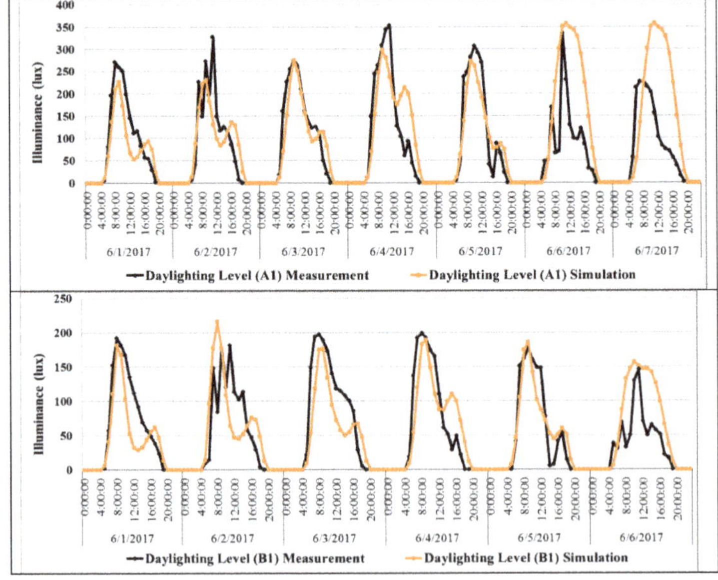

Figure 7. *Cont.*

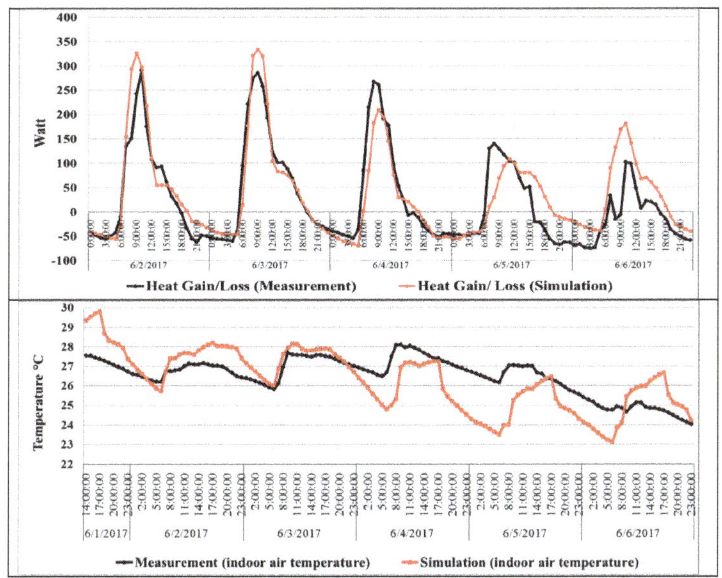

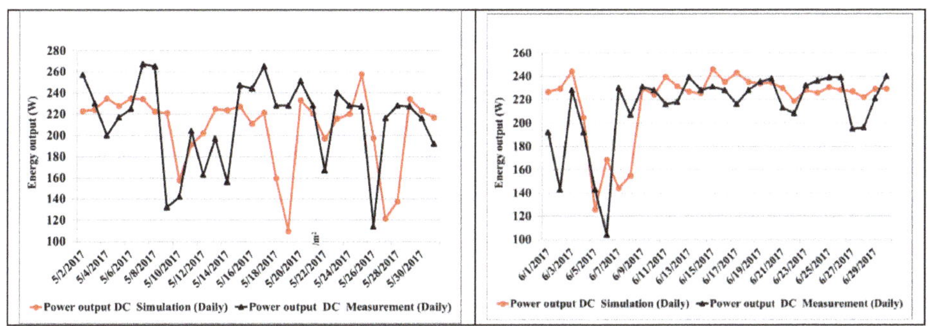

Figure 7. Comparison of the measured and simulated data (WPI) at the reference points (**A1,B1**); heat gain, heat loss, and indoor air temperature.

The daily energy output data were also compared with the simulated and tilted solar radiance for greater accuracy. A remarkably more fluctuating trend of power generation was observed in spring (May) than in the summer season due to the variations in sky conditions in the spring season, as shown in Figure 8. The monthly average of the energy output was between 213 Wh in May (at least) and 245 Wh in July as the maximum. However, the monthly total amount of energy output ranged up to 7 kWh per month, which is a considerable amount when compared to the energy consumption of a typical office. A validation of the energy output shows the perfect reliability of the model, where the mean bias error is between 0.48% to 2.21%, and the coefficient variation root mean square is between 11.95% in July because the sky conditions during this month are totally clear, while the coefficient variations is 22.7% in the spring season.

Figure 8. *Cont.*

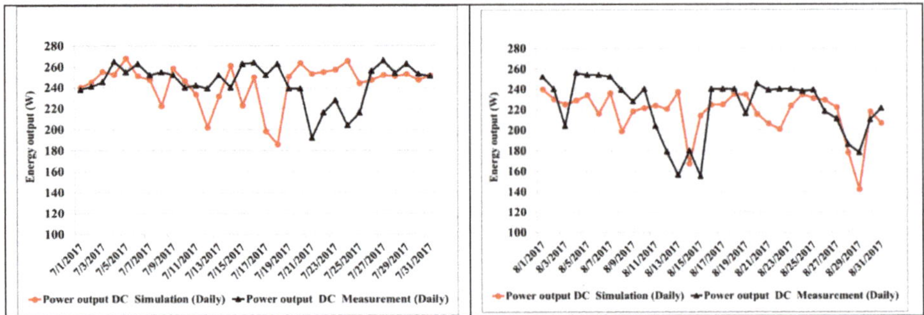

Figure 8. Comparison of the measured and simulated daily energy output of the BIPV window in the summer and spring season.

3.2. Evaluation of the energy output

The Figure 9 shows the significant differences in the energy output among the different months under the semi-arid climate in Algeria. The southern-oriented BIPV windows produced less energy in the summer compared to the winter months since the sun passes quite close to the zenith during the summer season (height 81°) and during the winter season, it passes at low latitudes. The maximum monthly energy output is around 5.95 kWh/m^2 in November, while the minimum energy output is around 2.8 kWh/m^2 in June. Further, north is seen as the worst orientation for energy output of BIPV windows. Similar results were achieved in the west and east facades due to the symmetry of the incident solar energy, where the maximum output energy loss from the east to west façade did not exceed 14%.

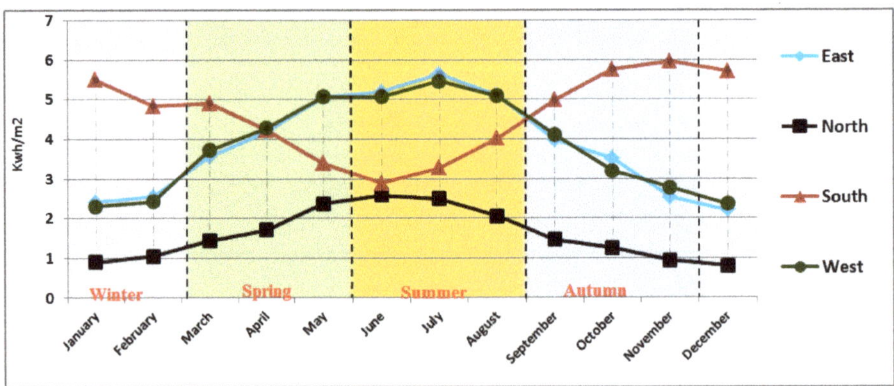

Figure 9. Monthly Distribution of the energy output PV module with conversion efficiency n = 4.

As shown in Figure 10, based on the annual data, it can be seen that the most effective energy output during the year was 111.112 kWh/m^2, which was obtained by facing the module toward the S–E/S/S–W facades, while the lowest energy output was obtained by facing the PV module to the north façade, producing an output that was about three times greater. Nevertheless, both the east and west facades were within the acceptable level of energy output. This indicates the significant role of the sun's path in each season throughout the year in the design of BIPV windows. The energy output increased dramatically from the 90 to the 30 degree slope and then decreased up to the horizontal 0 slope. The highest energy output was obtained at the 30 degree slope. However, the vertical BIPV windows only produced 61.24% of energy compared to those installed at the 30 degree slope. This particular result is consistent with previous research conducted in Beirut, Lebanon [41].

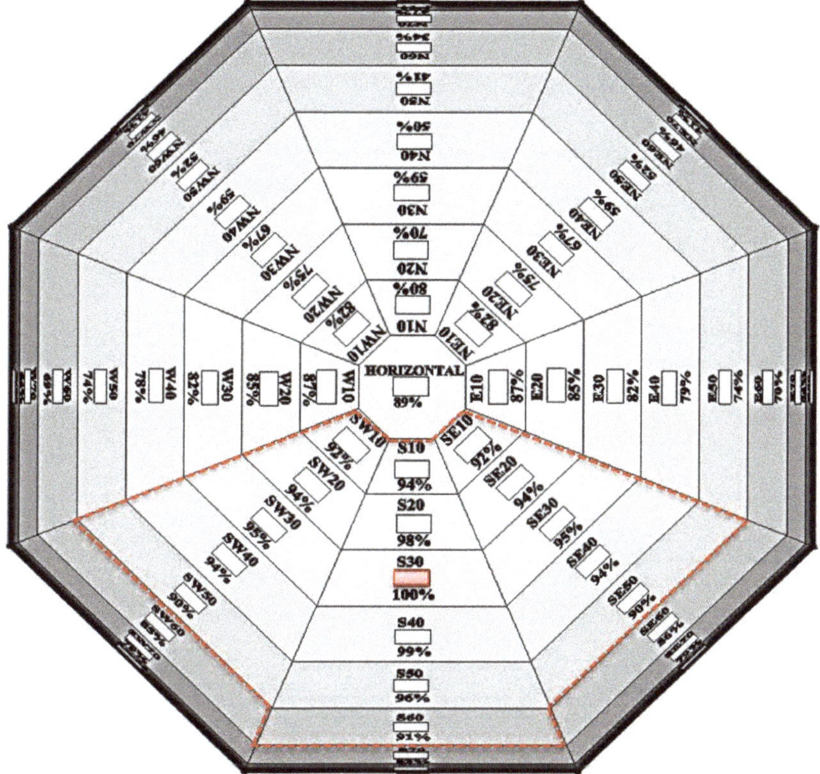

Figure 10. Percentage for the BIPV windows' energy output performance based on the tilt angle and orientations under the semi-arid region in Algeria.

Generally, the BIPV window facing the south façade achieved the highest energy output. However, this output decreased at the 30 degree slope by increasing the tilt angle in all orientations based on the ratio between the highest energy output facing the south slope (with 30 considered as 100%, with 178 kWh/y, and including the energy output of the other facades and tilted surface from the horizontal to the vertical façade). The tested PV module within the red line zone presents the best tilt angles and orientations for the PV module performance at higher than 90% of its capacity, mainly because this area receives the maximum amount of solar radiance. Moreover, this result revealed that a significant percentage of the 3D model can be used by architects and designers in Algeria.

On the other hand, the cell temperature and energy output of the thin film modules were investigated using the *equivalent one diode* model throughout the four design days (21st June, 21st September/March, and 21st December), thereby providing a general overview of the effects of cell temperature on the application of BIPV windows at different tilt angles (10°, 30°, 50°, 70°, and 90°) under semi-arid conditions.

The simulation of the solar cell temperature recorded the lowest value (7 °C) in the early winter morning, and the highest value was recorded to be 50 °C in the evening of the summer season. Accordingly, the PV module recorded the maximum energy output during summer. The graphs illustrated in Figure 11 demonstrate that by increasing the tilt angle from 10° to a vertical angle, using an interval of 20° towards the South facade, the cell temperature declined dramatically. Furthermore, during both the summer and winter afternoon, this method achieved a difference of 10 °C, particularly between 10 and 30 °C. A minimum reduction of at least 1 °C for the cell temperature was also recorded in the early morning between 8 a.m. and 9 a.m. These results further demonstrate that the effect on the energy output of increasing the cell temperature of the thin film modules was negligible compared to the solar radiance, due to the conversion efficiency of the solar cell line with the energy output. This result confirms that the use of thin film BIPV windows is appropriate under semi-arid climate conditions.

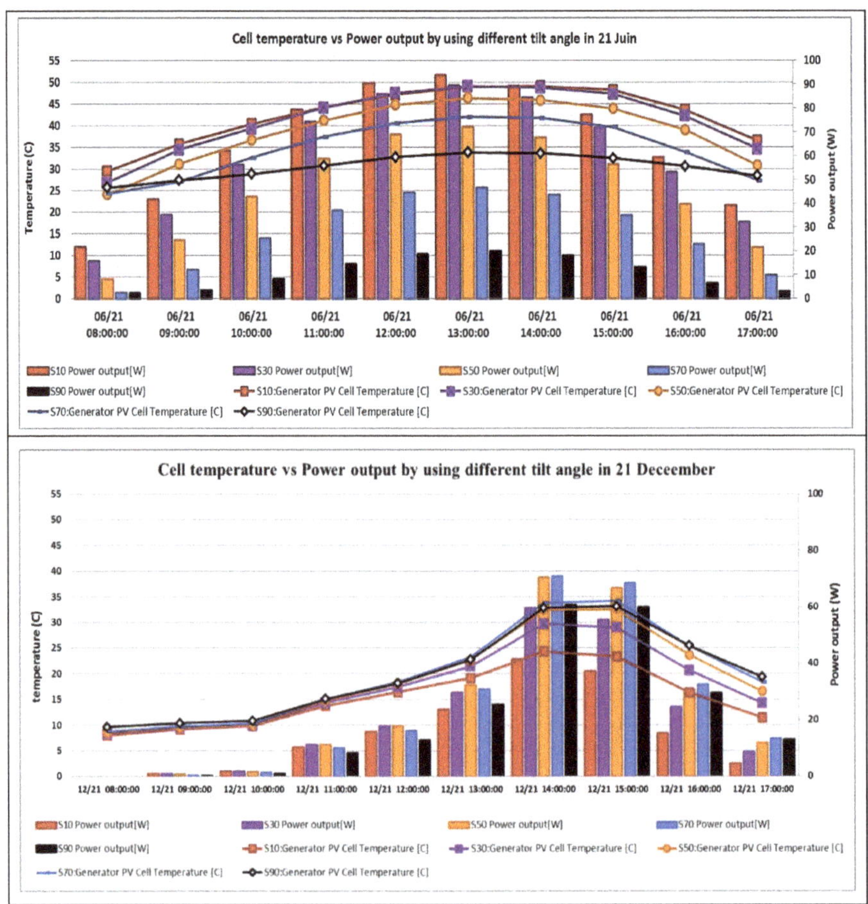

Figure 11. The effect of the cell temperature against the energy output with different tilt angles.

3.3. Evaluation of the Lighting Energy

In this study, thin-film BIPV window modules were treated as uniform optical properties. Three effective visible transmittance values of the BIPV window modules were simulated: 10%, 20%, and 30%. The minimum value of 10% was selected in order to ensure a certain minimum outside view. The graphs below describe the lighting electricity consumption as a function of solar cell transmittance and WWR. The annual total lighting electricity consumption reduced by increasing the solar cell transmittance. Meanwhile, this value decreased by increasing the WWR based on effective daylight availability. The results of lighting energy consumption can be summarised through two different scenarios:

- Scenario one involves an increase in the WWR of the BIPV window from 10% to 100%; meanwhile, the VLT is kept constant at 10%, 20%, or 30%. The graph below shows a very steep decline in lighting energy from 10% to 60% WWR. For example, the yearly lighting energy consumption in the south with 30% of VLT was 19.1k Wh/m^2, which diminished to only 1.8 kWh/m^2 per year, as shown in Figure 12. Then, the energy gradually decreased due to sufficient daylight in the work plane (refer to Figure A1 for the other orientations);

- Scenario two involves increasing the VLT by means of an interval of 10% and fixing the WWR. The decline percentage ranged between 6% and 10% in a small WWR, reaching up to 65%–80% with a large WWR of the PV modules; the maximum lighting energy reduction percentage was 80% and was achieved with a fully glazed PV module oriented toward the southern façade.

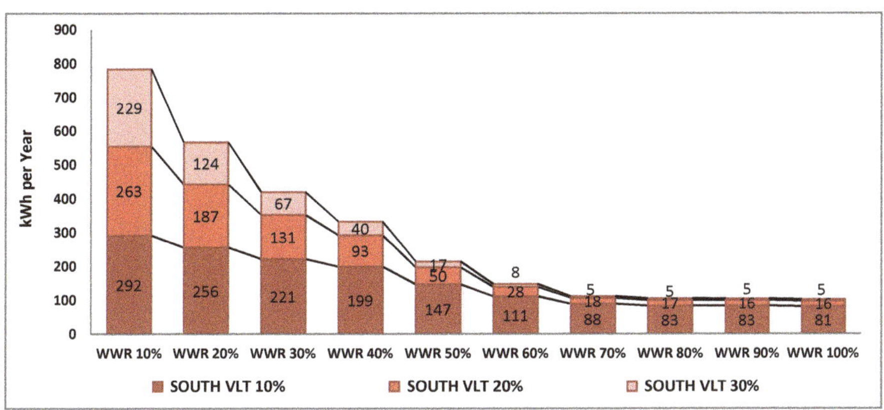

Figure 12. Lighting energy consumption using different VLT values for the PV modules on the south façade.

3.4. Evaluation of the Cooling and Heating Energy

The graph below depicts the variation in total cooling and heating energy consumption for the base-model of nine different BIPV modules as a function of the WWR in cardinal orientations. A positive correlation is observed between the WWR of the PV modules and the base-model with cooling energy consumption, where the larger WWR and the higher cooling load were caused by variations in the SHGC and U-values. The graphs in Figure 13 illustrate the three main divisions: (1) the double-glazed PV modules (A1, A2, A3, and C); (2) the single-glazed PV modules (B1, B2, B3, D, and E), and (3) the base-model. Consequently, the cooling loads changed considerably among these three divisions, where the result was very close, with a small WWR, and the difference between the three divisions was remarkably large for the WWR due to the increased solar heat gained by the PV module. Therefore, the best performance for energy savings based on the cooling load was achieved by the double-glazed PV modules, A1, A2, C, and A3. Moreover, the maximum value of the cooling load energy does not exceed 60 kWh/m^2 in the cardinal orientation. Meanwhile, all PV modules were found to have less energy consumption than the base model, where the cooling energy consumption increased to more than three times greater than the large WWR of the south, east, and west facades, and more than double that of the north facade. The energy savings accounted for by the cooling load were very significant, reaching up to 108, 100, 86, and 29 kWh/m^2 in the east, south, west, and north facades, respectively, compared to the double-glazed PV module. Moreover, approximately half of these percentages were acquired by applying a single PV module (B1, B2, B3, D, or E).

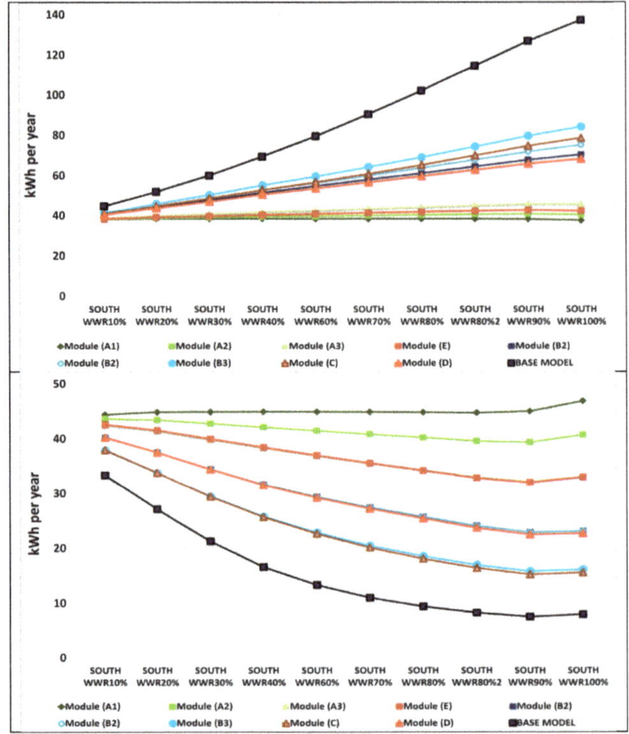

Figure 13. Annual cooling energy consumption of the BIPV modules compared to the base model on the South façade.

Inversely, the base-model consumed less heating energy than all PV modules in the cardinal orientations. Further, the rate of heating energy for the double-glazed PV modules was mostly constant, even with an increase in WWR. The heating energy consumption frequently decreased, except in the north façade due to the absence of solar transmission. The lowest value of the heating energy loads in the south facade, particularly in those with a large WWR, was due to the augmentation of the heat gain. Even though the east and west facades had the same performance, the west facade was slightly higher, with a large WWR. However, the peak heating energy consumption was achieved by the double-glazed PV modules (A1, A2, C, and A3), primarily due to their high insulation. Therefore, the energy savings of all PV modules were negative because the base-model had higher solar transmittance (SHGC). Refer to the Figures A2 and A3 for graphs of the other orientations.

3.5. Evaluation of the Optimum Overall Energy Consumption (OEC)

The Overall Energy Consumption (OEC) of the base-model trend highlights that, within a semi-arid climate, employing bigger windows is counterproductive, as large windows create bigger areas for heat transfer in winter and in numerous days during spring and autumn. A larger cooling demand with bigger south-facing window sizes, as well as increasing the WWR for north-facing windows, results in a lower heating demand. It is also observed that smaller window sizes cannot be reduced randomly because electric lighting consumption is an issue, predominantly with 10% WWR, as shown in Figure 14. The highest OEC of the base-model was observed at the eastern and western façades. This result agrees with the findings of [30] which is the only study conducted on the overall energy performance of a typical office. The lowest energy use of the base-model was identified at 20% WWR for cardinal alignments. At the same time, the solutions with the lowest total OEC were found to have the highest percentage of visual comfort or UDI300-750 lux, unless the east facade did not meet the minimum requirement of the UDI. For the base-model with medium and large window sizes, the use of artificial light was found to be mostly insignificant, but the risk of glare was extremely high and exceeded the shaded area.

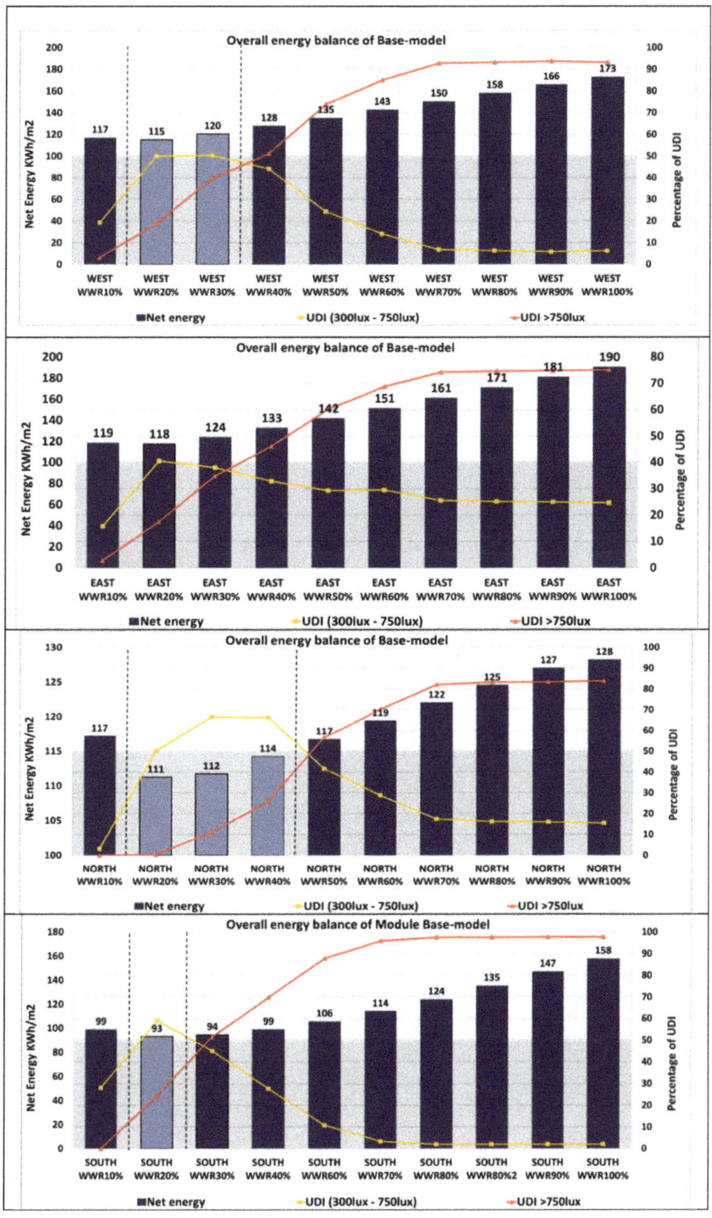

Figure 14. The optimum window design for the base-model in a cardinal orientation, west, east, north, and south.

As shown in the Figure 15, the PV Modules with 10% VLT or less indicate that optimising the WWR against the OEC hindered visual comfort. The PV modules A1, B1, C, D, and E that were optimised to reduce the OEC with respect to increasing the WWR, did not meet the visual comfort criteria because of the low light transmittance of these PV modules. Consequently, the office setting was controlled by high electric lighting use. At the same time, the PV modules with between 20% and 30% VLT (A2, A3, B2, and B3) showed a reduction in the OEC from a small to a medium WWR, and, between 70% and 100%, the OEC increased again, thereby overcoming the energy output against the total heating load and lighting energy, as shown in Figures A4–A6.

The results demonstrate that the OEC of the PV modules, compared with the base-model, is inverse for the WWR. The trend of the PV modules with 10% VLT and less A1, B1, C, D, and E reduced significantly by increasing the WWR due to the increment of the energy output. In contrast, the OEC of the base-model increased after 20% of WWR, due to the increment of the cooling load with no energy output. Moreover, the optimum design solutions limited the PV modules and WWR for this particular type of climate. In this study, only four PV modules among the nine could meet the targets of both visual comfort in the cardinal orientation and the specific WWR. The main reason for selecting these four PV modules (A2, A3, B2, and B3) was due to the required degree of transparency for the VLT to achieve the minimum requirement of visual comfort.

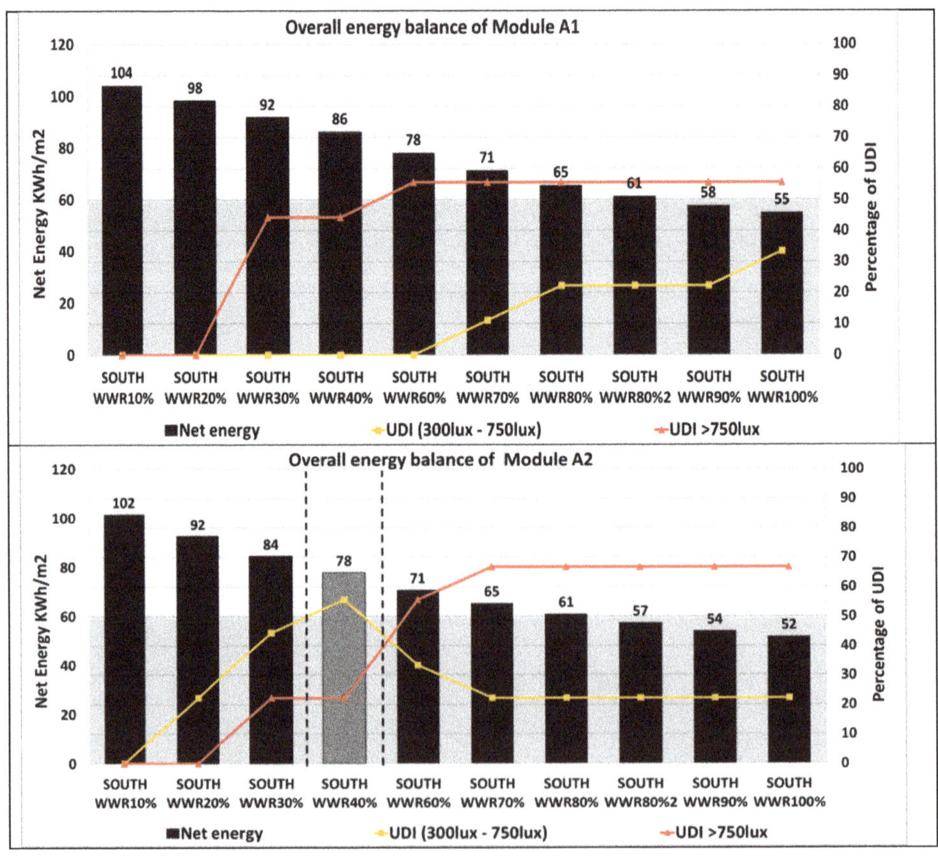

Figure 15. *Cont.*

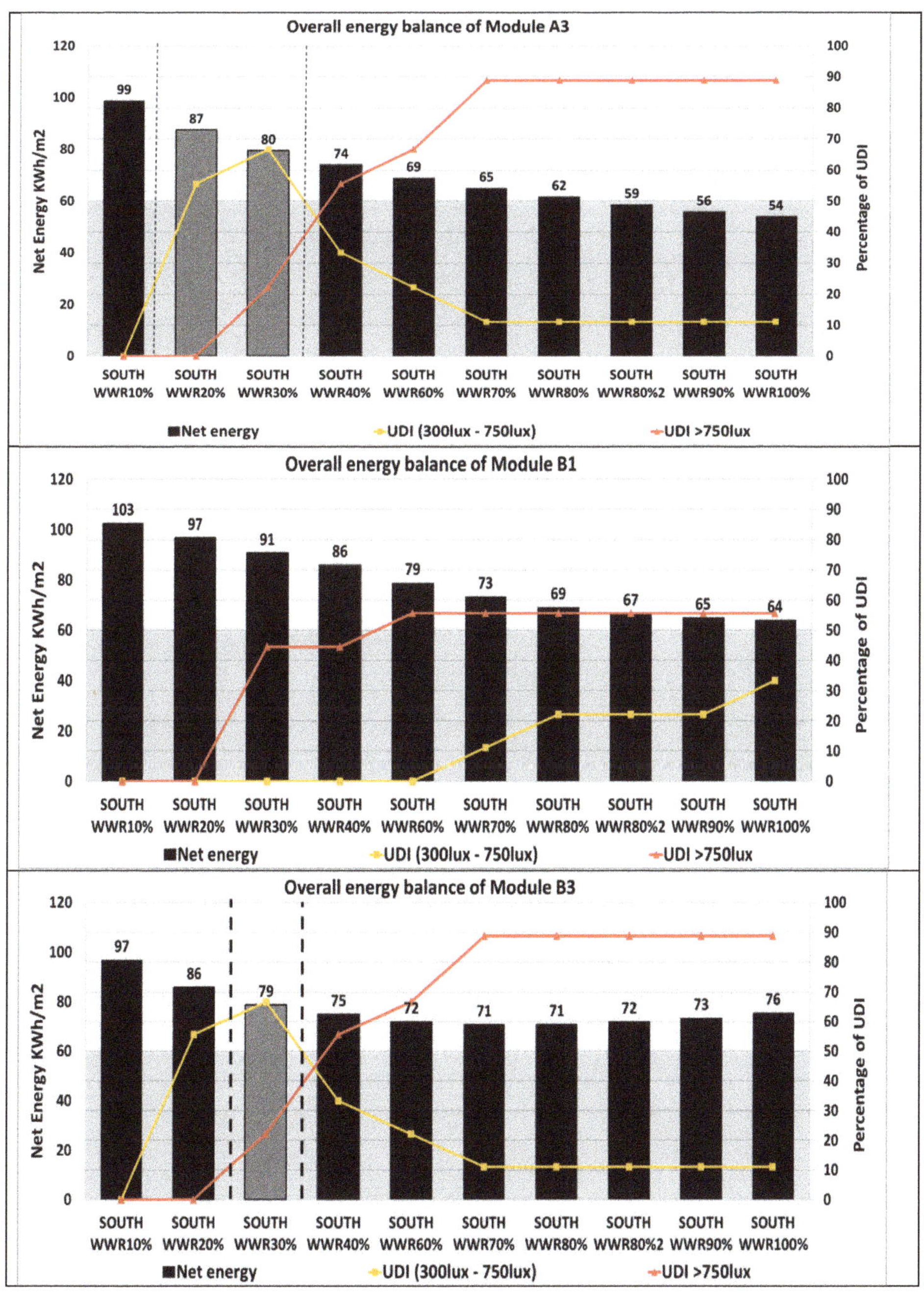

Figure 15. *Cont.*

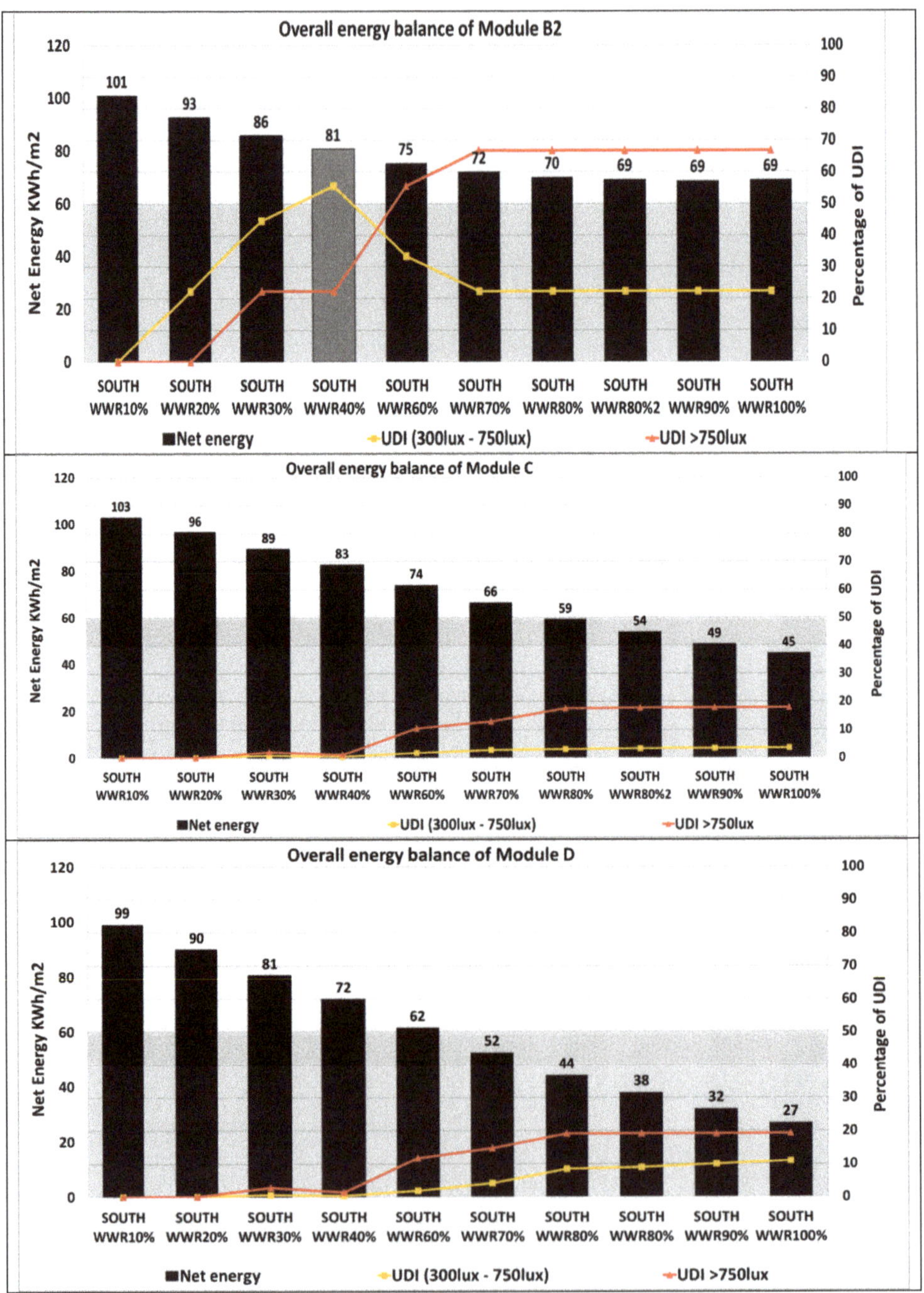

Figure 15. *Cont.*

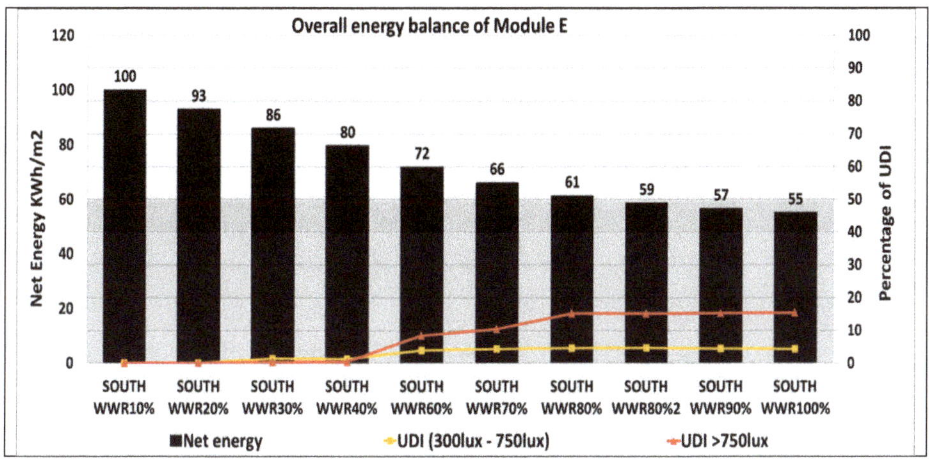

Figure 15. The optimum BIPV window design in the southern orientation.

3.6. Evaluation of Visual Comfort

To acquire more details about the quality of visual discomfort, specifically the uncomfortable glare issue that may be caused by the base-model or different configurations of PV modules, this study utilised CEI Glare Index (CGI) metrics to assess the glare status for each case, in addition to graphical presentations. Tables 6 and 7 indicate that the means of the CGI values of the base-model varied from 18.85 to 28.23 during the studied time and cardinal orientations; the means of CGI only obtained acceptable values during the winter solstice, while the remaining design days provided uncomfortable values. On the other hand, the mean CGI for the optimum PV modules selected in this research ranges from 13.48 to 24.2. As consequence, these results precipitate a sharp decrease in the mean CGI values throughout the year compared to the base model. The PV module only exceeds the limit of 22 in an east orientation during the morning period, when the office is exposed directly to sunlight. Thus, the base-model aggravates this condition, since the average CGI values in all PV modules in each orientation are lower than the CGI values for the base model because of the large differences in terms of visible transparency (VLT). In all cases, it is remarkable that there is a significant improvement in terms of visual comfort by reducing the means of the CEI glare index by at least 3.5 degrees. The means of CGI are barely perceptible in the south orientation, however, with 40% WWR. As result, the use of an optimally designed PV module strategy in the cardinal orientation could provide a significant reduction in glare.

Table 6. The International Commission on Illumination CIE glare index of the optimum PV modules in the cardinal orientation compared to the base-model during summer solstice, winter, and spring equinox.

Orientation		East		South		West	
Date & Time	WWR + VLT	Base model WWR 20%	PV module A3 WWR 50%	Base model WWR20%	PV module A2 WWR 40%	Base model WWR20%	PV module A3 WWR 60%
Summer solstice	9.00am	27.79	23.49	23.69	18.69	24.13	19.64
21 June	15.00 pm	23.95	19.98	23.87	18.45	26.18	21.88
Spring equinox	9.00 am	28.23	24.20	24.02	18.94	23.89	19.46
21 March	15.00 pm	23.83	20.66	25.05	19.82	26.14	21.94
Winter Solstice 21	9.00 am	18.85	15.02	18.85	13.48	18.85	14.84
December	15.00 pm	20.33	16.74	20.33	14.88	20.33	16.31

Table 7. Improvement in the Mean CEI Glare Index evaluation for optimum PV module designs.

Date Time		East	South	West
Summer solstice 21 June	9.00 am	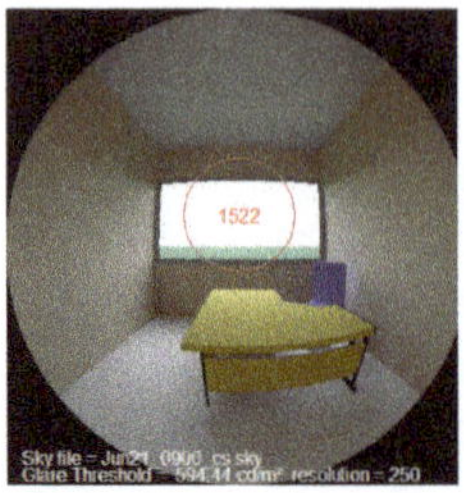 1522 Sky file = Jun21_0900_cs.sky Glare Threshold = 594.44 cd/m² resolution = 250	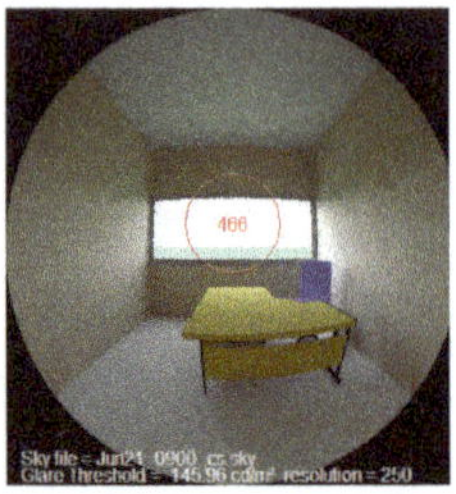 466 Sky file = Jun21_0900_cs.sky Glare Threshold = 145.96 cd/m² resolution = 250	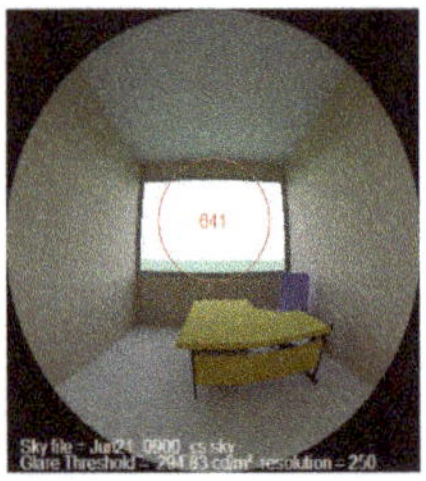 641 Sky file = Jun21_0900_cs.sky Glare Threshold = 794.83 cd/m² resolution = 250
	15.00 pm	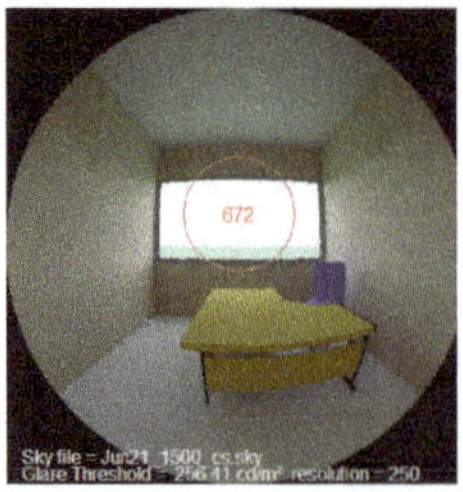 672 Sky file = Jun21_1500_cs.sky Glare Threshold = 256.41 cd/m² resolution = 250	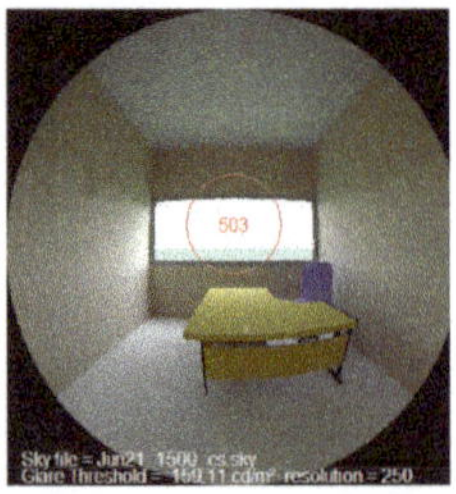 503 Sky file = Jun21_1500_cs.sky Glare Threshold = 169.11 cd/m² resolution = 250	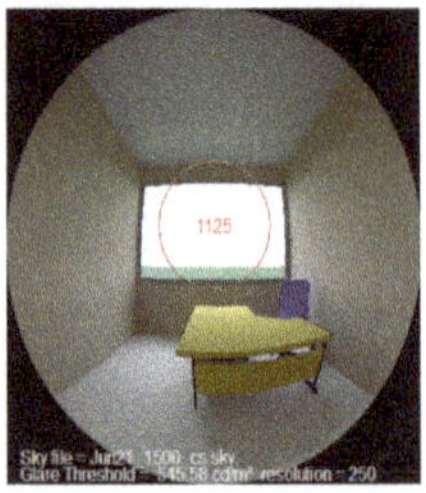 1125 Sky file = Jun21_1500_cs.sky Glare Threshold = 645.58 cd/m² resolution = 250
Spring equinox 21 March	9.00 am	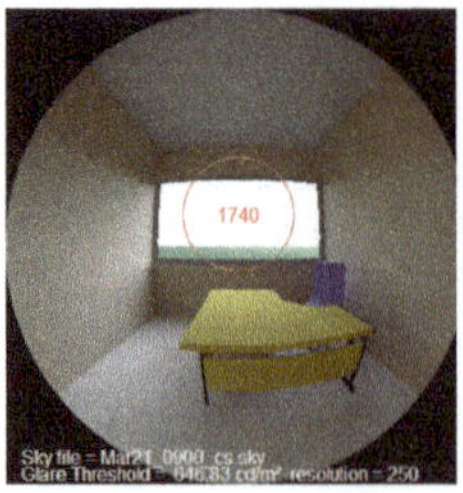 1740 Sky file = Mar21_0900_cs.sky Glare Threshold = 646.83 cd/m² resolution = 250	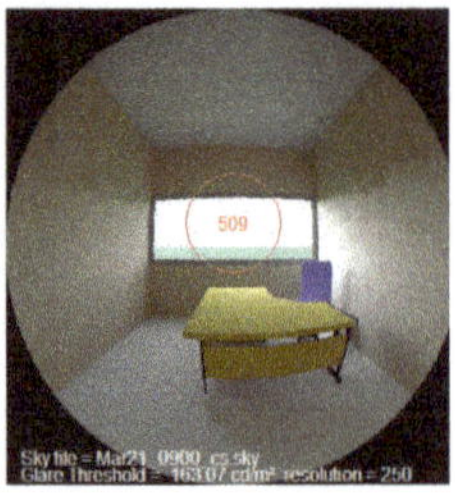 509 Sky file = Mar21_0900_cs.sky Glare Threshold = 163.07 cd/m² resolution = 250	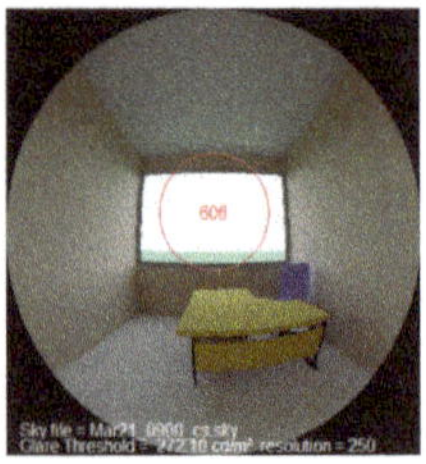 606 Sky file = Mar21_0900_cs.sky Glare Threshold = 272.18 cd/m² resolution = 250
	15.00 pm	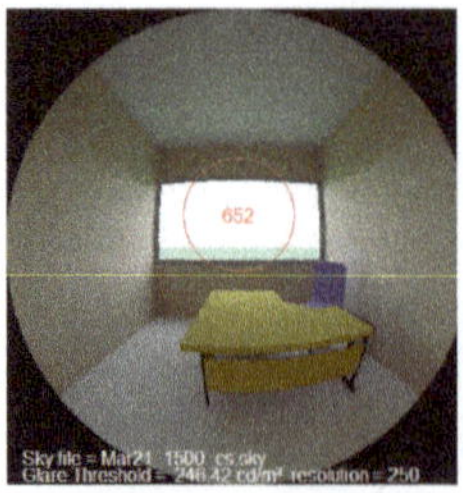 652 Sky file = Mar21_1500_cs.sky Glare Threshold = 246.42 cd/m² resolution = 250	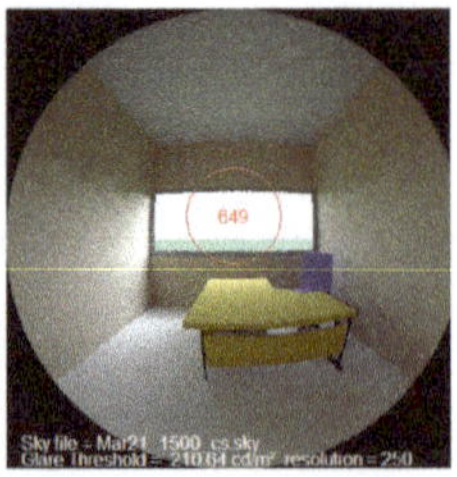 649 Sky file = Mar21_1500_cs.sky Glare Threshold = 210.64 cd/m² resolution = 250	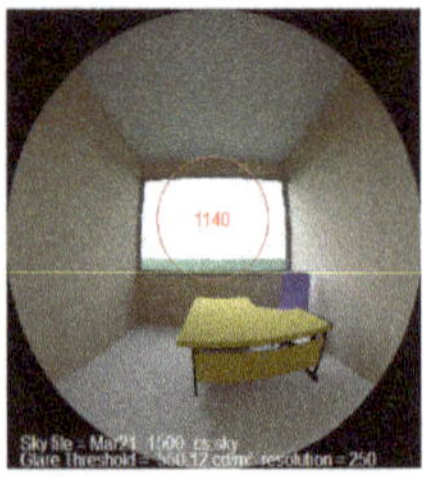 1140 Sky file = Mar21_1500_cs.sky Glare Threshold = 560.17 cd/m² resolution = 250
Winter solstice 22 December	9.00am	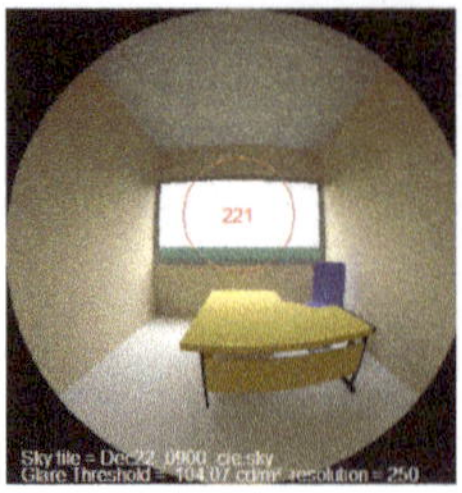 221 Sky file = Dec22_0900_cie.sky Glare Threshold = 104.07 cd/m² resolution = 250	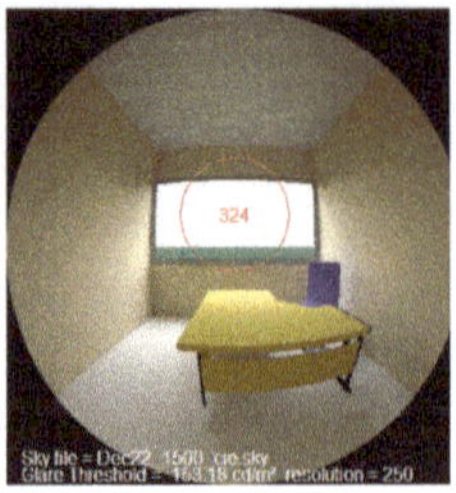 324 Sky file = Dec22_1500_cie.sky Glare Threshold = 163.18 cd/m² resolution = 250	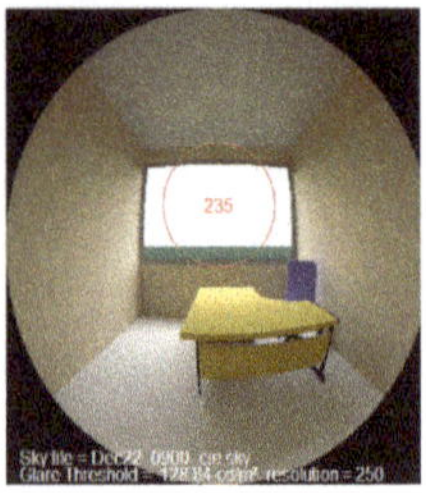 235 Sky file = Dec22_0900_cie.sky Glare Threshold = 128.84 cd/m² resolution = 250

Table 7. *Cont.*

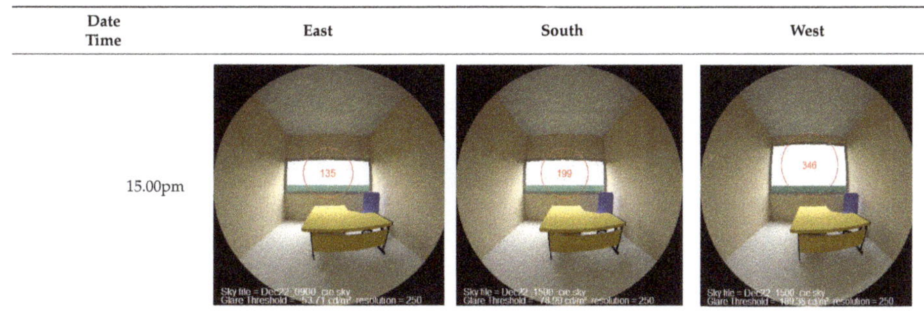

Date Time	East	South	West
15.00pm			

3.7. Energy Saving of the BIPV Window Modules Compared to the Base Model

The largest potential percentage savings that can be attained by accepting nine various PV modules in place of the more commonly employed clear glazing (see the base model) in cardinal alignments can be seen in Figure 16. These graphs showcase inconsistent savings in an approximate range of 1.29% to 60%, and, in some cases, the result was negative (no-savings) compared to the base model. A significant percentage of savings were achieved in the southern orientation by using module D, whose energy savings were estimated to be 60%, due to it having the highest conversion efficiency (n = 8) among the modules. Conversely, module D had a negative percentage in the northern orientation.

This result indicates that the conversion efficiency is not significant, due to the absence of solar radiance. The maximum savings percentage was only achieved by double-glazing the PV modules (A1, A2, A3, C). The remaining modules were all negative, and the results demonstrate that thermal performance was more important than conversion efficiency. Furthermore, the eastern and western orientations had approximately similar results, where PV modules B1, B2, and B3 presented a negative percentage mainly due to their weak thermal performance. This occurred because the conversion efficiency does not compensate for the high energy consumed by the cooling load, except in the southern orientation, with a WWR of 100%, which could be accomplished by 19.33%, 9.13%, and 14.02% energy savings. Consequently, even the low conversion efficiency of all the PV modules in a southern alignment is deemed to be comparatively more energy efficient than existing base model window technology.

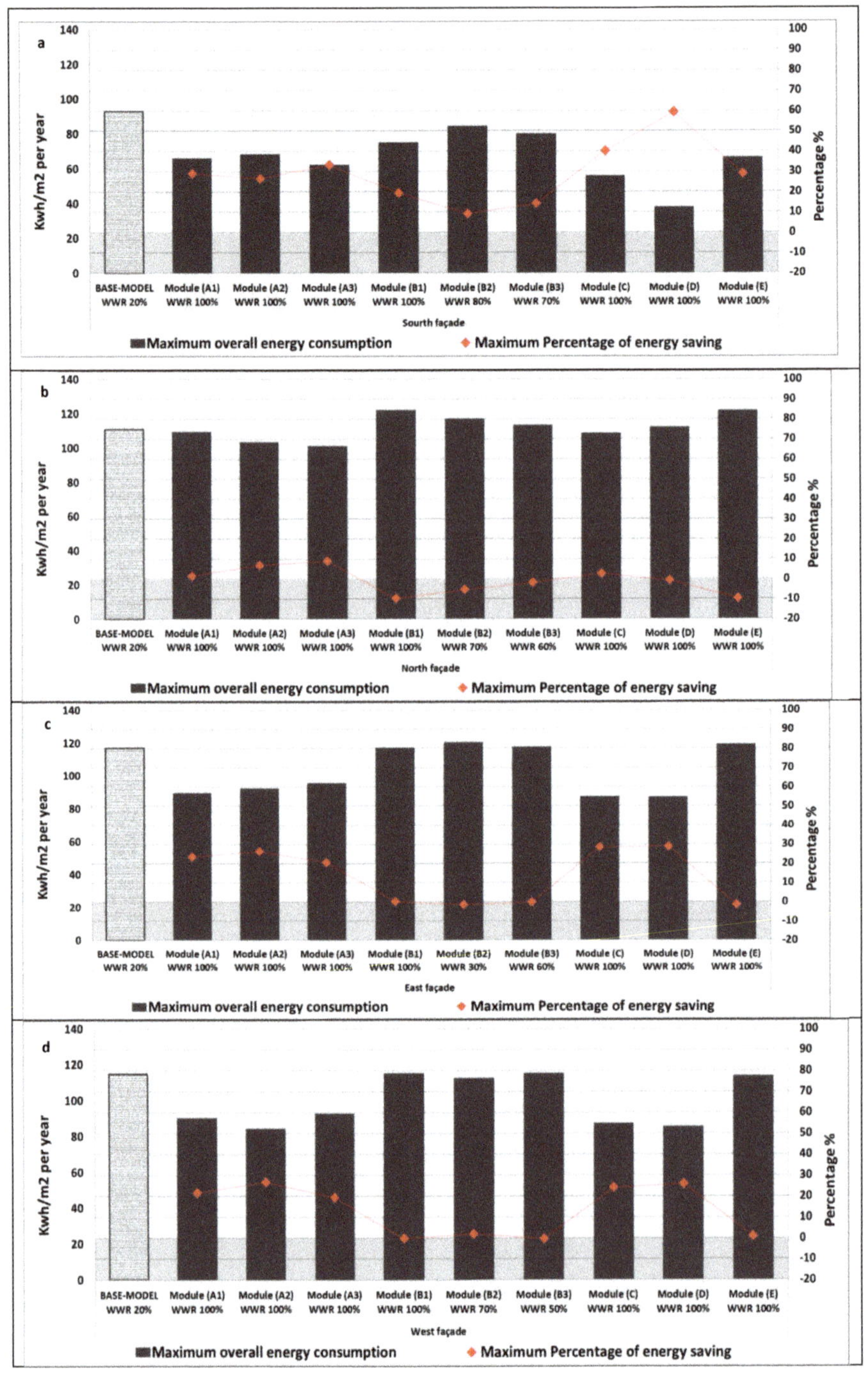

Figure 16. The maximum percentage of energy savings for BIPV modules compared with the base model in cardinal orientations: south, north, east, and west.

Figure 16 shows the lowest overall energy consumption of all PV modules and the base-model in the southern orientation compared to other orientations, demonstrating that the value of the overall energy consumption is less than 95 kWh/m2.yr for the base model, with 20% WWR. However, the PV module D reached 25 kWh/m2.yr, with 100% WWR, which is close to zero energy. The WWR of the PV modules achieved the highest percentage of energy savings with full PV glazing in most cases and orientations, except for PV modules B2 and B3 due to their U-values and SHGC being higher than those of the other PV modules.

In contrast, the overall energy savings of the optimum WWR of the PV modules were much lower than the maximum energy savings. Figure 17 shows that the energy savings ranged between 6% and 23%. Notably, these percentages fluctuate in cardinal orientations. For the western façade, the PV module (A3) with a WWR of 60% attained the highest percentage (23%) compared to the base-model, with a WWR of 20% considered to be the optimum WWR for the west, south, and north orientations. For the eastern orientation, the base-model could not meet the requirement of visual comfort. Therefore, the eastern façade base model cannot achieve the necessary target. Instead, two PV modules (A3, with a WWR of 50%, and B3, with a WWR of 50%) can replace the base model and act as a solution for the eastern façade. The northern orientation includes several alternate solutions. The common characteristics of these other PV modules include a peak transparency of 30%, with a large WWR ranging between 60% and 100%. The energy savings for the PV modules with double glazing (A3) ranged between 16.22% and 18.92%, while the energy savings of the single-glazed PV module (B3) were lower than those of the PV module (A3) by at last two-fold. Inversely, the southern orientation energy savings meet the required targets by using small WWR values of 20% and 30% due to the risk of glare and thermal discomfort in a façade with a large opening.

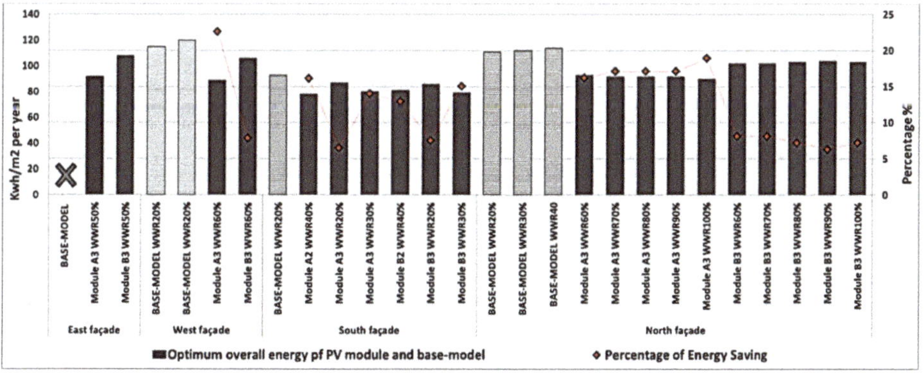

Figure 17. The optimum WWR of the PV modules and base-model in a cardinal orientation and its energy savings against the optimum base model.

Ultimately, the positive effects of the overall energy savings in the cardinal orientations within the studied semi-arid climate are consistent with past research results in different climates, although the percentages varied in every context due to the variety of configurations for the PV modules. For instance, the energy savings in Spain, a Mediterranean climate, comprise up to 59% of energy used. Meanwhile, in Japan, 55% of all energy is saved using a solar cell transmittance of 40% in comparison to a single-glazed façade (Wong et al., 2008). In Singapore and Brazil, with their tropical climates, energy is saved by 16.7% to 41.3% [42]. A recent study identified that the most energy saved by applying a CdTe-based PV module within an Indian climate was 60.4% [42].

4. Design Recommendations

The results from this research allow us to suggest the following design recommendations for the usage of BIPV windows technology in office buildings in the semi-arid region in Algeria:

i. In general, the adoption of BIPV window modules has a positive impact on the overall energy saving in an office building. however, care must be taken to select the adequate properties of PV modules in cardinal orientations;

-   North orientation: This orientation is not recommended for use in BIPV window applications. In this case, it is necessary to use a double-glazed window to overcome the thermal discomfort issue;
-   South orientation: This orientation is highly recommended, particularly for PV modules with high conversion efficiency;
-   East and west orientation: The application of a BIPV window is acceptable in these orientations, since both orientations produce almost the same results (higher than 70% of yearly energy output;

ii. The application of daylight control strategies with effective solar transparency and WWR is highly recommended;

iii. A lateral typology for a typical office building should be used to achieve optimal distribution and an adequate daylight uniformity of > 0.6;

iv. Based on the results of the base-model and BIPV window modules, the east–west axis was shown to consume higher overall energy than the south–north axis. Therefore, apart from directing the office buildings toward the south–north axis, vertical or horizontal louvers are also suggested for use with the east and west facades of the building;

v. As shown from this research, the optimum design of 20% WWR for the base model was found to produce the greatest energy savings and provided sufficient daylight for office buildings in cardinal orientations, unless the north facade adopted 30% of the WWR to provide sufficient visual comfort;

vi. Generally, it is recommended to use double-glazed PV modules rather than single-glazed PV modules in cardinal orientations;

vii. In this research, the optimum design of various BIPV windows is different for each orientation level as presented in Figure 18. The recommended orientations are as given as follows:

(a) For the East façade: PV modules (B3) with a WWR of 50%;
(b) For the South façade: PV modules (B2) with a WWR of 40%;
(c) For the West façade: PV modules (B3) with a WWR of 60%;
(d) For the North façade: PV modules (B3) with a WWR of 100%;

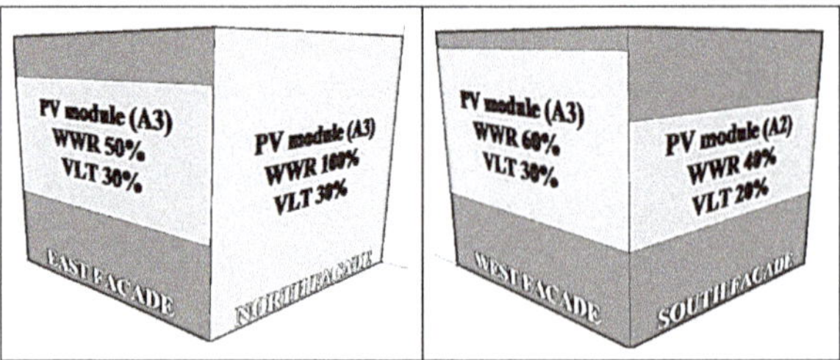

Figure 18. Optimum design for BIPV windows (Window Wall Ratios (WWR), Visible Light Transmittance (VLT), and conversion efficiency) in each cardinal orientation for office building in vertical facades.

viii.   It is recommended to use BIPV window modules with 10% VLT for locations that do not require visual comfort, such as archival rooms or resting areas;

ix.   As depicted in the 3D model in Figure 19 below, architects can use the output percentages obtained from the various tilt angles and orientations of the BIPV window in the early stages of design.

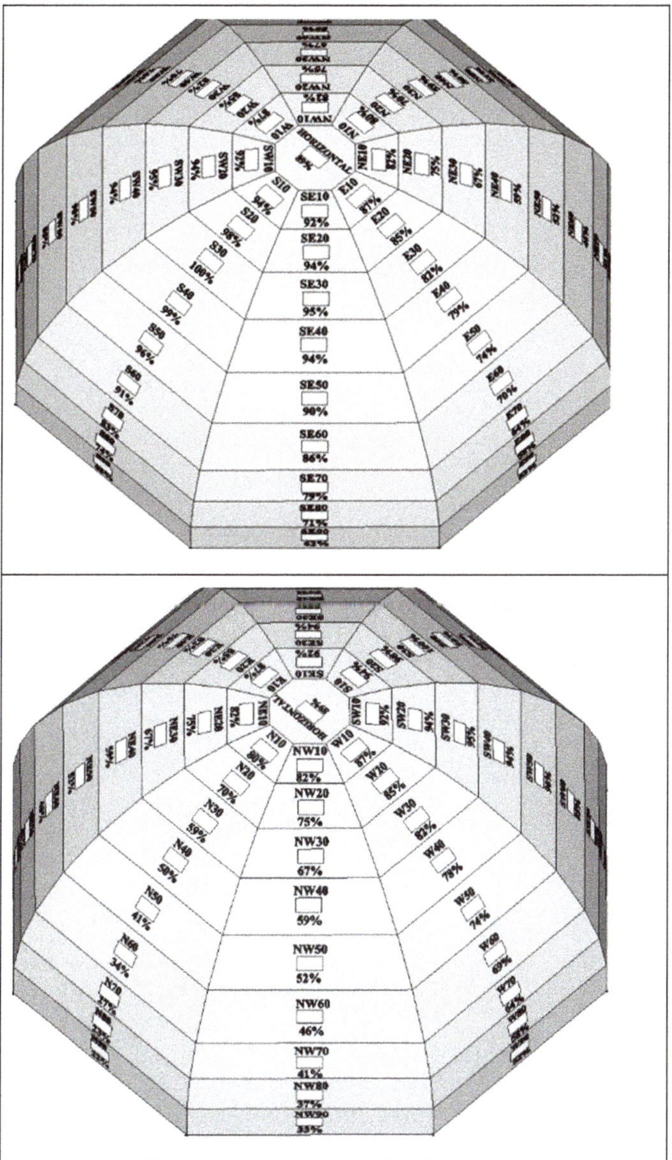

Figure 19. A 3D model guideline for the energy output percentages obtained from various tilt angles and orientations for architects and designers.

5. Conclusions

Apart from demonstrating the importance of appropriate BIPV window designs in the realization of Zero Energy Building, this study has also shown BIPV to be an energy efficient lighting design strategy that enhances the visual comfort of offices with windows. Policymakers and architects can also exploit the results of this research to retrofit conventional building designs with BIPV windows as well as in the implementation of new building designs in the semi-arid regions. However, further studies may address the inter-correlation of environmental factors and design studies to obtain more precise measurements on the BIPV windows' overall energy performance, as well as evaluate the return of investments (ROI) for BIPV on new buildings and its impact on Algeria's economy. In the end, this study contributes to better sustainable design research and practice and suggests the effective usage of BIPV windows in a cardinal orientation. Window size and various optical and thermal BIPV window configuration strategies should be included in the guidelines, with special reference to the Algerian climate, to maximize the energy savings by up to 23%. Meanwhile, this method provides visual comfort and helps prevent damage to the environment by significantly reducing CO_2 emissions and pollution.

Author Contributions: A.M. conceived the study, performed all measurements and simulations, interpreted the results, and wrote the manuscript. M.Z.K. supervised and helped each stage of the project in terms of data collection, analysis, and interpretation. G.A.A. Albaqawy interpreted, revised and finalized the manuscript. All authors have read and agreed to the published version of the manuscript.

Funding: This research received no external funding.

Acknowledgments: I would like to thanks Mesloub said and Mohd zin kandar for their extraordinary support to complete this project. Also the authors would thank the editor, the managing editor, the academic editor, and three anonymous reviewers for their helpful comments which greatly improved earlier versions of the manuscript.

Conflicts of Interest: The authors declare no conflict of interest.

Appendix A

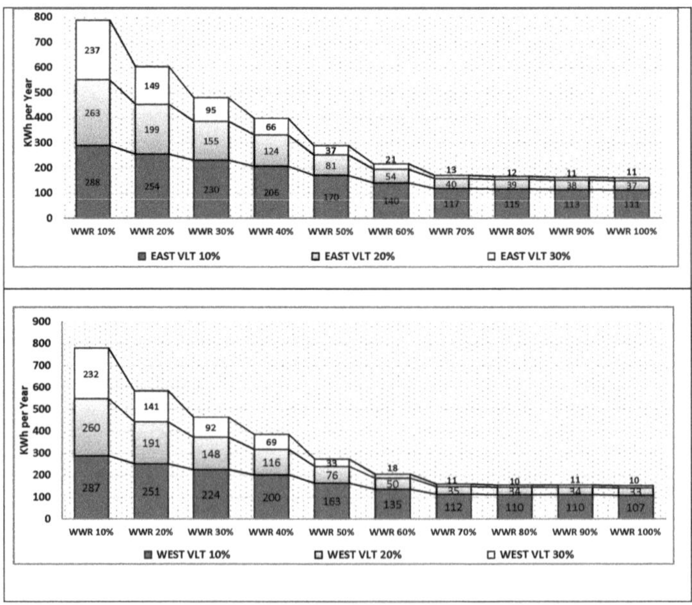

Figure A1. *Cont.*

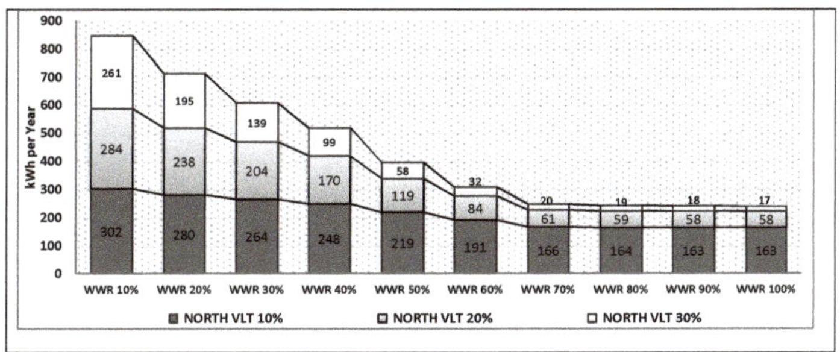

Figure A1. Lighting energy consumption when using different VLT values for the PV modules in cardinal orientations: east, west and north, respectively.

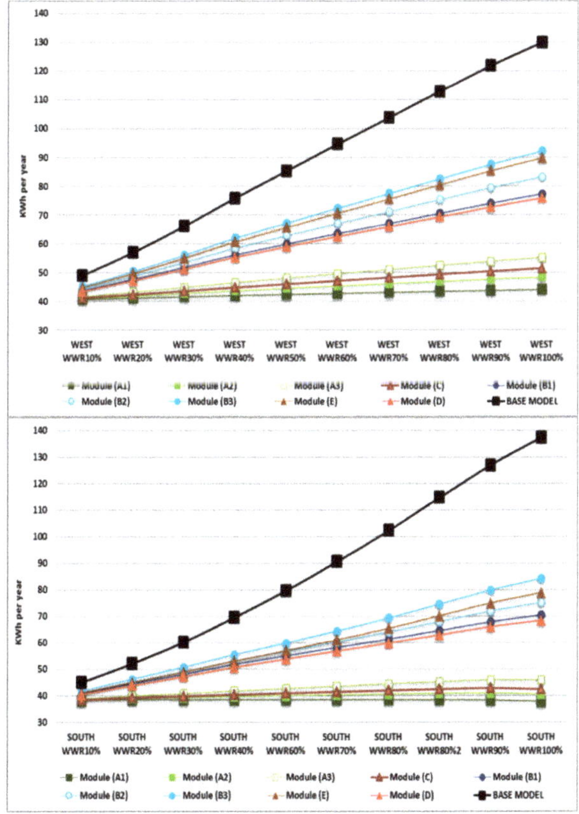

Figure A2. *Cont.*

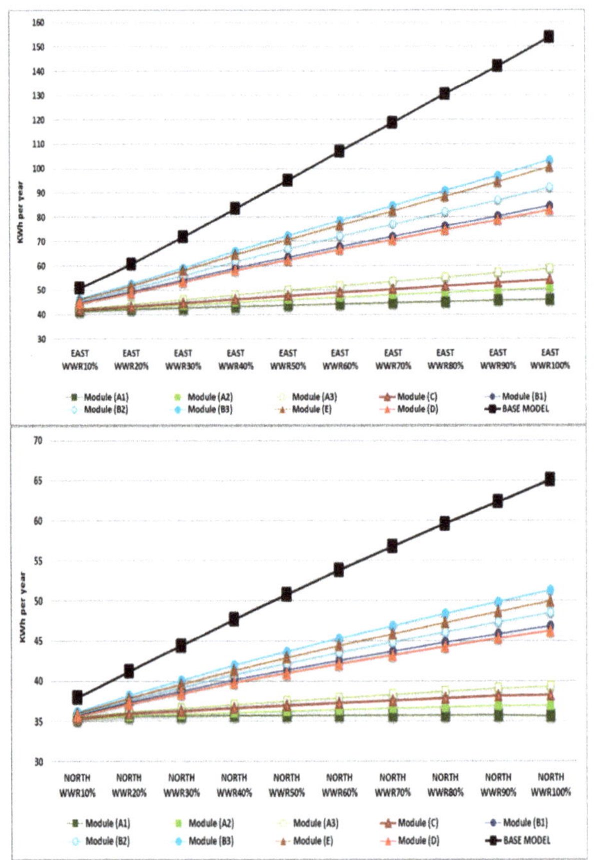

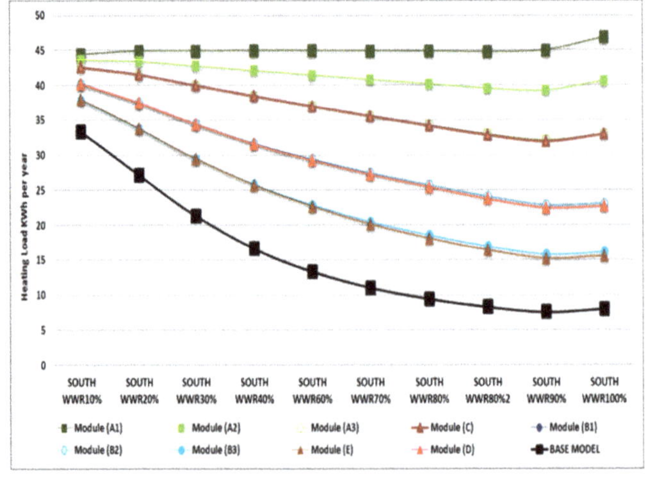

Figure A2. Annual cooling energy consumption of the BIPV modules against the base model.

Figure A3. *Cont.*

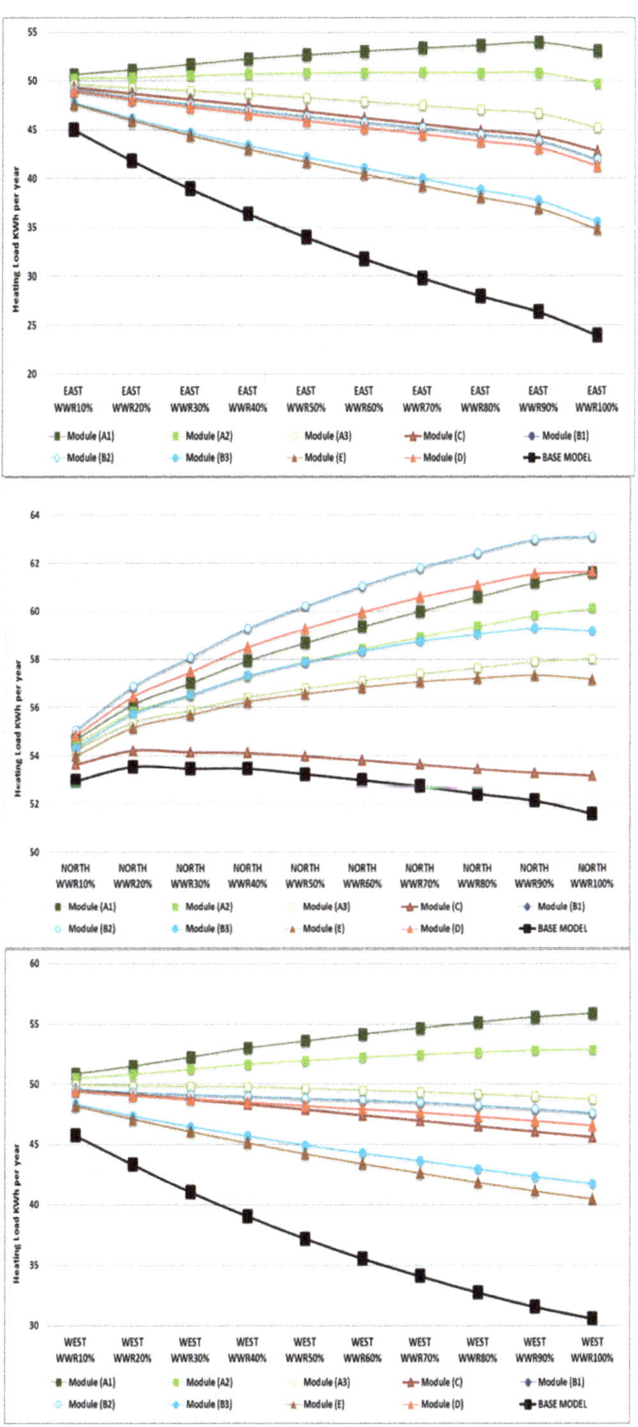

Figure A3. Annual heating energy consumption of the BIPV modules against the base model.

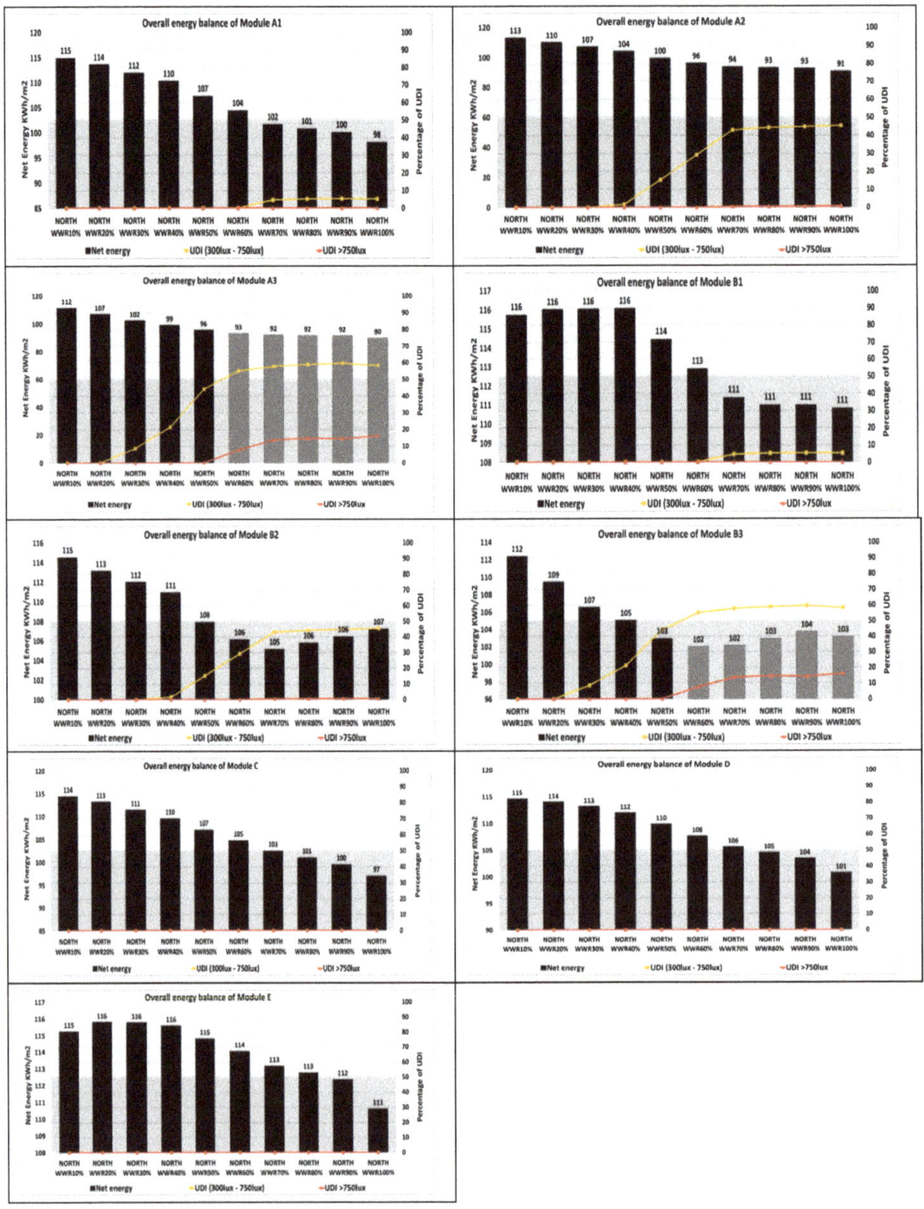

Figure A4. Overall energy performance and visual comfort of the PV modules to achieve the optimum design in the northern façade.

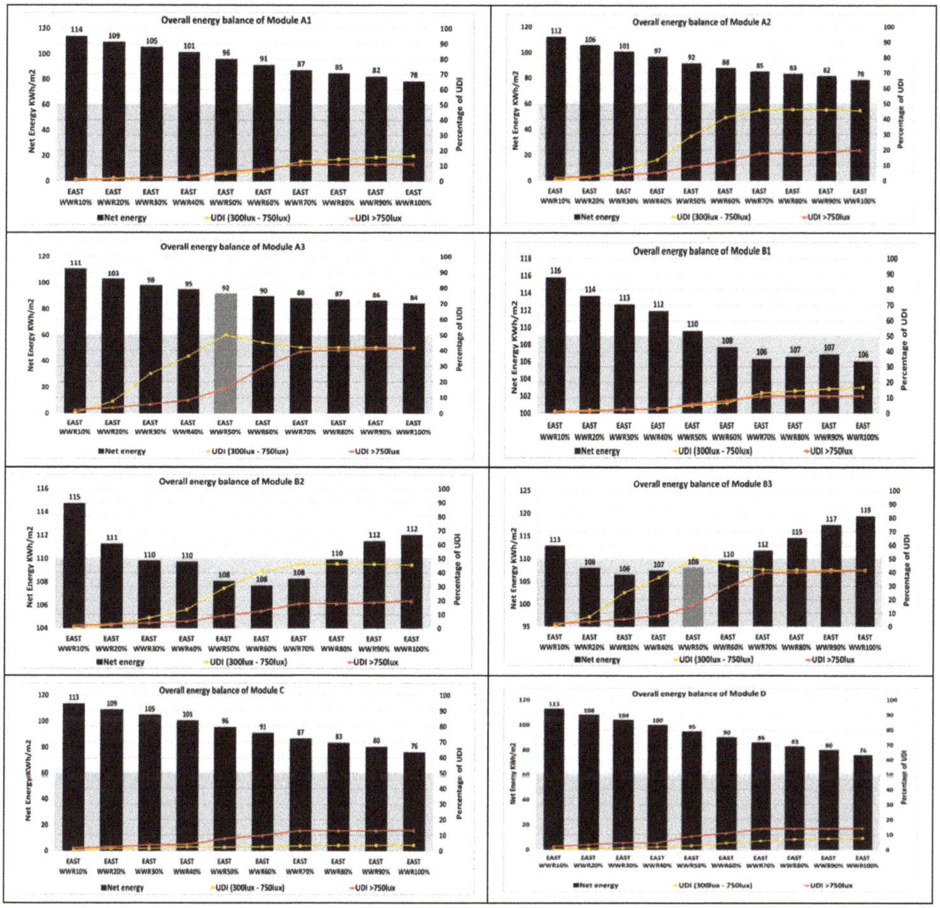

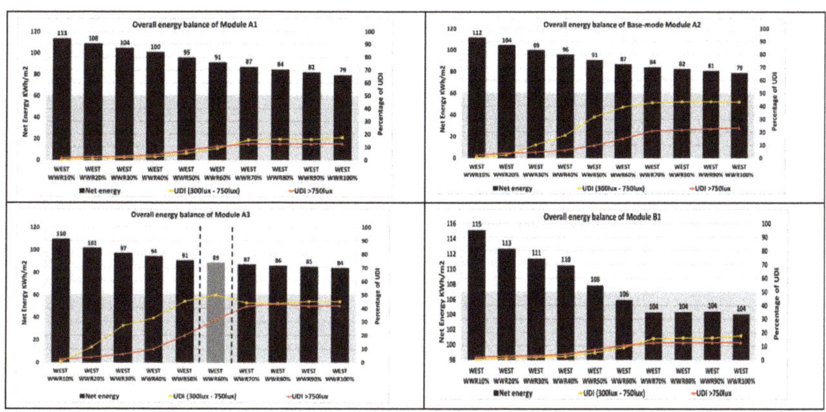

Figure A5. Overall energy performance and visual comfort of the PV modules to achieve the optimum design in the eastern façade.

Figure A6. *Cont.*

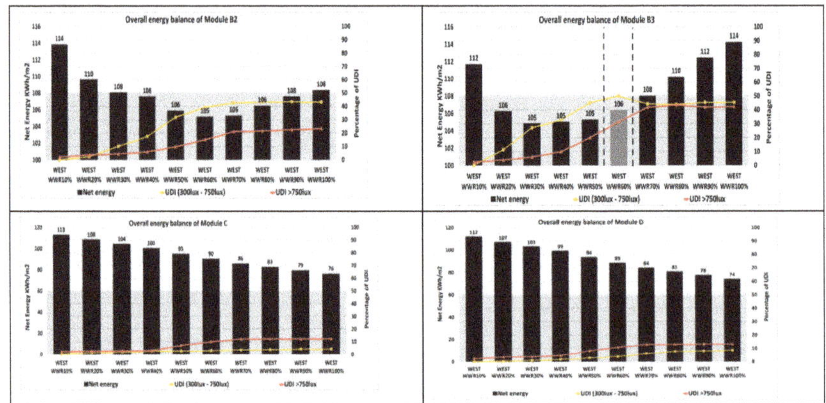

Figure A6. Overall energy performance and visual comfort of the PV modules to achieve the optimum design in the western façade.

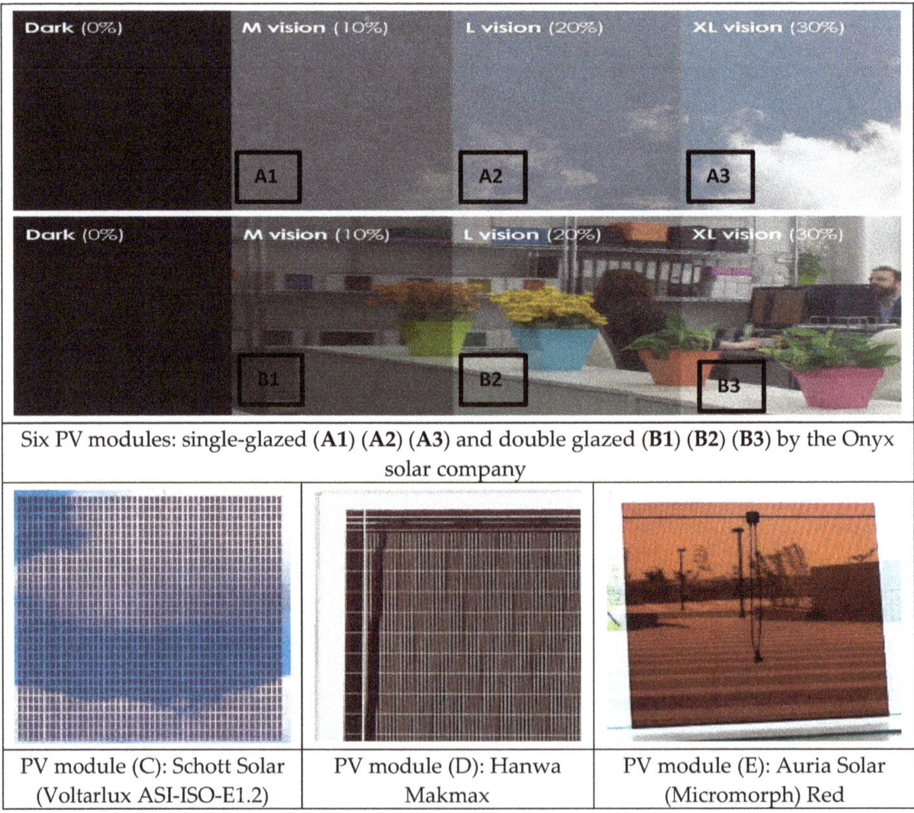

Six PV modules: single-glazed (**A1**) (**A2**) (**A3**) and double glazed (**B1**) (**B2**) (**B3**) by the Onyx solar company		
PV module (C): Schott Solar (Voltarlux ASI-ISO-E1.2)	PV module (D): Hanwa Makmax	PV module (E): Auria Solar (Micromorph) Red

Figure A7. The PV modules used in the simulations.

References

1. Peng, C.; Huang, Y.; Wu, Z. Building-integrated photovoltaics (BIPV) in architectural design in China. *Energy Build.* **2011**, *43*, 3592–3598. [CrossRef]
2. Stambouli, A.B.; Khiat, Z.; Flazi, S.; Kitamura, Y. A review on the renewable energy development in Algeria: Current perspective, energy scenario and sustainability issues. *Renew. Sustain. Energy Rev.* **2012**, *16*, 4445–4460. [CrossRef]
3. Hui, S.C. *From Renewable Energy to Sustainability: The Challenge for Hong Kong*; Hong Kong Institution of Engineers: Hong Kong, China, 1997; pp. 351–358.
4. Ferhat, S.; Boutrahi, M. The residential built heritage of Algiers and its energy behaviour: Case study of the building number 3 flats of the Aero-habitat district. *Wit Trans. Ecol. Environ.* **2014**, *186*, 39–53.
5. Belakehal, A.; Bennadji, A.; Aoul, K. Office Buildings Daylighting Design in Hot Arid Regions: Forms, Codes and Occupants' Point of View. In Proceedings of the International Engineering Conference on Hot Arid Regions (IECHAR 2010), Al-Ahsa, Saudi Arabia, 1 March 2010.
6. Amirat, M.; El Hassar, S. Economies d'energie dans le secteur de l'habitat consommation electrique des menages: Cas d'un foyer algérien typique en periode d'hiver. *Rev. Des. Énergies Renouvelables* **2005**, *8*, 27–37.
7. Himri, Y.; Malik, A.S.; Stambouli, A.B.; Himri, S.; Draoui, B. Review and use of the Algerian renewable energy for sustainable development. *Renew. Sustain. Energy Rev.* **2009**, *13*, 1584–1591. [CrossRef]
8. Li, D.H.; Lam, T.N.; Chan, W.W.; Mak, A.H. Energy and cost analysis of semi-transparent photovoltaic in office buildings. *Appl. Energy* **2009**, *86*, 722–729. [CrossRef]
9. Yoon, J.-H.; Song, J.; Lee, S.-J. Practical application of building integrated photovoltaic (BIPV) system using transparent amorphous silicon thin-film PV module. *Sol. Energy* **2011**, *85*, 723–733. [CrossRef]
10. Lu, L.; Law, K.M. Overall energy performance of semi-transparent single-glazed photovoltaic (PV) window for a typical office in Hong Kong. *Renew. Energy* **2013**, *49*, 250–254. [CrossRef]
11. Zhang, W.; Lu, L.; Peng, J.; Song, A. Comparison of the overall energy performance of semi-transparent photovoltaic windows and common energy-efficient windows in Hong Kong. *Energy Build.* **2016**, *128*, 511–518. [CrossRef]
12. Siong, C.T.; Janssen, P. Semi-Transparent Building Integrated Photovoltaic Facades–Maximise Energy Savings Using Evolutionary Multi-Objective Optimisation. In Proceedings of the 18th International Conference on Computer-Aided Architectural Design Research in Asia (CAADRIA 2013), Singapore, 15–18 May 2013; pp. 127–136.
13. Kapsis, K.; Dermardiros, V.; Athienitis, A. Daylight performance of perimeter office façades utilizing semi-transparent photovoltaic windows: A simulation study. *Energy Procedia* **2015**, *78*, 334–339.
14. Jakubiec, J.; Reinhart, C. DIVA-FOR-RHINO 2.0: Environmental parametric modeling in Rhinoceros/Grasshopper using RADIANCE, Daysim and EnergyPlus. In Proceedings of the Conference of Building Simulation, Sedney, Australia, 14–16 November 2011.
15. Reinhart, C. *Tutorial on the Use of Daysim Simulations for Sustainable Design*; Harvard Design School: Ottawa, ON, Canada, 2011.
16. López, C.S.P.; Sangiorgi, M. Comparison assessment of BIPV façade semi-transparent modules: Further insights on human comfort conditions. *Energy Procedia* **2014**, *48*, 1419–1428. [CrossRef]
17. Miyazaki, T.; Akisawa, A.; Kashiwagi, T. Energy savings of office buildings by the use of semi-transparent solar cells for windows. *Renew. Energy* **2005**, *30*, 281–304. [CrossRef]
18. Wong, P.W.; Shimoda, Y.; Nonaka, M.; Inoue, M.; Mizuno, M. Semi-transparent PV: Thermal performance, power generation, daylight modeling and energy saving potential in a residential application. *Renew. Energy* **2008**, *33*, 1024–1036. [CrossRef]
19. Mondol, J.D.; Yohanis, Y.G.; Norton, B. The impact of array inclination and orientation on the performance of a grid-connected photovoltaic system. *Renew. Energy* **2007**, *32*, 118–140. [CrossRef]
20. Haysom, J.E.; Hinzer, K.; Wright, D. Impact of electricity tariffs on optimal orientation of photovoltaic modules. *Prog. Photovolt. Res. Appl.* **2016**, *24*, 253–260. [CrossRef]
21. Chen, W.; Shen, H.; Liu, Y. Performance evaluation of PV arrays at different tilt angles and orientations in BIPV. Taiyangneng Xuebao. *Acta Energ. Sol. Sin.* **2009**, *30*, 206–210.
22. Yang, J.-H.; Mao, J.-J.; Chen, Z.-H. Calculation of solar radiation on variously oriented tilted surface and optimum tilt angle. *J. Shanghai Jiaotong Univ. Chin. Ed.* **2002**, *36*, 1032–1036.

23. Ng, P.K.; Mithraratne, N.; Kua, H.W. Energy analysis of semi-transparent BIPV in Singapore buildings. *Energy Build.* **2013**, *66*, 274–281. [CrossRef]
24. Wang, M.; Peng, J.; Li, N.; Yang, H.; Wang, C.; Li, X.; Lu, T. Comparison of energy performance between PV double skin facades and PV insulating glass units. *Appl. Energy* **2017**, *194*, 148–160. [CrossRef]
25. An, H.J.; Yoon, J.H.; An, Y.S.; Heo, E. Heating and cooling performance of office buildings with a-Si BIPV windows considering operating conditions in temperate climates: The case of Korea. *Sustainability* **2018**, *10*, 4856. [CrossRef]
26. Martinopoulos, G.; Serasidou, A.; Antoniadou, P.; Papadopoulos, A.M. Building integrated shading and building applied photovoltaic system assessment in the energy performance and thermal comfort of office buildings. *Sustainability* **2018**, *10*, 4670. [CrossRef]
27. Skandalos, N.; Tywoniak, J. Influence of PV facade configuration on the energy demand and visual comfort in office buildings. *J. Phys. Conf. Ser.* **2019**, *1343*, 012094. [CrossRef]
28. Ruiz, G.R.; Bandera, C.F. Validation of calibrated energy models: Common errors. *Energies* **2017**, *10*, 1587. [CrossRef]
29. Djamel, Z.; Noureddine, Z. The impact of window configuration on the overall building energy consumption under specific climate conditions. *Energy Procedia* **2017**, *115*, 162–172. [CrossRef]
30. Meteonorm. Handbook Part I and II. 2010. Available online: http://meteonorm.com/de/download (accessed on 18 November 2019).
31. Osman, M.M.; Alibaba, A.P.D.H.Z. Comparative Studies on Integration of Photovoltaic in Hot and Cold Climate. *Scientific Research Journal* **2015**, *3*, 48–60.
32. Peng, J.; Lu, L.; Yang, H.; Ma, T. Validation of the Sandia model with indoor and outdoor measurements for semi-transparent amorphous silicon PV modules. *Renew. Energy* **2015**, *80*, 316–323. [CrossRef]
33. Griffith, B.T.; Ellis, P.G. *Photovoltaic and Solar Thermal Modeling with the Energy Plus Calculation Engine*; National Renewable Energy Lab.: Golden, CO, USA, 2004.
34. Infield, D.; Eicker, U.; Fux, V.; Mei, L.; Schumacher, J. A simplified approach to thermal performance calculation for building integrated mechanically ventilated PV facades. *Build. Environ.* **2006**, *41*, 893–901. [CrossRef]
35. Fung, T.Y.; Yang, H. Study on thermal performance of semi-transparent building-integrated photovoltaic glazings. *Energy Build.* **2008**, *40*, 341–350. [CrossRef]
36. Lee, E.S.; DiBartolomeo, D.; Selkowitz, S. Daylighting control performance of a thin-film ceramic electrochromic window: Field study results. *Energy Build.* **2006**, *38*, 30–44. [CrossRef]
37. Reinhart, C.F.; Mardaljevic, J.; Rogers, Z. Dynamic daylight performance metrics for sustainable building design. *Leukos* **2006**, *3*, 7–31. [CrossRef]
38. Wah, W.P.; Shimoda, Y.; Nonaka, M.; Inoue, M.; Mizuno, M. Field study and modeling of semi-transparent PV in power, thermal and optical aspects. *J. Asian Archit. Build. Eng.* **2005**, *4*, 549–556. [CrossRef]
39. Olivieri, L.; Caamaño-Martin, E.; Olivieri, F.; Neila, J. Integral energy performance characterization of semi-transparent photovoltaic elements for building integration under real operation conditions. *Energy Build.* **2014**, *68*, 280–291. [CrossRef]
40. Salem, T.; Kinab, E. Analysis of Building-integrated Photovoltaic Systems: A Case Study of Commercial Buildings under Mediterranean Climate. *Procedia Eng.* **2015**, *118*, 538–545. [CrossRef]
41. Didoné, E.L.; Wagner, A. Semi-transparent PV windows: A study for office buildings in Brazil. *Energy Build.* **2013**, *67*, 136–142. [CrossRef]
42. Barman, S.; Chowdhury, A.; Mathur, S.; Mathur, J. Assessment of the efficiency of window integrated CdTe based semi-transparent photovoltaic module. *Sustain. Cities Soc.* **2018**, *37*, 250–262. [CrossRef]

Article

Optimal Placement and Sizing of Wind Generators in AC Grids Considering Reactive Power Capability and Wind Speed Curves

Walter Gil-González [1], Oscar Danilo Montoya [1,2], Luis Fernando Grisales-Noreña [3], Alberto-Jesus Perea-Moreno [4] and Quetzalcoatl Hernandez-Escobedo [5,*]

[1] Laboratorio Inteligente de Energía, Universidad Tecnológica de Bolívar, km 1 vía Turbaco, Cartagena 131001, Colombia; wjgil@utp.edu.co (W.G.-G.); o.d.montoyagiraldo@ieee.org or omontoya@utb.edu.co (O.D.M.)

[2] Facultad de Ingeniería, Universidad Distrital Francisco José de Caldas, Carrera 7 No. 40B-53, Bogotá D.C. 11021, Colombia

[3] Grupo GIIEN, Facultad de Ingeniería, Institución Universitaria Pascual Bravo, Campus Robledo, Medellín 050036, Colombia; luisgrisales@itm.edu.co

[4] Departamento de Física Aplicada, Universidad de Córdoba, ceiA3, Campus de Rabanales, 14071 Córdoba, Spain; g12pemoa@uco.es

[5] Escuela Nacional de Estudios Superiores, Campus Juriquilla, UNAM, Queretaro 3001, Mexico

* Correspondence: qhernandez@unam.mx

Received: 10 March 2020; Accepted: 2 April 2020; Published: 8 April 2020

Abstract: This paper presents an optimization model for the optimal placement and sizing of wind turbines, considering their reactive power capacity, wind speed, and demand curves. The optimization model is nonlinear and is focused on minimizing power losses in AC distribution networks. Also, paired wind turbine and power conversion systems are treated via chargeability factor η at the peak hour. This factor represents the percentage of usage of the power conversion system in the nominal wind speed conditions, and allows to support reactive power dynamically during all periods of the day as a function of the distribution system requirements. In addition, an artificial neural network is used for short-term forecasting to deal with uncertainties in wind power generation. We assume that the number of wind power distributed generators could be from zero to three generators integrated into the system, considering unit power factors and reactive power injections to follow up the effect of reactive power compensation in the daily operation. The General Algebraic Modeling System (GAMS) is employed to solve the proposed optimization model.

Keywords: wind power generation; artificial neural networks; chargeability factor; reactive power capacity; wind speed and demand curves

1. Introduction

Recently, the rapid growth of flexible AC distribution systems, smart grid, renewable energy sources, energy storage devices, and DC networks has led many researchers to study the optimal planning and operation of these systems [1]. This has been propelled mainly for the integration of renewable energy sources into the electrical power system around the world [2,3]. However, renewable generation sources have inherent technical and operational challenges, such as the need for appropriate integration without congesting the transmission lines, increasing energy losses, or voltage profiles degradation, among others [2,4]. Hence, the placement, sizing, and operation of renewable energies are important and play a fundamental role in the electrical system performance [5]. Due to these facts, it is necessary to propose strategies for optimal placement and sizing of renewable energies in order to reduce the network power losses without affecting performance and quality of service.

In this context, several methodologies have been proposed for optimal placement and sizing of renewable energy resources in the electrical distribution system. In [6], the optimal location of the distributed generators was introduced in order to reduce the power losses. In [7], a method for optimal sizing and optimal placement of renewable energy was presented, implementing an iterative search approach along with the Newton Raphson method. In [8], an adaptive quantum-inspired evolutionary algorithm was employed to locate and size distributed generators to reduce power losses in the networks. In [9], a multi-objective function was described to minimize the real power and reactive losses and to enhance the voltage profiles using the General Algebraic Modeling System (GAMS). In [10], an improved variant of the genetic algorithm has been proposed for optimal planning of wind power generators considering reactive power dispatch capabilities. In [2], a Bat algorithm for optimal placement and sizing of the distributed energy resources was developed, considering load variations to minimize power loss and enhance voltage profile. In [11], to solve this same problem, an invasive weed optimization algorithm was proposed. In [12], an analytical approach for optimal size and location of solar photovoltaic was proposed. The authors of [12] focus on reducing power losses and improving the voltage profiles. In [13], the optimal distributed generator placement in radial distribution networks based on a symbiotic organism's search algorithm was presented to reduce the network losses. Many other approaches have been developed based on evolutionary algorithms for the integration of renewable energy sources considering several aims, such as the harmony-based search algorithm (HSA) [14], artificial bee colony (ABC) [15], teaching and learning optimization method [16], particle swarm optimization [17], among others. Multiple hybrid approaches that combine two optimization techniques have also been introduced. In [18], ant colony optimization and fuzzy approaches were mixed. In [19], a combination of ant colony optimization and ABC algorithm was performed. A mixture of the particle artificial bee colony with the HSA algorithm was shown in [20]. In [21], incremental learning and PSO algorithms were mixed. The authors of [22] present a PSO algorithm combined with a feasible solution search to optimize the reactive power dispatch in a wind farm test system. Although there is plenty of research on the optimal location and size for wind power, all are focused on considering reactive power capacity, wind speed curves or load variation, but none of them take these problems simultaneously.

This study tackles the problem of optimal placement and sizing renewable energy resources based on wind turbines in AC distribution networks considering demand and load curves. The main contribution in regards with literature approaches lies in the proposal of a mixed-integer nonlinear programming (MINLP) model that includes variable reactive power capabilities in the power conversion system that composes the wind generation system. In this model, the chargeability factor in the voltage source converters is introduced as a function of the peak active power generation. This creates variable reactive power, making it possible to use the wind power system as a variable energy compensator that can operate with lagging or leading power factor depending on the grid requirements. The General Algebraic Modeling System (GAMS) is employed to solve the proposed MINLP model due to its excellent results in similar optimization problems [5,23]. In summary, the contributions of this paper are:

- A methodology for optimal placement and sizing of wind power generators considering reactive power capability and wind speed curves is described. In addition, the methodology also takes into account demand curves over a 24-h period.
- The proposed methodology is focused on minimizing power losses, which can be applied to radial and mesh networks.
- A chargeability factor η is proposed to represent the capacity to support the reactive power of a wind power system.

This study is organized as follows: Section 2 describes the mathematical modeling for the problem of the optimal location of wind power sources in AC distribution networks considering load variations and wind speed curves. Section 3 introduces the artificial neural network to forecast wind speed variations. Section 4 presents the main characteristics of the software implementation of the proposed mathematical model. In Section 5 the test system is presented, and Section 6 presents the numerical simulations. Lastly, the main concluding remarks derived from this research are shown in Section 7.

2. Mathematical Modeling

The problem of the optimal location of wind power sources in AC distribution networks considering load variations and wind speed curves is a nonlinear, non-differentiable, and non-convex optimization problem that combines continuous and discrete variables. In general terms, these characteristics generate an MINLP model, as reported in [5]. The main interest of this MINLP model is to minimize the total daily energy losses in all the branches of the AC grid. The complete mathematical model is presented below:

Objective function:

$$\min \ z = \sum_{t \in \Omega_T} \left[\sum_{i \in \Omega_N} V_{i,t} \left(\sum_{j \in \Omega_N} V_{j,t} Y_{ij} \cos \left(\theta_{i,t} - \theta_{j,t} - \phi_{ij} \right) \right) \right] \Delta_T, \tag{1}$$

where z is the value of the objective function related to the minimization of the daily energy losses, $V_{i,t}$ ($V_{j,t}$) is the voltage magnitude at node i (j) at the period of time t. Y_{ij} is the magnitude of the admittance that relates nodes i and j. $\theta_{i,t}$ ($\theta_{j,t}$) is the angle of the voltage at node i (j) at the period of time t; ϕ_{ij} is the angle of the admittance that relates nodes i and j. Observe that Δ_T is the period of time of the analysis (typically defined as 1 h for daily economic dispatch); and Ω_N and Ω_T are the sets that contain the nodes and the periods of time, respectively.

Set of constraints:

$$P_{i,t}^{CG} + y_i^{WT} P_{i,t}^{WT\text{nom}} - P_{i,t}^{D} = V_{i,t} \sum_{j \in \Omega_N} V_{j,t} Y_{ij} \cos \left(\theta_{i,t} - \theta_{j,t} - \phi_{ij} \right), \quad \{ \forall i \in \Omega_N, \forall t \in \Omega_T \} \tag{2}$$

$$Q_{i,t}^{CG} + Q_{i,t}^{WT} - Q_{i,t}^{D} = V_{i,t} \sum_{j \in \Omega_N} V_{j,t} Y_{ij} \sin \left(\theta_{i,t} - \theta_{j,t} - \phi_{ij} \right), \quad \{ \forall i \in \Omega_N, \forall t \in \Omega_T \} \tag{3}$$

$$V_i^{\min} \leq V_{i,t} \leq V_i^{\max}, \quad \{ \forall i \in \Omega_N, \forall t \in \Omega_T \} \tag{4}$$

$$0 \leq y_i^{WT} \leq P_i^{WT\max} x_i^{WT}, \quad \{ \forall i \in \Omega_N \} \tag{5}$$

$$-\frac{y_i^{WT}}{\eta} \sqrt{1 - \eta^2 \left(P_{i,t}^{WT\text{nom}} \right)^2} \leq Q_{i,t}^{WT} \leq \frac{y_i^{WT}}{\eta} \sqrt{1 - \eta^2 \left(P_{i,t}^{WT\text{nom}} \right)^2}, \quad \{ \forall i \in \Omega_N, \forall t \in \Omega_T \} \tag{6}$$

$$\sum_{i \in \Omega_N} x_i^{WT} \leq NG_{WT}^{\max}, \tag{7}$$

where $P_{i,t}^{CG}$ is the active power generation at node i at the period of time t by the conventional generator; y_i^{WT} is the decision variable that defines the size of the wind turbine located at node i; $P_{i,t}^{WT\text{nom}}$ is the nominal curve in per unit of the active power generated by the wind turbine connected at node i in the period of time t (this curve is obtained after applying the artificial neural network (ANN) forecasting methodology). $P_{i,t}^{D}$ is the active power consumption at node i in the period of time t. $Q_{i,t}^{CG}$ is the reactive power generation at node i at the period of time t by the conventional generator; $Q_{i,t}^{WT}$ is the reactive power generated by the wind turbine connected at node i in the period of time t (this value depends of the active power transported by the paired wind turbine–power converters). $Q_{i,t}^{D}$ is the reactive power consumption at node i in the period of time t. $V_i^{\min}$ and $V_i^{\max}$ are the minimum and maximum bounds for the voltage profile at node i. $P_i^{WT\max}$ is the maximum active power generation capability (size) of the wind turbine that can be connected at node i. x_i^{WT} is the binary variable related to the location of a wind turbine at node i, i.e., if $x_i^{WT} = 1$ the generator is located, and $x_i^{WT} = 0$ otherwise. The constant η corresponds to the chargeability factor of the paired wind turbine–power converter at the maximum active power point, i.e., $0 < \eta \leq 1$. Here we assume that this chargeability factor equals 90%. $NG_{WT}^{\max}$ is the maximum number of wind turbines available for location on the AC grid.

The mathematical model defined from (1) to (7) has the following interpretation: Equation (1) represents the objective function at it determines the total energy losses in an operation period of 24 h. In (2) and (3) are defined the active and reactive power balance equations; Expression (4) represents a box-type constraint that represents the minimum and maximum voltage regulation bounds in the AC grid. Equation (5) determines the size of the wind turbine that can be located at node i if the decision variable x_i^{WT} is activated. Equation (6) determines the maximum and minimum reactive power bounds of the wind turbine located at node i; note that this reactive power can be positive or negative, which implies that the wind turbine can work with lagging or leading power factor as variable energy compensator. In (7) the maximum number of wind turbines available for location in the AC grid is defined.

In Figure 1 the interconnection scheme of the wind turbine to the electrical AC network is presented. This connection is made via back-to-back voltage source converters. In this connection, via nonlinear control strategies such as passivity-based control [24,25], sliding mode control [26,27] or feedback linearization [28,29], among others, it is possible to control the active and reactive power flow exchanged between the AC grid and the wind turbine system independently.

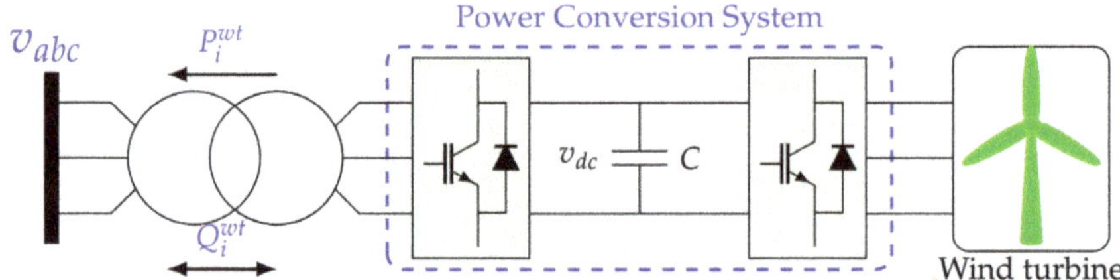

Figure 1. Electrical interconnection of a wind turbine to AC three-phase distribution network via voltage source converters using back-to-back configuration for type-D connections [30].

Based on Figure 1, and considering that the power losses in the power conversion system and transformer are negligible, the reactive power inequality constraint reported in (6) is reached as follows: first, we assume that when the maximum active power is obtained from the wind turbine system the power conversion system works at 90%, we name this factor as η, which implies that under this condition, if a wind turbine is located at node i, the per unit representation is as follows:

$$\max_{t \in \Omega_T} \left\{ P_{i,t}^{WT_{\text{nom}}} \right\} = \eta S_i^{WT_{\text{nom}}}. \tag{8}$$

Second, if we consider the reactive power variable $Q_{i,t}^{WT}$ in per unit representation, then, the following result yields:

$$Q_{i,t}^{WT} = \pm \sqrt{\left(S_i^{WT_{\text{nom}}} \right)^2 - \left(P_{i,t}^{WT_{\text{nom}}} \right)^2}. \tag{9}$$

Now, if we substitute (8) in (9), we obtain the result presented below:

$$Q_{i,t}^{WT} = \pm \frac{1}{\eta} \sqrt{\left(\max_{t \in \Omega_T} \left\{ P_{i,t}^{WT_{\text{nom}}} \right\} \right)^2 - \eta^2 \left(P_{i,t}^{WT_{\text{nom}}} \right)^2}. \tag{10}$$

Since our interest is to support reactive power dynamically, we replace the equality sign in (10) by an inequality constraint as reported in (6). Observe that in per unit representation $\max_{t \in \Omega_T} \left\{ P_{i,t}^{WT_{\text{nom}}} \right\} = 1$. Figure 2 illustrates the reactive power limits of wind energy for an active power delivered at the given time.

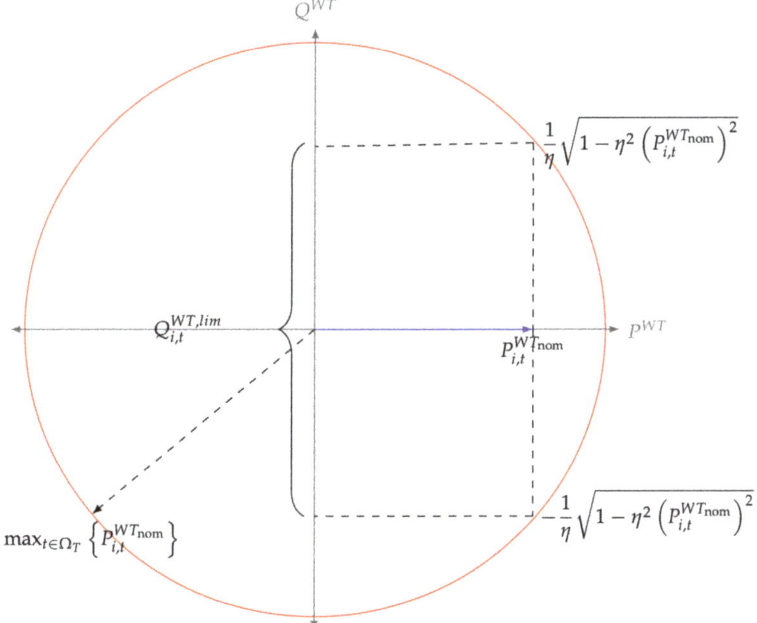

Figure 2. Example of reactive power limits for an active power delivered.

In the case of the real and imaginary power dispatches (control strategy), i.e., active and reactive power generation in WTs, it is important to mention that the control of these are made via nonlinear strategies applied to the power system conversion presented in Figure 1. This picture presents a type-D (or Type-4) WT connection, where the first voltage source converter controls the active power providing by the wind generator and supports all its reactive power requirements. While the second voltage source converter adjusts this active power to be transferred to the electrical grid with an adequate frequency at the same time that interchanges reactive power with this [30]. In addition, the voltage source converter also enables reactive power to be delivered or absorbed. In this part, we are interested in presenting the tertiary control stage in power system operation, i.e., the optimization stage in hierarchical control structures. The tertiary control stage is also called optimal power flow and it can be implemented as centralized or distributed [31]. This works if links are sent to each active device (e.g., power electronic converters) to provide references to primary and secondary controls [32].

3. Wind Power Forecasting

The integration of renewable energy sources is a great challenge, framed in the intermittence of their primary resources. Therefore, it is essential to consider these intermittences in the planning, and their potential for power generation. Additionally, improper placement and sizing of the wind powers may cause troubles, such as transmission line overload, voltage profiles, and increases in power losses [33]. Hence, the placement and sizing of these generators should take into account the high variability of the wind speed with the purpose of minimizing the forecasting errors and, therefore, avoiding the issues mentioned above.

Here we adopt the methodology proposed in [23], which is based on ANN. The ANN needs as input information the temperature, humidity, pressure, and time to properly estimate the wind speed. The historical wind speed data are depicted in Figure 3, which are taken from [33] and [34].

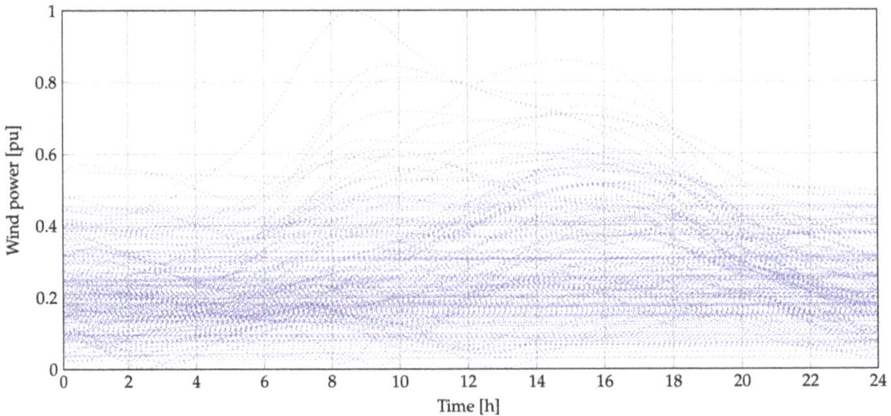

Figure 3. Historical wind speed data used for the artificial neural network (ANN) training process (adapted from [23,33]).

Note in Figure 3 that the illustrated wind speed comprises multiple days over the course of the 24-h period, which provides the ANN with sufficient information to make an adequate prediction. It is also important to mention although the plots are not shown for the temperature, humidity, pressure, and time, this information is also implemented for the ANN. These data are found in the Caribbean region of Colombia, and the availability of wind speed energy is measured over a year.

Artificial Neural Network

The ANN has implemented several topics, such as pattern classification, clustering, optimization, function approximation, and prediction [35]. The ANN is a mathematical tool that can store and remind the characteristics of a system and, therefore, obtain learning for the purpose of predicting future events or stochastic variables. The ANN has some advantages such as fast time in processing information and small data size, no complexity in pattern recognition tasks, and reprogramming is not required [36].

The training process of an ANN requires input and output data of the system to obtain a nonlinear mapping. The ANN is implemented when it does not have a system model in which the input and output data are related. The nonlinear learning rule is reached, as follows:

$$y(t) = f\left(y(t-1), ..., y(t-n_y), x(t-1), ..., x(t-n_x)\right) \tag{11}$$

where $x \in R^n$ and $y \in R$ are input and output data. n_y and n_x are the last values of the prediction and the input data, respectively. Observe that y also depends on the last n_y values of the variable under prediction.

The *ntstool* of MATLAB software was used in the training process of the ANN to predict the wind speed. The ANN configuration is composed of four inputs (temperature, humidity, pressure, and time), four delays, and 12 hidden neurons. We set up the ANN in the training, adjustment, and validation processes with a 70%, 15%, and 15% of the data, respectively. In Figure 4 the ANN scheme with the wind speed predictions implemented in MATLAB is shown.

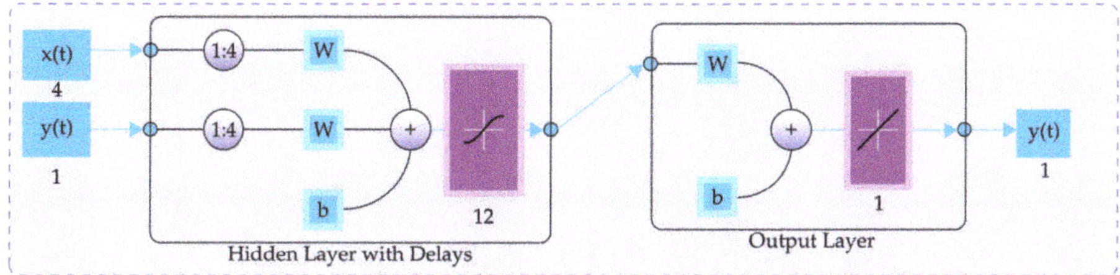

Figure 4. ANN scheme for wind speed prediction [23].

4. Solution Methodology

To solve the MINLP model that represents the optimal placement and sizing of wind turbines in AC distribution networks considering wind and demand curves, in this research the general algebraic modeling system is employed, widely known in specialized literature as GAMS. This optimization package has been successfully used for solving nonlinear large-scale optimization problems, such as optimal location and sizing of distributed generators in AC and DC distribution networks [5,37]; optimal operation of battery energy storage systems [23,38]; multi-objective optimization of the stack of a thermoacoustic engine [39] and optimization of pump and valve schedules in complex, large-scale water distribution systems [40], and so on. The main characteristics of GAMS implementation are listed below [23].

✓ Representation of the mathematical models by sets, which generates a compact formulation.
✓ The mathematical structure of the model, i.e., (1)–(7), is preserved via symbolic representation by using a plain text.
✓ This optimization software allows solving multiple optimization problems, such as linear programming, discrete models, and general non-convex formulations.
✓ Availability of a free version for demonstration, useful to introduce undergraduate students with mathematical optimization.
✓ Non-advanced programming skills are required for using GAMS, since it has an intuitive manner to implement a mathematical model using a basic plain text interface.

The basic elements for implementing a mathematical model in the GAMS interface are presented in Algorithm 1.

Algorithm 1: Main features for implementing a mathematical optimization model in GAMS

Data: Define the nature of the optimization problem and select the test system.
Sets Definition of sets, parameters (constant vectors), scalars (constant number), and tables (constant matrices).
Variables: Determine the type of variables, e.g., integer, continuous or binary.
Equations: Write the optimization problem, i.e., the set of Equations (1)–(7).
Solution: Select a MINLP solver to reach the solution via minimization of the objective function.
Visualization: Extract the variables of interest, i.e., location of the generators and their sizes, the value of the objective function, etc.
Result: Optimal location and sizing of wind turbines in AC distribution networks under daily operative scenarios.

General characteristics about GAMS implementations, such as reserve words and specialized functions can be consulted in [41].

5. Test Systems

To validate the proposed mathematical model for optimal location and placement of renewable energy sources such as wind power considering reactive power capabilities in AC distribution networks, we consider two test systems with a radial structure composed of 27 and 69 nodes, respectively. These systems are presented below.

5.1. 27-Node Test System

The 27-node test system is a radial test system with similar characteristics of the Colombian distribution networks located in the Caribbean region [37]. The branch information and nodal demands at the peak hour are reported in Table 1, and the electrical configuration of this test feeder is depicted in Figure 5. For this test system, we employ 13.8 kV and 1000 kW as voltage and power bases, respectively. To evaluate the daily operation of this system, including PV systems, we employ the demand variation and the wind generation capacity reported in Figure 6. Note that in this test feeder, the maximum size of each wind turbine is left free to verify the effect of the renewable injection to reduce the daily energy losses.

Table 1. Branch and load information of the 27-node test system.

Node		R_{ij}	X_{ij}	P_j	Q_j	Node		R_{ij}	X_{ij}	P_j	Q_j
i	j	[Ω]	[Ω]	[kW]	[kW]	i	j	[Ω]	[Ω]	[kW]	[kW]
1	2	0.15208	0.19855	0	0	14	15	0.87630	0.41330	106.3	65.8
2	3	0.65805	0.59745	0	0	15	16	0.87630	0.41330	25	158
3	4	0.19742	0.17924	297.5	184.4	3	17	0.87630	0.41330	255	158
4	5	0.43848	0.26038	0	0	17	18	0.52578	0.24798	127.5	79
5	6	0.48720	0.28931	255	158	18	19	0.78867	0.37197	297.5	184.4
6	7	0.48197	0.22732	0	0	19	20	0.83248	0.39263	340	210.7
7	8	0.87630	0.41330	212.5	131.7	20	21	0.87630	0.41330	85	52.7
8	9	1.09540	0.51663	0	0	4	22	0.87630	0.41330	106.3	65.8
9	10	0.87630	0.41330	266.1	164.9	5	23	0.87630	0.41330	55.3	34.2
2	11	0.87630	0.41330	85	52.7	6	24	0.35052	0.16532	69.7	43.2
11	12	1.07780	0.50836	340	210.7	8	25	0.52578	0.24798	256	158
12	13	0.65722	0.30998	297.5	184.4	8	26	0.52578	0.24798	63.8	39.5
13	14	0.49073	0.23145	191.3	118.5	26	27	0.70104	0.33064	170	105.4

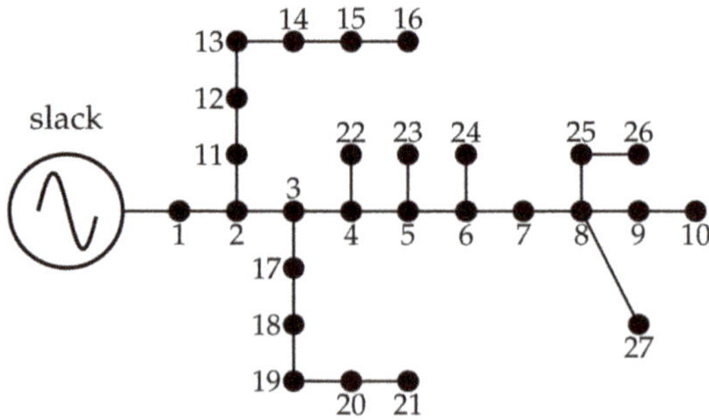

Figure 5. Electrical configuration of the 27-node test system.

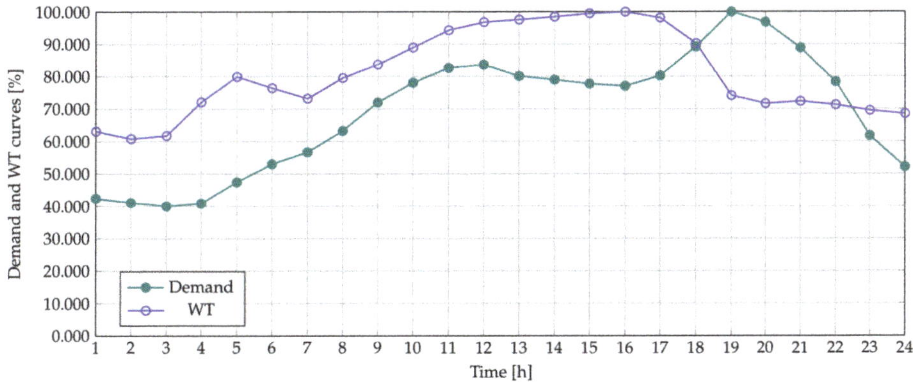

Figure 6. Percentage of power consumption and availability on a typical sunny day in the Caribbean region of Colombia [37].

5.2. 69-Node Test Feeder

This test feeder illustrated in Figure 7 is widely known in specialized literature as the Baran and Wu test system with 69 nodes and 68 branches with 12.66 kV of operating voltage [21]. This test feeder has 3890.7 kW and 2693.6 kVAr of total active and reactive power demand. The initial active energy losses of this system equals 3525.7520 kWh/day. For this test system, we also considered the possibility of installing 3 wind turbines, limited from 0 kW to 2000 kW each. In addition, we also considered 12.66 kV and 1000 kW as voltage and power base values, respectively.

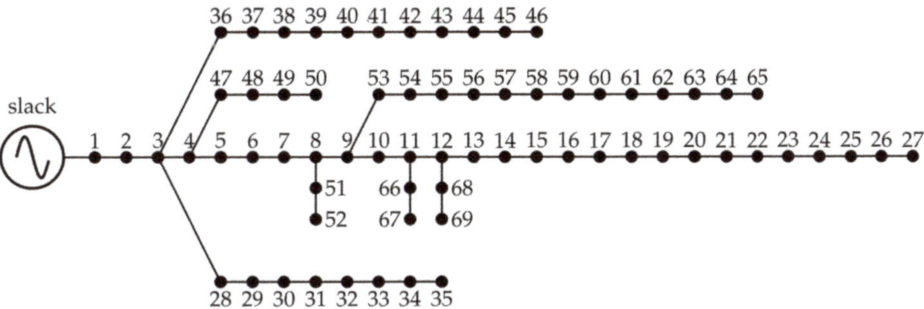

Figure 7. Electrical configuration of the 69-node test system.

The information of the branches and the load consumption of the 69-node test feeder is presented in Table 2. Additionally, the demand and wind turbine curves are the same presented in Figure 6.

Table 2. Parameters of the 69-node test feeder.

Node		R_{ij}	X_{ij}	P_j	Q_j	Node		R_{ij}	X_{ij}	P_j	Q_j
i	j	[Ω]	[Ω]	[kW]	[kW]	i	j	[Ω]	[Ω]	[kW]	[kW]
1	2	0.0005	0.0012	0	0	3	36	0.0044	0.0108	26	18.55
2	3	0.0005	0.0012	0	0	36	37	0.0640	0.1565	26	18.55
3	4	0.0015	0.0036	0	0	37	38	0.1053	0.1230	0	0
4	5	0.0251	0.0294	0	0	38	39	0.0304	0.0355	24	17
5	6	0.3660	0.1864	2.6	2.2	39	40	0.0018	0.0021	24	17
6	7	0.3811	0.1941	40.4	30	40	41	0.7283	0.8509	102	1
7	8	0.0922	0.0470	75	54	41	42	0.3100	0.3623	0	0
8	9	0.0493	0.0251	30	22	42	43	0.0410	0.0478	6	4.3
9	10	0.8190	0.2707	28	19	43	44	0.0092	0.0116	0	0
10	11	0.1872	0.0619	145	104	44	45	0.1089	0.1373	39.22	26.3
11	12	0.7114	0.2351	145	104	45	46	0.0009	0.0012	39.22	26.3
12	13	1.0300	0.3400	8	5	4	47	0.0034	0.0084	0	0
13	14	1.0440	0.3450	8	5	47	48	0.0851	0.2083	79	56.4
14	15	1.0580	0.3496	0	0	48	49	0.2898	0.7091	384.7	274.5
15	16	0.1966	0.0650	45	30	49	50	0.0822	0.2011	384.7	274.5
16	17	0.3744	0.1238	60	35	8	51	0.0928	0.0473	40.5	28.3
17	18	0.0047	0.0016	60	35	51	52	0.3319	0.1140	3.6	2.7
18	19	0.3276	0.1083	0	0	9	53	0.1740	0.0886	4.35	3.5
19	20	0.2106	0.0690	1	0.6	53	54	0.2030	0.1034	26.4	19
20	21	0.3416	0.1129	114	81	54	55	0.2842	0.1447	24	17.2
21	22	0.0140	0.0046	5	3.5	55	56	0.2813	0.1433	0	0
22	23	0.1591	0.0526	0	0	56	57	1.5900	0.5337	0	0
23	24	0.3463	0.1145	28	20	57	58	0.7837	0.2630	0	0
24	25	0.7488	0.2475	0	0	58	59	0.3042	0.1006	100	72
25	26	0.3089	0.1021	14	10	59	60	0.3861	0.1172	0	0
26	27	0.1732	0.0572	14	10	60	61	0.5075	0.2585	1244	888
3	28	0.0044	0.0108	26	18.6	61	62	0.0974	0.0496	32	23
28	29	0.0640	0.1565	26	18.6	62	63	0.1450	0.0738	0	0
29	30	0.3978	0.1315	0	0	63	64	0.7105	0.3619	227	162
30	31	0.0702	0.0232	0	0	64	65	1.0410	0.5302	59	42
31	32	0.3510	0.1160	0	0	11	66	0.2012	0.0611	18	13
32	33	0.8390	0.2816	10	10	66	67	0.0047	0.0014	18	13
33	34	1.7080	0.5646	14	14	12	68	0.7394	0.2444	28	20
34	35	1.4740	0.4873	4	4	68	69	0.0047	0.0016	28	20

6. Computational Validation

The solution of the general MINLP model defined from (1) to (7) for the optimal placement and sizing of wind turbines considering reactive power capabilities is made using the GAMS optimization package with the CONOPT solver in a desktop computer with an INTEL(R) Core(TM) i5-3550 3.5 GHz processor and 8 GB of RAM running a 64-bit version of Windows 7 Professional [37].

In order to verify the effectiveness of using wind turbines considering the reactive power capability in the converters, we solve the problem from zero to three possible distributed generators considering unity power factor and reactive power injections; these cases allow to detect the effect of reactive power compensation in the daily operative behavior of the AC distribution systems in regards to the energy losses minimization [5].

6.1. 27-Node Test Feeder

For this test system we evaluate the the operation of the paired wind turbine power electronic converter with unity and variable power factor. In Table 3 the solution achieved by GAMS considering unity power factor is reported and in Table 4 the solution with variable reactive power compensation is presented. Note that in these simulation results, the chargeability factor is selected as $\eta = 0.9$ p.u. In this test system the daily energy losses is 1652.0190 kWh/day.

Table 3. Solution reached by General Algebraic Modeling System (GAMS) considering unity power factor for the 27-node test system.

Number of WTs	Location [Node]	Size [kW]	Energy Losses [kWh/Day]
1	18	1547.17725	1185.164
2	{7, 14}	{1477.5895, 878.1173}	823.6474
3	{8, 15, 20}	{1019.7799, 650.5079, 680.1683}	628.4598

Table 4. Solution reached by GAMS considering variable factor for the 27-node test system.

Number of WTs	Location [Node]	Size [kW]	Energy Losses [kWh/Day]
1	8	1432.89260	827.9621
2	{9, 13}	{1100.6703, 1069.0459}	582.6569
3	{9, 13, 19}	{777.5496, 767.6551, 798.0454}	283.1916

Results in Tables 3 and 4 show the positive effect of including renewable generation in the daily operation of AC distribution networks. It is important to mention that depending on the operative consign of the wind turbine, i.e., unity power factor or variable power factor, the location and size of the generators are susceptible to change; for example in the case of three WTs when power factor is unitary, the location of the generators is at nodes 8, 15, and 20 with a total active power generation at the peak hour about 2350.4561 kW; while in the case of the variable power factor the location of the generators is at nodes 9, 13 and 19, with a total active power generation at the peak hour about 2343.2501 kW. Note that nodes suffer slight variations in their locations for both scenarios, nevertheless, with a little reduction of the total power capability (7.2060 kW) the variable power factor approach allows a reduction in about 345.2682 kWh/day passing from 628.4598 kWh/day to 283.1916 kWh/day when unitary power factor changes from being unitary to variable. The previous results confirm the conclusions reached by authors of [6], where both cases were analyzed for AC distributed networks considering only the peak hour in their analysis.

Results in Figure 8 confirm the positive effect of using variable power factor in wind turbine applications via power conversion system to dynamically support active and reactive power to the AC grid. In this sense, for one WT the reduction in the daily energy losses is about 21.62%, for two WTs is about 14.59%, and for three wind turbines is about 20.90% when compared unitary and variable power factor approaches.

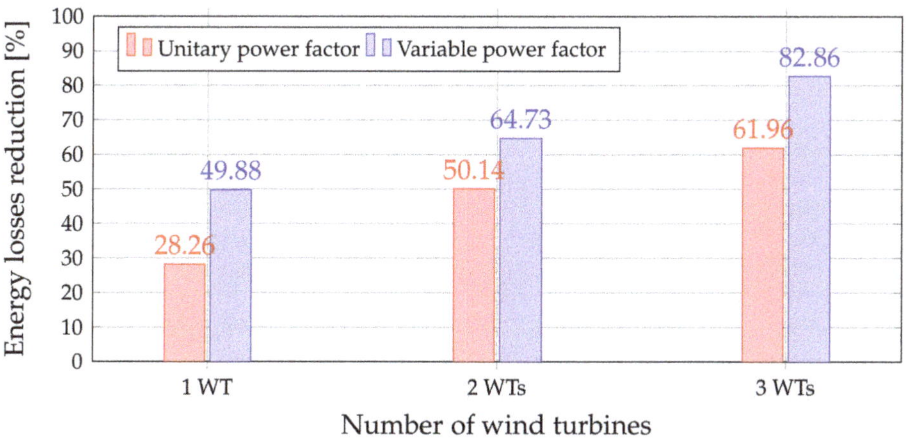

Figure 8. Energy losses reduction in the 27-node test system.

An important fact in the results presented in Figure 8 is that there is no linear relation between the number of renewable sources placed in the distribution system and the energy losses reduction, since the effect of an additional renewable source is less in comparison with the previous case, as concluded in [37] for photovoltaic plants. This behavior is expected, since the objective function (1) is nonlinear and non-convex, which implies that linear tendencies cannot be extrapolated from it.

It is worthy to mention that for this test feeder when unitary power factor is considered the GAMS package takes about 25 s to solve the problem, while in the variable power factor case this time is increased to 45 s. This increment is caused by the fact that in the variable power factor scenario appears the amount of reactive power as a variable, which introduces complexity to the optimization model and the solution space increases substantially with a rate of about 24 additional variables per WT.

6.2. 69-Node Test Feeder

Tables 5 and 6 present the optimal location and sizing of wind turbines in the 69-node test feeder considering unity and variable power factor, respectively.

In the case of the unitary power factor approach reported in Table 5 we can observe node 60 with a strong effect in the energy losses reduction, since in all the three cases, it appears with important power injections. Here it is important to mention that for one WT, the daily energy losses are about 1179.3480 kWh/day; while for three WT units this number is reduced to 949.2134 kWh/day. This implies that the nonlinear relation between number of renewable sources and energy losses is strongly demonstrated in the 69-node test feeder, since the difference in both cases is only 230.1346 kWh/day passing from 1639.2766 kW to 1993.7893 kW in the peak hour regarding the amount of power installed, i.e., 354.5127 kW of additional power.

Table 5. Solution reached by GAMS considering unity power factor for the 69-node test system.

Number of WTs	Location [Node]	Size [kW]	Energy Losses [kWh/Day]
1	60	1639.2766	1179.3480
2	{17, 60}	{450.1575, 1554.3188}	1048.7920
3	{17, 60, 61}	{450.1243, 189.2822, 1354.3828}	949.2134

Table 6. Solution reached by GAMS considering variable factor for the 69-node test system.

Number of WTs	Location [Node]	Size [kW]	Energy Losses [kWh/Day]
1	62	1597.3204	380.2232
2	{63, 69}	{1472.3336, 562.0680}	268.0452
3	{18, 61, 64}	{456.7767, 1280.2409, 256.9057}	159.6235

When variable power factor is analyzed the effect of reactive power injection is evident (see Table 6), since the daily energy losses present strong reduction for all the combinations of wind turbines. For example, if we observe the case of two WTs, then, the daily energy losses pass from 1048.7920 kWh/day to 268.0452 kWh/day by injecting at the peak hour a total of 2034.4016 kW; which is similar to the 2004.4763 kW in the unitary power factor scenario. Figure 9 confirms that the variable power factor approach increases by about 30% the daily losses reduction when compared to unitary power factor approach. This result indicates that distributed generators with dynamic reactive power compensation can be a powerful alternative to improve the electrical performance of the AC distribution networks, only by using adequate control schemes in the power conversion system presented in Figure 1.

Similarly, as shown by the 27-node test system, in the 69-node case the number of generators and their impact in the total energy losses reduction is strongly nonlinear, since for two and three WTs we observe a clear saturation in the objective function performance, with a small increment of about 5%. This implies that for utilities it is important to consider these nonlinear phenomena in their grids, since large penetrations of renewable energy can affect negatively their electric power system performances.

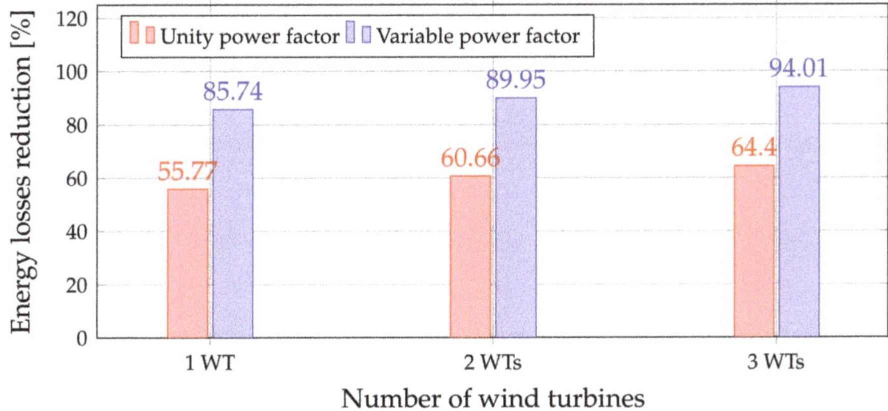

Figure 9. Energy losses reduction in the 69-node test system.

The processing times required in the case of the 69-node test feeder when the unity power factor was used was about 120 s, while in the case of the variable power factor, it increases to 450 s. This increment has the same explanation presented for the 27-node test feeder, and it is related to the increment in the number of variables in the optimization problem.

6.3. Large-Scale Power System Evaluation

To evaluate the applicability of the proposed mathematical model for optimal location and sizing of wind turbines in electrical networks with reactive power capabilities via power electronic converters, we implement a large-scale power system composed of 24 nodes with 38 interconnections via transmission lines and transformers [42,43]. This system has a total of active and reactive power demands of about 2850 MW and 580 MVAr, respectively. In addition, these power demands are fed by 10 conventional generators and 1 reactive power compensator. The electrical configuration of this test feeder is presented in Figure 10 and the parameters are reported in Tables 7–9, respectively (the voltage and power bases for this test feeder are 100 MW and 220 kV). For this test system, we assume that the slack node corresponds to the generator located at node 13, which is set with voltage output of 1.050 p.u. It is important to mention that the rest of the generators are free for generating the required power to minimize the total daily energy losses, and the voltage regulation bounds for this power system are 0.900 p.u. and 1.100 p.u., respectively.

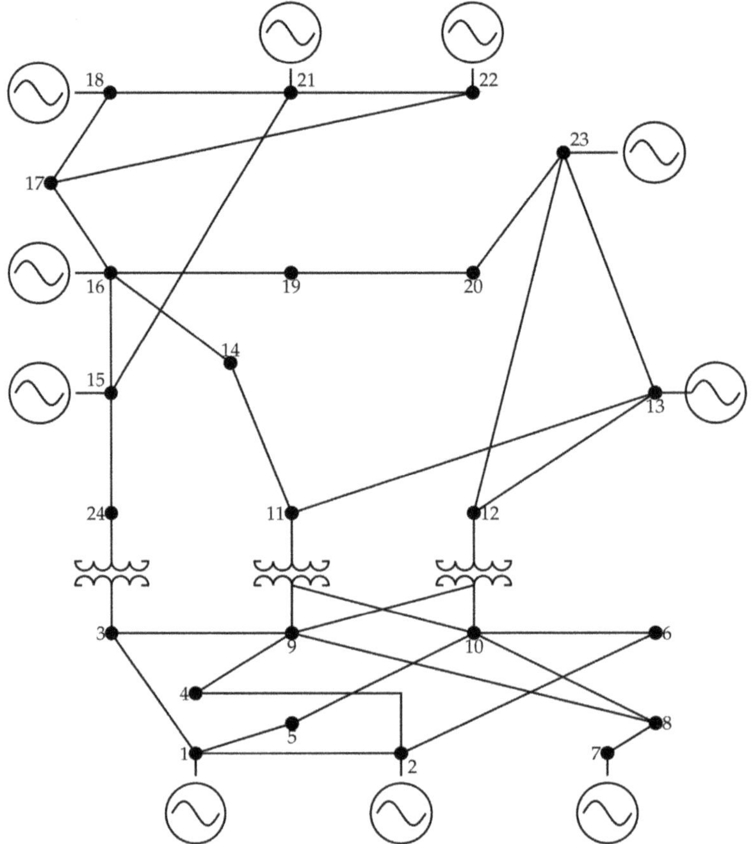

Figure 10. Electrical configuration of the IEEE 24-node test system.

Table 7. Conventional generator capabilities.

Node	P_g^{min}	P_g^{min}	Q_g^{min}	Q_g^{max}	Node	P_g^{min}	P_g^{min}	Q_g^{min}	Q_g^{max}
1	0.500	2.500	−0.750	0.750	16	0.150	1.350	−0.500	0.750
2	0.250	1.850	−0.900	0.700	16	0.150	1.350	−0.500	0.750
7	0.400	3.000	−0.800	1.200	18	0.500	4.000	−1.850	2.000
13	0.000	6.000	−5.000	5.000	21	0.450	4.500	−1.500	1.500
14	0.000	0.000	−1.500	1.500	22	0.250	2.800	−1.000	1.000
15	0.250	1.500	−0.750	0.800	23	0.750	6.000	−1.650	1.750

All data is per unit.

Table 8. Demand information for the 24-node test feeder.

Node	P_d	Q_d	Node	P_d	Q_d	Node	P_d	Q_d	Node	P_d	Q_d
1	1.080	0.220	7	1.250	0.250	13	2.650	0.540	19	1.810	0.370
2	0.970	0.200	8	1.710	0.350	14	1.940	0.390	20	1.280	0.260
3	1.800	0.370	9	1.750	0.360	15	3.170	0.640	21	0.000	0.000
4	0.740	0.150	10	1.950	0.400	16	1.000	0.200	22	0.000	0.000
5	0.710	0.140	11	0.000	0.000	17	0.000	0.000	23	0.000	0.000
6	1.360	0.280	12	0.000	0.000	18	3.330	0.680	24	0.000	0.000

All data is per unit.

Table 9. Branch information for the 24-node test feeder (all data in p.u.).

Node i	Node j	r_{ij}	x_{ij}	b_j	Tap	Node i	Node j	r_{ij}	x_{ij}	b_j	Tap
1	2	0.0026	0.0139	0.4611	0	12	13	0.0061	0.0476	0.0999	0
1	3	0.0546	0.2112	0.0572	0	12	23	0.0124	0.0966	0.2030	0
1	5	0.0218	0.0845	0.0229	0	13	23	0.0111	0.0865	0.1818	0
2	4	0.0328	0.1267	0.0343	0	14	16	0.0050	0.0389	0.0818	0
2	6	0.0497	0.1920	0.0520	0	15	16	0.0022	0.0173	0.0364	0
3	9	0.0308	0.1190	0.0322	0	15	21	0.0063	0.0490	0.1030	0
3	24	0.0023	0.0839	0	1.015	15	21	0.0063	0.0490	0.1030	0
4	9	0.0268	0.1037	0.0281	0	15	24	0.0067	0.0519	0.1091	0
5	10	0.0228	0.0883	0.0239	0	16	17	0.0033	0.0259	0.0545	0
6	10	0.0139	0.0605	2.4590	0	16	19	0.0030	0.0231	0.0485	0
7	8	0.0159	0.0614	0.0166	0	17	18	0.0018	0.0144	0.0303	0
8	9	0.0427	0.1651	0.0447	0	17	22	0.0135	0.1053	0.2212	0
8	10	0.0427	0.1651	0.0447	0	18	21	0.0033	0.0259	0.0545	0
9	11	0.0023	0.0839	0	1.030	18	21	0.0033	0.0259	0.0545	0
9	12	0.0023	0.0839	0	1.030	19	20	0.0051	0.0396	0.0833	0
10	11	0.0023	0.0839	0	1.015	19	20	0.0051	0.0396	0.0833	0
10	12	0.0023	0.0839	0	1.015	20	23	0.0028	0.0216	0.0455	0
11	13	0.0061	0.0476	0.0999	0	20	23	0.0028	0.0216	0.0455	0
11	14	0.0054	0.0418	0.0879	0	21	22	0.0087	0.0678	0.1424	0

To simulate the 24-node test feeder, we consider the case that reactive power is available to be injected into the power system since results for 27- and 69-node test feeders have demonstrated that variable power is more efficient regarding energy losses reduction than unity power factor. In Table 10 the optimal solutions reached by the GAMS package is reported on the high-voltage meshed power system depicted in Figure 10.

Table 10. Solution reached by GAMS considering variable factor for the 24-node test system.

No. of WTs	Loc. [Node]	Size [MW]	Energy Losses [MWh/Day]	Reduction [%]
1	6	208.9887	225.8110	45.0384
2	{6, 11}	{180.2410, 396.0309}	169.4930	58.7460
3	{3, 6, 11}	{167.1043, 183.3276, 358.0971}	131.8012	67.9201

From the results in Table 10, we can observe that:

- To reach a total daily energy losses reduction of about 45.0384% it is required to install a WT at node 6 with a nominal rate of 208.9887 MW; and to increase the daily energy reduction by 13.7075%, 576.2719 MW is required to be installed, which implies more than 100% of additional power injection. This result implies that energy losses formulated in (1) have a strong nonlinear relation between the amount of power injection by renewable energy resources and their location in regards with the objective function minimization. Note that this behavior was also observed in both radial test feeders previously reported.
- The proposed mathematical model is suitable to be applied for both radial medium voltage and high-voltage meshed networks, which confirms that it is general and scalable for multiple grid topologies.

On the other hand, Figure 11 presents the behavior of the reactive power for the case of the installation of one wind turbine at node 6.

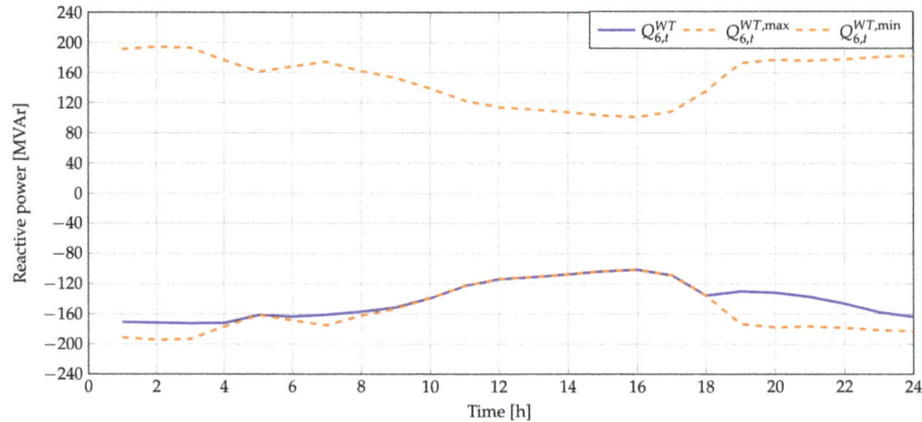

Figure 11. Reactive power generation for the case of one WT installed at node 6.

Observe that positive and negative reactive power bounds depicted in Figure 11 correspond to the graphical behavior of constraint (6) when y_6^{WT} is 208.9887 MW. It is important to highlight that these bounds are variable as a function of the active power generation since the power electronic converter that interfaces this WT has limited apparent power capabilities and function of the chargeability factor η, as demonstrated in Section 2. Finally, it is worth mentioning that in the case of the one wind turbine, the power electronic converter works in the second quadrant, because it provides active power at the same time that absorbs reactive power. This implies that this converter is absorbing the excessive reactive energy supplied by all the conventional power sources.

6.4. Assessment of ANN Performance

To demonstrate the efficiency of the ANN approach for predicting the renewable energy availability in wind turbines, we fix the locations reported in Table 4 for the 27-node test feeder and Table 6 for the 69-node test feeder for the WTS to evaluate the objective function when using the real wind energy curve (the real wind power curve can be consulted in [23]). In Figure 12 the absolute error regarding the objective function between the real and estimated curves for both test systems is presented. Note that the error reported in this picture is calculated, as follows:

$$\epsilon = \frac{|v_r - v_e|}{v_r},\tag{12}$$

where ϵ is the estimation error, v_r is the real value of the objective function and v_e the estimation reached with the ANN prediction.

Figure 12 reports the estimation errors reached when the ANN prediction is compared with the real generation curve. This picture allows to conclude that the maximum error for both test feeder does not overpass 1.35% for different possibilities of wind turbine locations. This result confirms that the ANN is a powerful tool for renewable generation prediction, as previously published in [5,23] for wind and photovoltaic applications.

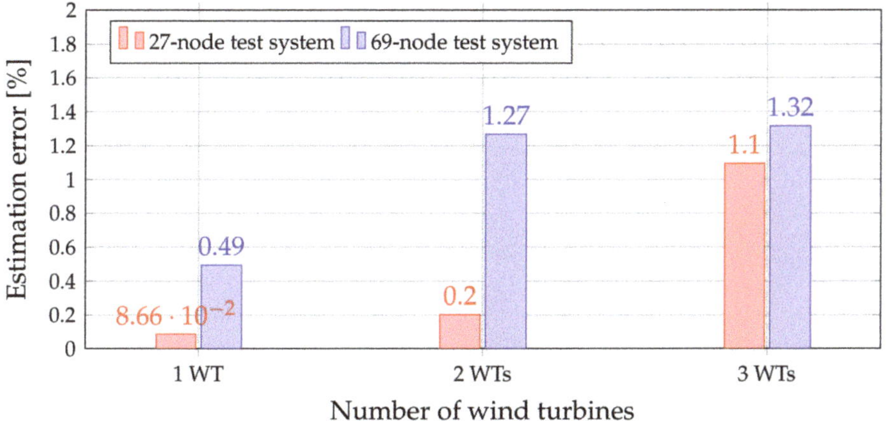

Figure 12. Error between the real generation curve for wind turbines and the predicted curve provided by the ANN approach.

7. Conclusions and Future Works

The variable energy compensation problem in AC distribution networks has been analyzed in this paper via optimal placement and sizing of wind turbines, considering daily wind and demand curves. The main contribution of this study is the nonlinear formulation related to the paired wind turbine and power conversion system via chargeability factor η at the peak hour. This factor represents the percentage of usage of the power conversion system in the nominal wind speed conditions, and it allows supporting reactive power dynamically during all periods of the day as a function of the distribution system requirements.

The effect of the variable power factor in comparison to the unitary scenario was evident for both test feeders since the reduction of the energy losses was at least 20% superior by using practically the same installed capability regarding active power injection at the peak hour. These results confirm that power conversions systems presented in the integration of renewable can be used to replace classical approaches, such as capacitor banks with fixed of variable steps. Power conversion systems can be operated with lagging or leading power factors, which is a definite advantage in variable demand scenarios.

Artificial neural networks employed for renewable generation forecasting evidenced estimation errors lower than 1.35% when real and predicted curves are compared in terms of the objective function estimation, which allows to demonstrate the efficiency and robustness of this tool for economic and optimal power dispatch problems in the presence of renewable uncertainties.

As future work, it will be possible to solve the proposed optimization problem via metaheuristic techniques by using a master-salve structure with genetic algorithms and vortex optimizers. Additionally, it will also be possible to include the chargeability factor η as an optimization variable. This will allow to identify the optimal size of the power conversion system. An additional future work might be related to the inclusion of contingency analysis in meshed power systems in order to identify the best location and size of the renewable source to make the entire system more reliable and secure under fault conditions.

Author Contributions: Conceptualization and writing—Review and editing, W.G.-G., O.D.M., L.F.G.-N., A.-J.P.-M. and Q.H.-E. All authors have read and agreed to the published version of the manuscript.

Funding: This work was partially supported by the National Scholarship Program Doctorates of the Administrative Department of Science, Technology, and Innovation of Colombia (COLCIENCIAS) by calling contest 727-2015.

Conflicts of Interest: The authors declare no conflicts of interest.

References

1. Mazhari, S.M.; Monsef, H.; Romero, R. A multi-objective distribution system expansion planning incorporating customer choices on reliability. *IEEE Trans. Power Syst.* **2015**, *31*, 1330–1340. [CrossRef]
2. Essallah, S.; Khedher, A.; Bouallegue, A. Integration of distributed generation in electrical grid: Optimal placement and sizing under different load conditions. *Comput. Electr. Eng.* **2019**, *79*, 106461. [CrossRef]
3. Mohseni-Bonab, S.M.; Rabiee, A. Optimal reactive power dispatch: A review, and a new stochastic voltage stability constrained multi-objective model at the presence of uncertain wind power generation. *IET Gener. Transm. Distrib.* **2017**, *11*, 815–829. [CrossRef]
4. Strunz, K.; Abbasi, E.; Huu, D.N. DC microgrid for wind and solar power integration. *IEEE Trans. Emerg. Sel. Top. Power Electron.* **2013**, *2*, 115–126. [CrossRef]
5. Montoya, O.D.; Grisales-Noreña, L.F.; Gil-González, W.; Alcalá, G.; Hernandez-Escobedo, Q. Optimal Location and Sizing of PV Sources in DC Networks for Minimizing Greenhouse Emissions in Diesel Generators. *Symmetry* **2020**, *12*, 322. [CrossRef]
6. Kaur, S.; Kumbhar, G.; Sharma, J. A MINLP technique for optimal placement of multiple DG units in distribution systems. *Int. J. Electr. Power Energy Syst.* **2014**, *63*, 609–617. [CrossRef]
7. Ghosh, S.; Ghoshal, S.P.; Ghosh, S. Optimal sizing and placement of distributed generation in a network system. *Int. J. Electr. Power Energy Syst.* **2010**, *32*, 849–856. [CrossRef]
8. Manikanta, G.; Mani, A.; Singh, H.; Chaturvedi, D. Placing distributed generators in distribution system using adaptive quantum inspired evolutionary algorithm. In Proceedings of the Second International Conference on Research in Computational Intelligence and Communication Networks (ICRCICN), Kolkata, India, 23–25 September 2016; pp. 157–162.
9. Babu, P.V.; Singh, S. Optimal Placement of DG in Distribution network for Power loss minimization using NLP & PLS Technique. *Energy Procedia* **2016**, *90*, 441–454.
10. Meena, N.K.; Swarnkar, A.; Gupta, N.; Niazi, K. Wind Power Generation Planning by Utilizing the Reactive Power Capabilities of Generators. In Proceedings of the Asian Conference on Energy, Power and Transportation Electrification (ACEPT), Singapore, 24–26 October 2017; pp. 1–6.
11. Prabha, D.R.; Jayabarathi, T. Optimal placement and sizing of multiple distributed generating units in distribution networks by invasive weed optimization algorithm. *Ain Shams Eng. J.* **2016**, *7*, 683–694. [CrossRef]
12. Jamil, M.; Anees, A.S. Optimal sizing and location of SPV (solar photovoltaic) based MLDG (multiple location distributed generator) in distribution system for loss reduction, voltage profile improvement with economical benefits. *Energy* **2016**, *103*, 231–239. [CrossRef]
13. Das, B.; Mukherjee, V.; Das, D. DG placement in radial distribution network by symbiotic organisms search algorithm for real power loss minimization. *Appl. Soft Comput.* **2016**, *49*, 920–936. [CrossRef]
14. Parizad, A.; Khazali, A.H.; Kalantar, M. Sitting and sizing of distributed generation through Harmony Search Algorithm for improve voltage profile and reducuction of THD and losses. In Proceedings of the CCECE 2010, Calgary, AB, Canada, 2–5 May 2010; pp. 1–7.
15. Abu-Mouti, F.S.; El-Hawary, M. Optimal distributed generation allocation and sizing in distribution systems via artificial bee colony algorithm. *IEEE Trans. Power Deliv.* **2011**, *26*, 2090–2101. [CrossRef]
16. García, J.A.M.; Mena, A.J.G. Optimal distributed generation location and size using a modified teaching–learning based optimization algorithm. *Int. J. Electr. Power Energy Syst.* **2013**, *50*, 65–75. [CrossRef]
17. Cheng, S.; Chen, M.Y.; Fleming, P.J. Improved multi-objective particle swarm optimization with preference strategy for optimal DG integration into the distribution system. *Neurocomputing* **2015**, *148*, 23–29. [CrossRef]

18. Tolabi, H.B.; Ali, M.H.; Rizwan, M. Simultaneous reconfiguration, optimal placement of DSTATCOM, and photovoltaic array in a distribution system based on fuzzy-ACO approach. *IEEE Trans. Sustain. Energy* **2014**, *6*, 210–218. [CrossRef]

19. Muthukumar, K.; Jayalalitha, S. Optimal placement and sizing of distributed generators and shunt capacitors for power loss minimization in radial distribution networks using hybrid heuristic search optimization technique. *Int. J. Electr. Power Energy Syst.* **2016**, *78*, 299–319. [CrossRef]

20. Kefayat, M.; Ara, A.L.; Niaki, S.N. A hybrid of ant colony optimization and artificial bee colony algorithm for probabilistic optimal placement and sizing of distributed energy resources. *Energy Convers. Manag.* **2015**, *92*, 149–161. [CrossRef]

21. Grisales-Noreña, L.F.; Gonzalez Montoya, D.; Ramos-Paja, C.A. Optimal sizing and location of distributed generators based on PBIL and PSO techniques. *Energies* **2018**, *11*, 1018. [CrossRef]

22. Martinez-Rojas, M.; Sumper, A.; Gomis-Bellmunt, O.; Sudrià-Andreu, A. Reactive power dispatch in wind farms using particle swarm optimization technique and feasible solutions search. *Appl. Energy* **2011**, *88*, 4678–4686. [CrossRef]

23. Montoya, O.D.; Gil-González, W.; Grisales-Noreña, L.; Orozco-Henao, C.; Serra, F. Economic Dispatch of BESS and Renewable Generators in DC Microgrids Using Voltage-Dependent Load Models. *Energies* **2019**, *12*, 4494. [CrossRef]

24. Cisneros, R.; Mancilla-David, F.; Ortega, R. Passivity-Based Control of a Grid-Connected Small-Scale Windmill With Limited Control Authority. *IEEE J. Emerg. Sel. Top. Power Electron.* **2013**, *1*, 247–259. [CrossRef]

25. Namazi, M.M.; Nejad, S.M.S.; Tabesh, A.; Rashidi, A.; Liserre, M. Passivity-Based Control of Switched Reluctance-Based Wind System Supplying Constant Power Load. *IEEE Trans. Ind. Electron.* **2018**, *65*, 9550–9560. [CrossRef]

26. Barambones, O. Sliding Mode Control Strategy for Wind Turbine Power Maximization. *Energies* **2012**, *5*, 2310–2330. [CrossRef]

27. Lee, S.W.; Chun, K.H. Adaptive Sliding Mode Control for PMSG Wind Turbine Systems. *Energies* **2019**, *12*, 595. [CrossRef]

28. Alrifai, M.; Zribi, M.; Rayan, M. Feedback Linearization Controller for a Wind Energy Power System. *Energies* **2016**, *9*, 771. [CrossRef]

29. Li, P.; Wang, J.; Xiong, L.; Wu, F. Nonlinear Controllers Based on Exact Feedback Linearization for Series-Compensated DFIG-Based Wind Parks to Mitigate Sub-Synchronous Control Interaction. *Energies* **2017**, *10*, 1182. [CrossRef]

30. Salgado-Herrera, N.M.; Medina-Ríos, A.; Tapia-Sánchez, R.; Anaya-Lara, O. Reactive power compensation through active back to back converter in type-4 wind turbine. In Proceedings of the IEEE International Autumn Meeting on Power, Electronics and Computing (ROPEC), Ixtapa, Mexico, 9–11 November 2016; pp. 1–6.

31. Murillo-Yarce, D.; Garcés-Ruiz, A.; Escobar-Mejía, A. Passivity-Based Control for DC-Microgrids with Constant Power Terminals in Island Mode Operation. *Revista Facultad de Ingeniería Universidad de Antioquia* **2018**, *86*, 32–39. [CrossRef]

32. Choi, Y.H.; Cho, Y.S. Multiple-Point Voltage Control to Minimize Interaction Effects in Power Systems. *Energies* **2019**, *12*, 274. [CrossRef]

33. Gil-González, W.; Montoya, O.D.; Holguín, E.; Garces, A.; Grisales-Noreña, L.F. Economic dispatch of energy storage systems in dc microgrids employing a semidefinite programming model. *J. Energy Storage* **2019**, *21*, 1–8. [CrossRef]

34. Data, S.S.R. Time Series of Solar Radiation Data. Available online: http://www.soda-pro.com/ (accessed on 5 July 2019).

35. Koo, J.; Han, G.D.; Choi, H.J.; Shim, J.H. Wind-speed prediction and analysis based on geological and distance variables using an artificial neural network: A case study in South Korea. *Energy* **2015**, *93*, 1296–1302. [CrossRef]

36. Demuth, H.B.; Beale, M.H.; De Jess, O.; Hagan, M.T. *Neural Network Design*, 2nd ed.; Martin Hagan: Stillwater, OK, USA, 2014.

37. Montoya, O.D.; Gil-González, W.; Grisales-Noreña, L. An exact MINLP model for optimal location and sizing of DGs in distribution networks: A general algebraic modeling system approach. *Ain Shams Eng. J.* **2019**. [CrossRef]

38. Montoya, O.D.; Grajales, A.; Garces, A.; Castro, C.A. Distribution Systems Operation Considering Energy Storage Devices and Distributed Generation. *IEEE Lat. Am. Trans.* **2017**, *15*, 890–900. [CrossRef]

39. Tartibu, L.; Sun, B.; Kaunda, M. Multi-objective optimization of the stack of a thermoacoustic engine using GAMS. *Appl. Soft Comput.* **2015**, *28*, 30–43. [CrossRef]

40. Skworcow, P.; Paluszczyszyn, D.; Ulanicki, B.; Rudek, R.; Belrain, T. Optimisation of Pump and Valve Schedules in Complex Large-scale Water Distribution Systems Using GAMS Modelling Language. *Procedia Eng.* **2014**, *70*, 1566–1574, doi:10.1016/j.proeng.2014.02.173. [CrossRef]

41. Giraldo, O.D.M. Solving a Classical Optimization Problem Using GAMS Optimizer Package: Economic Dispatch Problem Implementation. *Ing. Y Cienc.* **2017**, *13*, 39–63. [CrossRef]

42. Soroudi, A. *Power System Optimization Modeling in GAMS*; Springer International Publishing: Cham, Switzerland, 2017.

43. Martínez-Lacañina, P.J.; Marcolini, A.M.; Martínez-Ramos, J.L. DCOPF contingency analysis including phase shifting transformers. In Proceedings of the Power Systems Computation Conference, Wroclaw, Poland, 18–22 August 2014; pp. 1–7.

sustainability

Article

Multi-Objective Stochastic Optimization for Determining Set-Point of Wind Farm System

Van-Hai Bui [1,2], Akhtar Hussain [1,2], Thai-Thanh Nguyen [3] and Hak-Man Kim [1,2,*]

[1]   Department of Electrical Engineering, Incheon National University, 12-1 Songdo-dong, Yeonsu-gu, Incheon 406-840, Korea; buivanhaibk@inu.ac.kr (V.-H.B.); hussainakhtar@inu.ac.kr (A.H.)

[2]   Research Institute for Northeast Asian Super Grid, Incheon National University, 119 Academy-ro, Yeonsu-gu, Incheon 22012, Korea

[3]   Department of Electrical and Computer Engineering, Clarkson University, 8 Clarkson Ave., Potsdam, NY 13699, USA; ntthanh@inu.ac.kr

*   Correspondence: hmkim@inu.ac.kr; Tel.: +82-32-835-8769; Fax: +82-32-835-0773

Abstract: Due to the uncertainty in output power of wind farm (WF) systems, a certain reserve capacity is often required in the power system to ensure service reliability and thereby increasing the operation and investment costs for the entire system. In order to reduce this uncertainty and reserve capacity, this study proposes a multi-objective stochastic optimization model to determine the set-points of the WF system. The first objective is to maximize the set-point of the WF system, while the second objective is to maximize the probability of fulfilling that set-point in the real-time operation. An increase in the probability of satisfying the set-point can reduce the uncertainty in the output power of the WF system. However, if the required probability increases, the set-point of the WF system decreases, which reduces the profitability of the WF system. Using the proposed method helps the WF operator in determining the optimal set-point for the WF system by making a trade-off between maximizing the set-point of WF and increasing the probability of fulfilling this set-point in real-time operation. This ensures that the WF system can offer an optimal set-point with a high probability of satisfying this set-point to the power system and thereby avoids a high penalty for mismatch power. In order to show the effectiveness of the proposed method, several case studies are carried out, and the effects of various parameters on the optimal set-point for the WF system are also analyzed. According to the parameters from the transmission system operator (TSO) and wind speed profile, the WF operator can easily determine the optimal set-point using the proposed strategy. A comparison of the profits that the WF system achieved with and without the proposed method is analyzed in detail, and the set-point of the WF system in different seasons is also presented.

Keywords: energy management systems; multi-objective function; optimal set-points; stochastic optimization; wind farm operation

Citation: Bui, V.-H.; Hussain, A.; Nguyen, T.-T.; Kim, H.-M. Multi-Objective Stochastic Optimization for Determining Set-Point of Wind Farm System. *Sustainability* **2021**, *13*, 624. https://doi.org/10.3390/su13020624

Received: 18 December 2020
Accepted: 8 January 2021
Published: 11 January 2021

Publisher's Note: MDPI stays neutral with regard to jurisdictional clai-ms in published maps and institutio-nal affiliations.

1. Introduction

Wind energy, along with other renewable energy sources, is expected to grow substantially in the coming decades and play an important role in fulfilling future world energy needs as well as contributing to reducing global warming. The International Energy Agency (IEA) estimates that the annual wind power could increase to more than 2180 TWh by 2030, which is seven times higher than accumulative wind power production up to 2009 [1,2].

In order to convert wind energy into electricity, a vast number of wind turbine generators (WTGs) and the WF systems have been under construction recently and injecting huge amounts of power into the power system. However, due to the rapid increase in the penetration of wind power, future power systems may face numerous challenges from the supply variability and uncertainty in the output power of WF systems. For small WF systems, this uncertainty can be neglected because the total output power of the WF system is small compared with the power system capacity. Recently, however, WF systems are designed with a huge installed capacity of up to several GW [3]. Therefore, the uncertainty in such large WF systems cannot be neglected, which adversely affects the operation of the power system in terms of power quality, system security, and system stability [4,5]. Various methods have been proposed to handle the uncertainty in the output power of the WF system in the operation of the power system [6,7].

The most common approach to reducing the effect of the uncertainty in the output power of the WF system is to use auxiliary supplies or reserve capacity, such as battery energy storage system (BESSS) [8,9], power-to-hydrogen-to-power system [10], power-to-gas energy storage [11], controllable distributed generators [12], etc. The optimal control of these auxiliary systems can reduce the effect of wind power curtailment by peak shaving as well as by compensating for the power mismatch by the uncertainty in WF's output power. However, the operation and investment costs for this reserve capacity are quite expensive due to the installation of additional controllable distributed sources. In order to reduce these costs, the optimal scheduling and sizing of the reserve capacity are required, considering the uncertainty in the output power of the WF system.

There are several optimization algorithms and strategies for the operation of power systems that have been proposed for optimal sizing and scheduling of the reserve capacity using robust optimization [13], stochastic optimization [14,15], dynamic programming [16], and reinforcement learning (RL)/deep RL [17]. The authors in [13] have proposed a two-stage distributed robust optimization model to investigate the optimization scheduling for the multi-energy coupled system, considering the uncertainty in wind power. This model aims to minimize the expectation of the operation cost under the worst-case condition. The authors in [14,15] have developed a two-stage stochastic programming model for optimal unit commitment and dispatch decisions. The authors in [16] have proposed a capacity sizing method for wind power–energy storage systems using dynamic programming. The authors in [17] have developed a double deep Q-learning-based distributed operation strategy for a BESS considering the uncertainty in the output of wind power. However, these studies in [13–17] only focus on the optimal operation of the power system with a certain uncertainty in the output power of the WF system. The transmission system operators (TSOs) attempt to optimize the operation scheduling of resources outside the WF system, such as power plants and BESSs, to ensure the service reliability in the worst-case (i.e., the output power of the WF is at the lowest bound). A large uncertainty bound can lead to a significant increase in the operation and investment costs due to the requirement of the huge amount of reserve capacity in the power system.

In order to reduce the amount of reserve capacity in the system, the uncertainty in the output power of the WF system should be decreased by optimizing the set-point of the WF system. Various methods have been proposed to determine the set-points of WTGs and the WF system with different objectives [18–20]. The authors in [18] have proposed an operation strategy to optimize the set-point of each WTGs for maximizing the total output power of the WF system. The authors in [19] have investigated the wind power smoothing effect considering the different number of WTGs and the operation of WTGs in the WF system. The authors in [20] have developed an operational strategy to minimize the power deviation in the WF system by optimizing the set-point for each WTGs. However, most studies have focused on maximizing the output power of a WF system [18], smoothing wind power output [19], or minimizing power deviation in the WF system [20]. Determining the set-point of the WF system to reduce the uncertainty of the output power has not been considered in the literature.

Therefore, this study mainly focuses on developing a strategy for the WF operator to determine the optimal set-point of the WF system to reduce the uncertainty in the output power. In the proposed strategy, a multi-objective stochastic optimization model is developed based on mixed-integer linear programming (MILP). The multi-objective function consists of two single objectives; the first objective is to maximize the set-point of the WF system, while the second objective is to maximize the probability of fulfilling the set-point of the WF system in real-time operation. As aforementioned, the set-point of the WF system helps TSO in determining the optimal scheduling for all external resources to fulfill the electric demands. In order to reduce the uncertainty of the output power, the WF operators need to assure that they can satisfy the set-point in the real-time operation and inject it into the power system. Any power mismatch between the actual output power and the committed power may result in a high penalty for the WF operator. To increase the profit of the WF by selling power to the grid, it is easy to observe that the WF should inform a high set-point to the TSO. However, if the set-point for WF systems increases, the probability of fulfilling such high set-point decreases. By using the proposed method, the WF operator is able to determine the set-point for the WF system by deciding a trade-off between maximizing the set-point and increasing the probability of fulfilling that set-point. A high probability of fulfilling the set-point helps the WF system avoiding a penalty for power mismatch between the actual output power and the set-point of output power and also reduces the uncertainty of the output power of the WF system. This helps the TSO to significantly reduce the reserve capacity and thereby reducing the investment and operation costs for the whole system. The effect of the ratio of weight factors and the minimum probability requirement on the set-point of the WF system are analyzed in detail in the simulation section. In addition, a comparison of the profits that the WF system achieved with and without the proposed method is analyzed in detail and the set-point of the WF system is also presented with different wind speed profiles for the four seasons in a year. The major contributions of this study are listed as follows:

- A multi-objective stochastic optimization model is developed to determine the optimal set-point of WF with different wind probability density functions. This helps to reduce the uncertainty of the output power of WF and thereby to reduce the requirement of reserve capacity;
- A novel algorithm is proposed for a trade-off between maximizing the set-point of WF and increasing the probability of satisfying this set-point in real-time operation. With any input information, the WF operator is able to find out the optimal set-point with a required probability;
- By increase, the probability of fulfilling the set-point in real-time operation, the uncertainty of the output power of WF can be decreased. This results in the reduction of operation cost of the whole system.

This paper is arranged as follows: In Section 2, the system configuration and operation of the system are presented. In Section 3, the detailed strategy for determining the optimal set-point of the system WF is presented. In Section 4, a MILP-based mathematical model for multi-objective stochastic optimization is formulated. In Section 5, the numerical results are analyzed, and the comparison on the set-point of the WF system is also presented. The conclusion of this study is summarized in Section 6.

2. System Configuration

Figure 1 depicts a typical WF system, which is connected to the power system to supply electric demands. The whole system is operated by a transmission system operator (TSO). The TSO's primary task is to determine the optimal scheduling for supply resources (i.e., power plants and renewable energies sources) and manage the operation of the entire system in real-time operation. To optimize the scheduling for power plants, TSO requires the set-point from the WF system. The WF system normally operates by the WF operator, and this management system is also responsible for determining the optimal set-point of the WF system and informing the TSO. This set-point of the WF system plays a vital role in the optimal scheduling of other resources. Therefore, the WF system must be able to fulfill its set-point in real-time operation. Any power mismatch between actual output and committed power results in a massive penalty for the WF system from TSO. Therefore, this study mainly focuses on determining the optimal set-point of a WF by a trade-off between maximizing the set-point and increasing the probability of satisfying the set-point in the real-time operation. The operation strategy for the WF system is presented in detail in the next section.

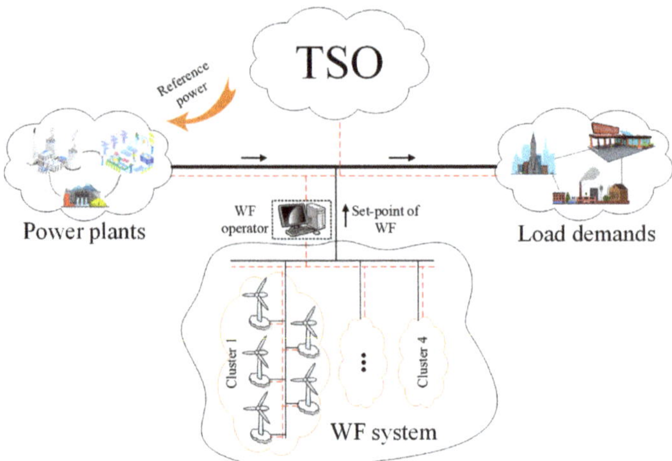

Figure 1. A typical wind farm (WF) system configuration.

3. A Strategy for Determining the Optimal Set-Point of WF System

In this section, we present a strategy for the WF operator to determine the optimal set-point of the WF system, as shown in Figure 2. First, the wind speed parameter is assumed to comply with the Weibull distribution, and the detailed information about the Weibull parameters (i.e., Weibull shape and scale) are taken as input data. Based on the probability density function (PDF) of wind speed data, numerous scenarios for wind speed at each interval is generated to ensure the accuracy of the proposed method. However, a large number of scenarios significantly increases the computation burden for the simulation system. Therefore, a scenario reduction algorithm was developed to merge similar scenarios, as shown in detail in Algorithm 1. After merging all similar scenarios, the output capacity of each WTGs is calculated using (7) with the corresponding wind speed in each scenario. The total output power of the WF system in each scenario is used to determine the optimal set-point of the WF system by solving a multi-objective stochastic optimization model. The first objective is to maximize WF's profitability by selling power to the power system (i.e., maximizing the set-point). However, the WF operator cannot always ensure that the WF system always meets the maximum set-point in real-time operation. Therefore, the WF operator may try to make a trade-off between maximizing the set-point of the WF system and increasing the probability of fulfilling that set-point considering the ratio of weight factors and the minimum probability requirement. In order to determine the actual probability for each set-point of the WF system, we also developed Algorithm 2, and the detailed explanation for Algorithm 2 is presented in Section 4.

As stated previously, Algorithm 1 was developed to reduce the number of scenarios by merging similar scenarios in the scenario set. First, the Kantorovich distances are calculated for each pair of scenarios in the scenario set, and then a similar pair of scenarios (k, s) is determined, as shown in Algorithm 1. Because the two similar scenarios often do not contribute much in evaluating the proposed method, one scenario can be omitted, and the probability for the other is updated simply by the sum of the probabilities of both scenarios [21,22]. This process is repeated until the number of scenarios reduced to the minimum scenario requirement.

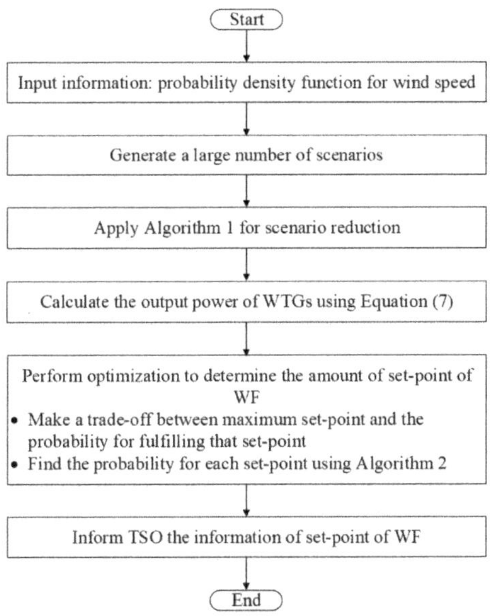

Figure 2. The strategy for determining the set-point of the WF system.

Algorithm 1: Scenario Reduction

Generate S scenarios

Scenarios i: $s(i) = \left\{ \lambda_1^i, \lambda_2^i, \lambda_n^i, ..., \lambda_N^i \right\}$

while $S < S_{req.}$ **do:**

    for s = 1 to S **do:**

        for k = 1 to S **do:**

            //Calculate Kantorovich distance

$$d(s,k) = \left(\sum_{n=1}^{N} \left(\lambda_n^s - \lambda_n^k \right)^2 \right)^{1/2}$$

        end

    end

    //determine scenario s that is reduced

$$\min_{s \in S} p(s) \cdot \left\{ \min_{\substack{k \neq s \\ k \in S}} \left(p(k) \cdot d(s,k) \right) \right\}$$

    //update number of scenarios

$$S \leftarrow S - 1$$

    //determine scenario k that nearest the reducing scenario s

$$k = \arg s \min_{\substack{k \neq s \\ k \in S}} \left\{ p(s) \cdot p(k) \cdot d(s,k) \right\}$$

    //change probability of scenario k

$$p(k) \leftarrow p(k) + p(s)$$

end

In the next section, a detailed mathematical model is developed to determine the set-point of the WF system with different input data.

4. Mathematical Model

In this section, a mathematical model is developed based on mixed-integer linear programming (MILP) to determine the optimal set-point of the WF system. This optimal set-point is determined by making a trade-off between maximizing the set-point of WF and increasing the probability of fulfilling this set-point in real-time operation. Suppose the WF operator informs a high set-point, which is more profitable; however, the probability of fulfilling that set-point may significantly reduce. Hence, it is important having a trade-off between these two factors (i.e., maximizing the set-point and increasing the probability of fulfilling that set-point). The following mathematical model is developed to analyze the effects of different parameters on determining the set-point of the WF system.

First, in order to evaluate the effectiveness of the proposed method, we assume that the wind speed at WTGs follows Weibull distribution during each season as in [22,23]. The probability density function (PDF) and cumulative distribution functions (CDF) of the Weibull distribution are shown in (1) and (2), respectively. The Weibull shape (k) and Weibull scale (λ) are taken as input parameters in different seasons, and these parameters are taken from [23].

$$f(v) = \frac{k}{\lambda}\left(\frac{v}{\lambda}\right)^{k-1} \exp\left[-\left(\frac{v}{\lambda}\right)^k\right] \tag{1}$$

$$F(v) = 1 - \exp\left[-\left(\frac{v}{\lambda}\right)^k\right] \tag{2}$$

To ensure accuracy in determining the optimal set-point of a WF system, numerous scenarios (S) needs to be generated using PDFs and CDFs. Each scenario is a row vector Vs consisting of the wind speed at each interval of the day from $v_{1,s}$ to $v_{T,s}$, as shown in (3). The probability of each scenario ($prob_s$) is determined by multiplying the probability of each time interval having a certain wind speed $v_{s,t}$, as shown in (4). In this study, we assume that the number of scenarios is large enough, and therefore the total probability of occurrence of the entire scenario set (S) is 1, as shown in (5).

$$V_s = (v_{s,1}, v_{s,2}, \ldots v_{s,t}, \ldots, v_{s,T}) \quad \forall s \in S \tag{3}$$

$$prob_s = \prod_{t=1}^{T}(p_{s,t}) \qquad \forall s \in S \tag{4}$$

where: $p_{s,t}$ is the probability of interval t having wind speed $v_{s,t}$

$$\sum_{s=1}^{S} prob_s = 1 \tag{5}$$

The total output power of the WF system is determined by the total output power of each WTGs, as shown in (6), where the amount of output power of each WTG is calculated by (7) for each corresponding input of wind speed. In order to determine the optimal set-point of the WF system, a multi-objective function is developed, as shown in (8). The first part of (8) represents the normalization of the set-point of the WF system, where the minimum and maximum value of the WF system's output power is determined using (9) and (10), respectively. The second part of (8) is the probability of satisfying the set-point in real-time operation for the WF system. The weight factors α and β show the importance of every single objective in the multi-objective function. The constraints (11) and (12) show the relationship between the weight factors α and β, the values of α and β must be in the range (0, 1) and their sum needs to be 1. If the value of α is close to 1, the WF operator is more concerned with maximizing the set-point of the WF system. On the contrary, if the value of α is close to 0, the WF operator is more concerned about the possibility that the

WF system can satisfy the set-point in real-time operation. Constraints (13), (14) represent the bound of the set-point of the WF system and constraint (15) represents the minimum probability requirement for satisfying the set-point of the WF system in real-time operation.

$$P_{WF,s,t}^{Out} = \sum_{n=1}^{N} P_{n,s,t}^{WTG} \qquad \forall s \in S, t \in T \tag{6}$$

$$P_{n,s,t}^{WTG} = \begin{cases} 0 & v_{n,s,t} < v_{cut-in} \ or \ v_{n,s,t} > v_{cut-out} \\ \frac{1}{2} C_p(\beta,\lambda)\rho\pi R^2 v_{n,s,t}^3 & v_{cut-in} \leq v_{n,s,t} < v_{rate} \\ P_{n,rate}^{WTG} & v_{rate} \leq v_{n,s,t} \leq v_{cut-out} \end{cases} \tag{7}$$
$$\forall n \in N, s \in S, t \in T$$

$$Max\left\{ \alpha \cdot \left(\frac{P_{WF}^{Sch} - P_{WF,min}^{Out}}{P_{WF,max}^{Out} - P_{WF,min}^{Out}} \right) + \beta \cdot \left(prob\left(P \geq P_{WF}^{Sch} \right) \right) \right\} \tag{8}$$

$$P_{WF,min}^{Out} = \min\left(\sum_{t=1}^{T} P_{WF,s,t}^{Out} \right) \qquad \forall s \in S \tag{9}$$

$$P_{WF,max}^{Out} = \max\left(\sum_{t=1}^{T} P_{WF,s,t}^{Out} \right) \qquad \forall s \in S \tag{10}$$

$$\alpha + \beta = 1 \tag{11}$$

$$0 \leq \alpha, \beta \leq 1 \tag{12}$$

$$0 \leq P_{WF}^{Sch} \leq T \cdot P_{WF,rate}^{Out} \tag{13}$$

$$P_{WF,rate}^{Out} = \sum_{n=1}^{N} P_{n,rate}^{WTG} \tag{14}$$

$$prob\left(P > P_{WF}^{Sch} \right) \geq prob_{req} \qquad \forall t \in T \tag{15}$$

The probability of fulfilling the set-point of WF in the left-side of constraint (15) is determined by Algorithm 2. This algorithm helps the WF operator determine the probability that the WF system can meet the set-point in real-time operation. Algorithm 2 checks the output power of the WF system in each scenario and compares it with a certain set-point (P_{WF}^{Sch}) of WF. Suppose the output power of a scenario is greater than P_{WF}^{Sch}, the probability for satisfying the set-point is updated by adding that scenario's probability. After checking all scenarios, the WF operator can determine the probability of satisfying the set-point.

Algorithm 2: Determining probability of a set-point

Input: all N_s scenarios

for s = 1 to N_s **do:**

$\quad$ **if** $\sum_{t=1}^{T} P_{WF,s,t}^{Out} \geq P_{WF}^{Sch}$ **do:**

$\quad\quad prob\left(P \geq P_{WF}^{Sch} \right) \leftarrow prob\left(P \geq P_{WF}^{Sch} \right) + prob_s$

$\quad$ **end**

end

In the next sections, the optimal set-point of the WF system is presented in detail with different PDFs of wind speed. Furthermore, the effects of various parameters on the optimal set-point of the WF system is analyzed in detail.

5. Numerical Results

In this section, different probability density functions (PDFs) are presented for the four seasons in a year, respectively. In each season, the optimal set-point is analyzed in detail based on the minimum probability requirements and the ratio of weight factors in the objective function (8).

5.1. Input Data

As stated earlier, wind speed follows the Weibull distribution. In this study, we analyze the changes in the set-point of the WF system during different seasons in a year. Each season has different parameters for the Weibull distribution. PDFs and CDFs of wind speed are shown in Figure 3a,b for different seasons, respectively. It can be observed the average wind speed in fall and summer is higher than in spring and fall. This means that the set-point in fall and summer is usually higher in spring and winter. Detailed parameters for Weibull shape and scale are tabulated in Table 1 for different seasons.

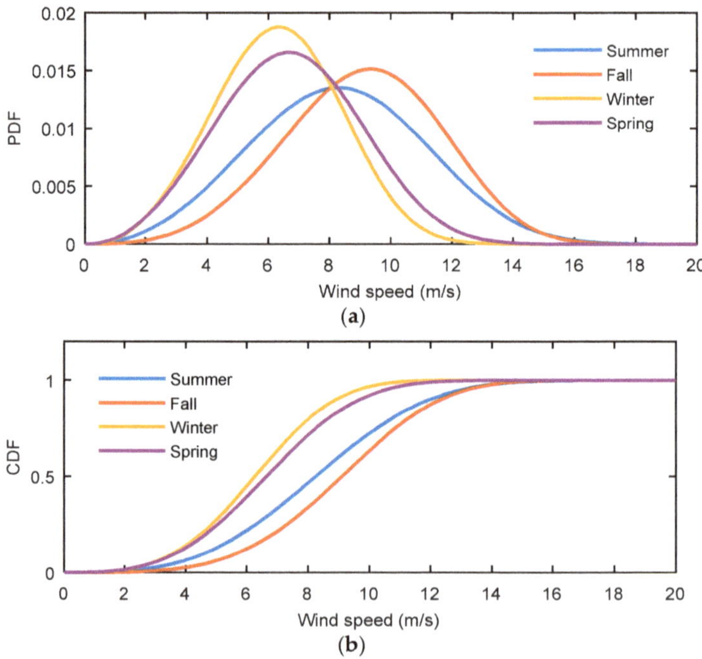

Figure 3. Weibull distribution model of wind speed in different seasons: (**a**) probability density function; (**b**) cumulative distribution function.

Table 1. Detailed parameters for Weibull distribution in different seasons [23].

Seasons	Parameters for Weibull Distribution	
	Weibull Shape (-)	Weibull Scale (m/s)
Spring	3.2	7.5
Summer	3.24	9.29
Fall	3.99	10.04
Winter	3.61	7.03

The test WF system consists of 20 WTGs, and the close WTGs are grouped to form a cluster. In this study, we assume that 20 WTGs are grouped into 4 clusters, and each cluster has 5 WTGs, as shown in Figure 1. All WTGs in the WF system has the same configuration, and detailed information for WTGs is presented as follows [24].

- The rated power is 10 MW;
- The minimum operation point is 10% of the rated power, i.e., 1 MW;
- The maximum ramp-up/ramp-down is 20% of the rated power, i.e., 2 MW.

To determine the optimal set- point of a WF system and analyze the effectiveness of the proposed method, the multi-objective stochastic optimization model is implemented in Visual Studio C++ integrated with IBM ILOG CPLEX 12.6 [25].

5.2. Determine Optimal Set-Point of WF in Spring with a Large Scenario Set

In this section, a detailed analysis of the optimal set-point of the WF system is presented with wind speed data in spring. The optimal set-point of the WF system is the total energy that the WF system injects into the power system during a day with a wind speed profile in spring. The scheduling horizon is a day, and each interval is set to 1 h. The effects of minimum probability requirement and ratio of weight factors on the set-point of WF are also presented in detail.

In order to ensure the accuracy of the proposed method, we generate 10,000 scenarios. However, a large number of scenarios increases the computation burden for the simulation system. Therefore, Algorithm 1 is used to reduce the number of scenarios to 1000. In the first case study, the ratio of weight factors (α/β) is fixed to $1/1$, and the minimum probability requirement in constraints (15) is varied from 0.1 to 0.95. The optimal set-point of the WF system is shown in Figure 4. It is easy to observe that the set-point of WF decreases if the minimum probability requirement increases. The set-point is nearly 1400 MWh if the minimum probability requirement is 0.1. This means that the WF can only guarantee to satisfy the set-point (i.e., 1400 MWh) with a probability of 10% in real-time operation. However, if the set-point reduces to nearly 950 MWh, the WF can guarantee to satisfy this set-point with a probability of up to 95% in real-time operation. The actual probability of fulfilling a given set-point is shown in detail in the second axes of Figure 4. It requires a trade-off between maximizing the set-point of WF and maximizing the probability of satisfying that set-point. This is because the WF operator may face a massive penalty for the power mismatch between the actual output power and the set-point of output power during the real-time operation of the power system. Therefore, it can be concluded that the set-point should be set at around 1000 MWh in this season, and the WF can guarantee to fulfill this set-point with a probability of up to 85%.

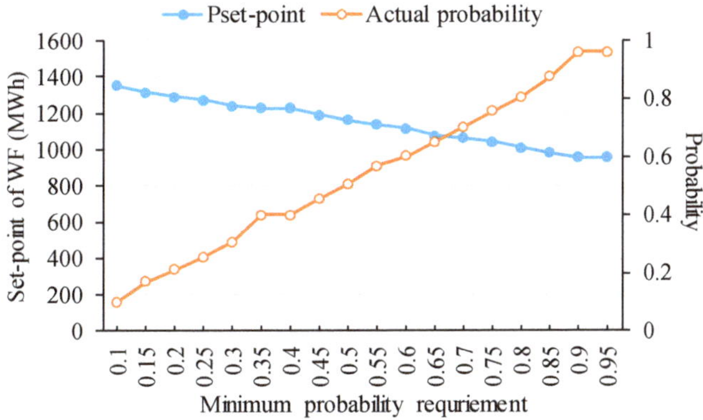

Figure 4. The set-point of WF with different values of minimum probability requirement.

In the second case study, the minimum probability requirement is set to 0.85 to avoid the penalty for power mismatch, while the weight factor α is varied from 0 to 1. The value of α close to 1, the WF operator tended to pay more attention to the maximum the set-point of the WF system. By contrast, if the value of α close to 0, the WF operator tends to pay more attention to the high probability of fulfilling this set-point in real-time operation (i.e., reduce the uncertainty of the output power of WF). Depending on information from TSOs, such as the selling price and the penalty for mismatch power, the WF operator will determine the ratio of weight factors (α/β) to take a trade-off between the profits from selling wind power and the possible penalty of power mismatch. It can be observed from Figure 5 that the set-point of the WF system is determined with different values of the weight factor α. In order to ensure the probability of satisfying the set-point from 90% in real-time operation, the set-point of the WF system should be set in a range from 800 MWh to around 1000 MWh, which corresponds to the value of the α weight factor from 0.05 to 0.8.

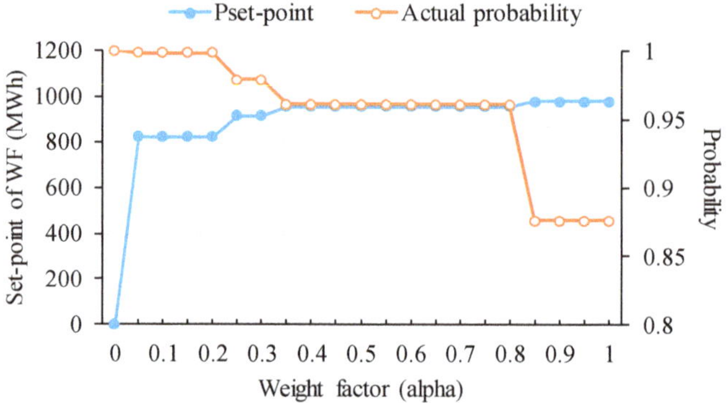

Figure 5. The set-point of WF with different values of weight factor (α).

Finally, the effects of the value of weight factor α and the minimum probability requirement on determining the set-point of the WF system are shown in Figure 6. In this case study, the value of weight factor α was varied from 0.05 to 1, and the minimum probability requirement is varied from 0.5 to 0.9. It can be observed from Figure 6 that the effect of the minimum probability requirement on the set-point of the WF system is negligible, especially in the case of the small value of α, while the value of α has a high effect on the set-point of the WF system. The maximum set-point of the WF system is 1160 MWh, corresponding to a value of α of 1 and the minimum probability requirement of 0.5. However, as mentioned earlier, the value of α and the minimum probability requirement is determined based on a trade-off between the profits from selling wind power and the penalty of power mismatch between the actual output and the committed power of the WF system. Based on the above-detailed analysis, the WF operator can easily determine the optimal set-point with any value of α and the minimum probability requirement.

5.3. Comparison of the Optimal Set-Point with and Without the Proposed Method

To show the effectiveness of the proposed method, a detailed comparison of the set-point of the WF system will be presented using the proposed method and not using the proposed method. As stated in Section 4, the proposed method is to determine the optimal set-point of a WF system by making a trade-off between maximizing the output power of the WF system and maximizing the probability of satisfying this set-point in real-time operation. This can reduce the penalty for mismatch power between the set-point and the actual output power. Without the proposed method, the WF operator usually sets the set-point based on the history data (i.e., PDF). However, this method can lead to the following two problems, (1) a low set-point with a high probability and (2) a high set-point with a low probability. Both cases can reduce the profit of the WF system.

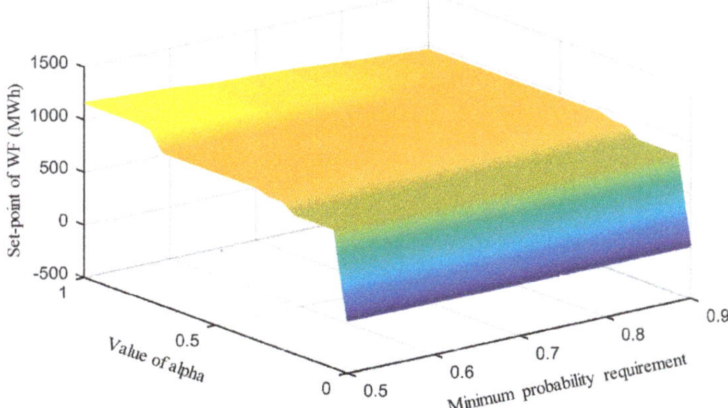

Figure 6. The set-point of WF with different values of weight factor (α) and minimum probability requirement.

Therefore, in this section, we analyze the effect of the set-point on the profit of a WF system using the wind speed profile in spring. Without the proposed method, we assume that the set-point of a WF system is 750 MWh and 1000 MWh. Based on the probability density function in Section 5.1, the corresponding probability to satisfy each set-point in real-time operation is easily determined. The detailed set-points of the WF system and the probability of fulfilling these set-points are tabulated in Table 2.

Table 2. Set-point of WF with and without the proposed method.

Without Proposed Method		With Proposed Method	
Set-Point (MWh)	**Probability**	**Set-Point (MWh)**	**Probability**
750	0.99	980	0.87
1000	0.81	-	-

To calculate the profit of the WF system, we assume that the selling price is 100 KRW/kWh, and the penalty for the mismatched power is 500 KRW/kWh. The profit of the WF system is calculated based on the set-point and the amount of mismatch power between the set-point and the actual output power in real-time operation, as shown in Table 3. When the set-point is small (i.e., 750 MWh), the penalty for the mismatched power decreases significantly because the WF system can ensure to meet this set-point with the probability of 0.99. However, the amount of selling power to the power system is also small and thus significantly reducing the profitability of the WF system. Conversely, when the set-point increases (i.e., 1000 MWh), the probability of fulfilling the set-point in real-time operation is only 0.81. Therefore, the WF system often faces a high penalty due to mismatched power. That is the main reason why we proposed a new algorithm to determine the optimal set-point of the WF system to maximize the total profit for the WF system. It can be seen that the optimal set-point is 980 MWh obtained using the proposed method, which provides the highest profit with a different amount of mismatch power between the set-point and the actual output power.

Table 3. Profit of WF with and without the proposed method ($\times 10^3$ KRW).

Set-Point (MWh)	Possible Power Mismatch (MWh)				
	10	20	30	40	50
750	74,200	74,150	74,100	74,050	74,000
1000	80,050	79,100	78,150	77,200	76,250
980 (optimal case)	84,610	83,960	83,310	82,660	82,010

5.4. Optimal Set-Point of WF in Different Seasons

In the previous sections, the effects of the various parameters on determining the set-point of the WF system were analyzed in detail using the wind speed data in spring. In this section, the set-point of the WF system is determined with different input parameters of wind speed for other seasons (i.e., summer, fall, and winter), and the minimum probability requirement is varied from 0.1 to 0.95.

The set-point of the WF system is shown in detail in Figure 7a–c for summer, fall, and winter, respectively. Similar to the discussion in Section 5.2, the set-point of the WF system will decrease if the minimum probability requirement increases. To ensure power supply reliability (i.e., reducing the power mismatch between the set-point of WF's output power and the actual output power), the minimum probability requirement is usually set greater than or equal to 0.85. If the minimum probability requirement is varied from 0.85 to 0.95, it can be seen from Figure 7a–c that the set-point changes from 1710 MWh to 1760 MWh for summer, from 2180 MWh to 2430 MWh for fall, and from 840 MWh to 870 MWh for winter, respectively. The change in the set-point of the WF system is reasonable with the given input data of wind speed in Table 1. The average value of the wind speed in fall and summer is much larger than that in winter. Therefore, although the minimum requirement probability is the same, the WF operator could determine a high set-point for the WF system during summer and fall, while this value usually decreases significantly during spring and winter. A detailed comparison of the set-point of the WF system is analyzed in detail in the next section.

5.5. Comparison of the Optimal Set-Point of WF among the Four Seasons

In this section, the set-point of the WF system and the actual probability to satisfy each set-point in real-time operation are presented and compared among the four seasons in a year with different wind speed parameters. In this case study, the weight factor α is varied from 0 to 1, and the minimum probability requirement is set to 0.85 to avoid a penalty for power mismatch. If the value of α is 0, the WF operator is only interested in maximizing the probability of satisfying the set-point of a WF system. Therefore, the set-point is set to 0, and the probability of fulfilling this set-point is 1. On the contrary, if the value of α varies from 0.4 to 1, the set-point of the WF system does not change much, as shown in Figure 8a. Therefore, the set-point of WF systems can be set at 980 MWh, 1760 MWh, 2430 MWh, 870 MWh for spring, summer, fall, and winter, respectively. It can be observed that the set-point of WF is largest in the fall and the smallest in winter. Corresponding to each set-point of the WF system, the WF operator always ensures that the probability of fulfilling the set-point in real-time operation is greater than or equal to 0.85, as shown in Figure 8b.

To show more clearly the difference in the set-point of the WF system during different seasons in a year, the detailed set-points of the WF system are tabulated in Table 4 with the optimum value in spring as a reference case. It can be observed that the set-points of the WF system increase significantly during summer and fall. If the value of α varies from 0 to 1, the increase in the set-point of WF could be from 72% to 100% in summer and 140% to 153% in fall compared with the set-point in spring, while the set-point in winter is slightly lower than in spring (i.e., from −7% to −12%).

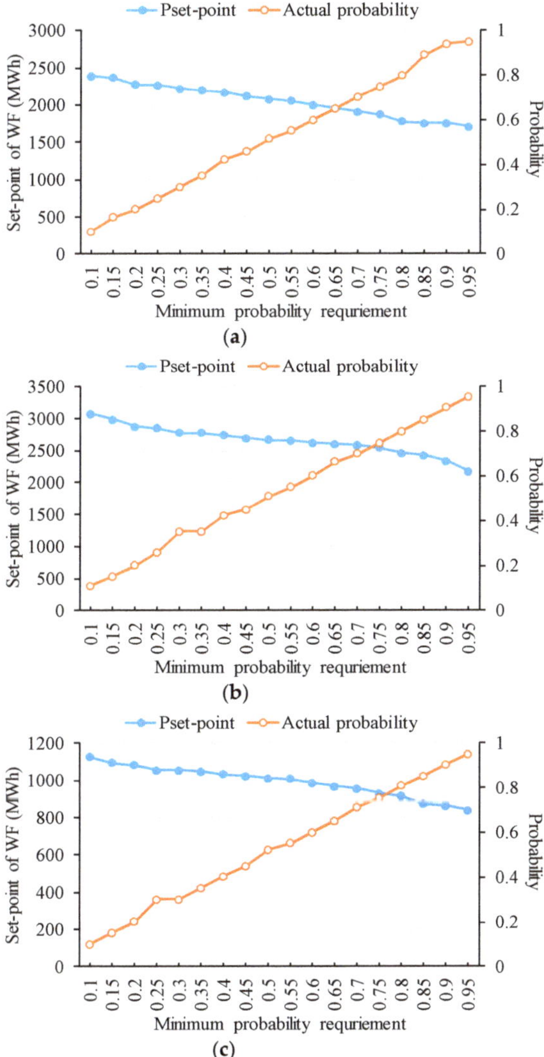

Figure 7. The set-point of WF with a different value of minimum probability requirement: (**a**) summer; (**b**) fall; (**c**) winter.

In this study, a detailed analysis of the set-point of the WF system was presented with different weight factors in the multi-objective function and wind speed profiles. It can be seen that the proposed method plays an important role in determining the optimal set-point for the WF system. This enables the WF operator to maintain a high profit by avoiding a penalty for any mismatch power between the set-point and the actual output power in real-time operation. The proposed method can be integrated into the energy management system of the WF system, and the optimal set-point is updated with any input information, such as wind speed profile and weight factors. In this study, the proposed method was tested with seasonal input data, and the optimal set-point is determined for a day in the season. However, the proposed method is also applicable to the different time scheduling horizons (e.g., an hour, a day, a week, etc.) with the corresponding probability density functions.

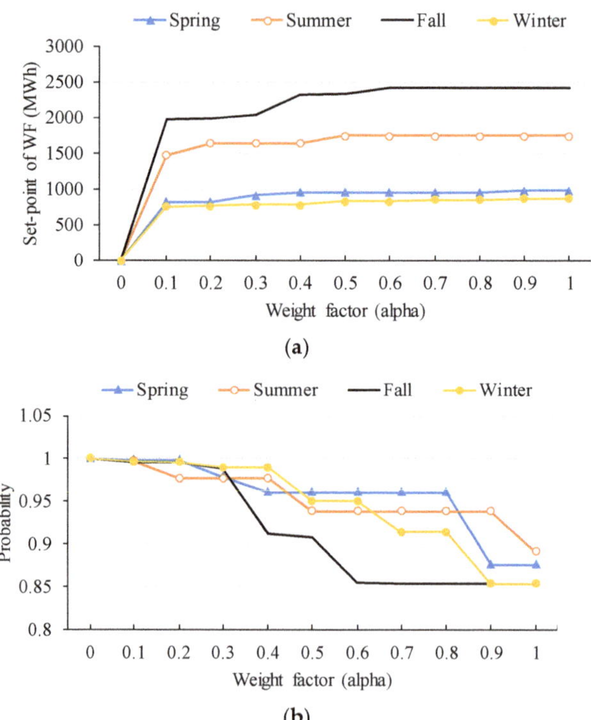

Figure 8. Comparison of the optimal results among different seasons: (**a**) the set-point of WF; (**b**) the probability of fulfilling the optimal set-point.

Table 4. Set-point of WF in different reasons and value of weight factors.

Value of Alpha	Increase in the Set-Point of WF			
	Spring (%)	Summer (%)	Fall (%)	Winter (%)
0	0.00	0.00	0.00	0.00
0.1	0.00	79.03	140.46	−7.64
0.2	0.00	100.11	141.34	−7.16
0.3	0.00	80.02	122.18	−14.33
0.4	0.00	72.10	143.30	−18.10
0.5	0.00	83.10	144.43	−12.61
0.6	0.00	83.10	153.06	−12.61
0.7	0.00	83.10	153.19	−10.02
0.8	0.00	83.10	153.19	−10.02
0.9	0.00	78.72	147.14	−11.21
1	0.00	78.83	147.14	−11.21

6. Conclusions

In this study, a multi-objective stochastic optimization model was proposed to determine the set-point for a WF system. The first objective is to maximize the set-point of the WF system, while the second objective is to maximize the probability of fulfilling that set-point in real-time operation. The proposed strategy mainly focuses on determining the set-point of the WF system by a trade-off between these two objectives considering the ratio of the weight factors in the multi-objective function and the minimum probability requirement. A comparison of the profit of the WF system using the proposed method and not using the proposed method were analyzed in detail. The results indicate that the WF system can always ensure the maximum profit at the optimal set-point achieved by the proposed method. Using the proposed method not only maintains a high set-point for the WF system but also ensures a high probability for satisfying this set-point in real-time operation. According to the wind speed profile in spring, the set-point of the WF system is set from 800 MWh to 1000 MWh with the value of α from 0.05 to 0.8 to ensure the probability of satisfying the set-point greater than or equal to 0.95 in real-time operation. A similar analysis also has been carried out with different wind data for four seasons, and the set-point of WF systems should be set at 980 MWh, 1760 MWh, 2430 MWh, 870 MWh for spring, summer, fall, and winter, respectively. It can be observed that the set-point of the WF system is largest in fall and is lowest in winter. With these set-points of WF in the four seasons, the WF operator always ensures that the probability of fulfilling these set-points in the real-time operation is greater than or equal to 0.85.

Author Contributions: V.-H.B. conceived and designed the experiments; V.-H.B., A.H., and T.-T.N. performed the experiments and analyzed the data; H.-M.K. revised and analyzed the results; V.-H.B. wrote the paper. All authors have read and agreed to the published version of the manuscript.

Funding: This research was supported by Korea Electric Power Corporation. (Grant number: R18XA03).

Institutional Review Board Statement: Not applicable.

Informed Consent Statement: Not applicable.

Data Availability Statement: Not applicable.

Conflicts of Interest: The authors declare no conflict of interest.

Nomenclatures:

Sets

T	Scheduling horizon
S	Set of scenarios
N	Set of WTGs

Indices

t	Index of time intervals
s	Index of scenarios
n	Index of WTGs

Parameters

$f(v), F(v)$	Probability density function and cumulative distribution function of wind speed
k, λ	Weibull shape and scale parameters
$v_{s,t}$	Wind speed at t in scenario s
V_s	Wind speed vector in scenario s
$prob_s$	Probability of scenario s
$P_{n,s,t}^{WTG}$	Output power of WTG n at t in scenario s
$P_{n,rate}^{WTG}$	Rated output power of WTG n
$v_{cut-in}, v_{cut-out}$	Cut-in, cut-out wind speed
v_{rate}	Rated wind speed of WTGs
$P_{WF,s,t}^{Out}$	Output power of the WF system at t in scenario s
P_{WF}^{Sch}	Optimal set-point of the WF system
$prob\left(P \geq P_{WF}^{Sch}\right)$	Probability of fulfilling the set-point in real-time operation
$prob_{req}$	Minimum required probability of fulfilling the set-point in real-time operation
α, β	Weigh factors of different objective
$P_{WF,min}^{Out}$	Minimum set-point of the WF system
$P_{WF,max}^{Out}$	Maximum set-point of the WF system
$P_{WF,rate}^{Out}$	Rated output power of the WF system

References

1. He, C.; Zhang, X.; Liu, T.; Wu, L.; Shahidehpour, M. Coordination of interdependent electricity grid and natural gas network—A review. *Curr. Sustain. Renew. Energy Rep.* **2018**, *5*, 23–36. [CrossRef]
2. Mirzaei, M.A.; Sadeghi-Yazdankhah, A.; Mohammadi-Ivatloo, B.; Marzband, M.; Shafie-khah, M.; Catalão, J.P. Integration of emerging resources in IGDT-based robust scheduling of combined power and natural gas systems considering flexible ramping products. *Energy* **2019**, *189*, 116195. [CrossRef]
3. Kim, J.H.; Nam, J.; Yoo, S.H. Public acceptance of a large-scale offshore wind power project in South Korea. *Mar. Policy* **2020**, *120*, 104141. [CrossRef]
4. Abbas, S.R.; Kazmi, S.A.A.; Naqvi, M.; Javed, A.; Naqvi, S.R.; Ullah, K.; Khan, T.U.R.; Shin, D.R. Impact analysis of large-scale wind farms integration in weak transmission grid from technical perspectives. *Energies* **2020**, *13*, 5513. [CrossRef]
5. Fernández-Guillamón, A.; Gómez-Lázaro, E.; Muljadi, E.; Molina-García, Á. Power systems with high renewable energy sources: A review of inertia and frequency control strategies over time. *Renew. Sustain. Energy Rev.* **2019**, *115*, 109369. [CrossRef]
6. Deng, X.; Lv, T. Power system planning with increasing variable renewable energy: A review of optimization models. *J. Clean. Prod.* **2020**, *246*, 118962. [CrossRef]
7. Mohandes, B.; El Moursi, M.S.; Hatziargyriou, N.; El Khatib, S. A review of power system flexibility with high penetration of renewables. *IEEE Trans. Power Syst.* **2019**, *34*, 3140–3155. [CrossRef]
8. Dui, X.; Zhu, G.; Yao, L. Two-stage optimization of battery energy storage capacity to decrease wind power curtailment in grid-connected wind farms. *IEEE Trans. Power Syst.* **2017**, *33*, 3296–3305. [CrossRef]
9. Jiang, X.; Nan, G.; Liu, H.; Guo, Z.; Zeng, Q.; Jin, Y. Optimization of battery energy storage system capacity for wind farm with considering auxiliary services compensation. *Appl. Sci.* **2018**, *8*, 1957. [CrossRef]
10. Zhang, Y.; Wang, L.; Wang, N.; Duan, L.; Zong, Y.; You, S.; Maréchal, F.; Yang, Y. Balancing wind-power fluctuation via onsite storage under uncertainty: Power-to-hydrogen-to-power versus lithium battery. *Renew. Sustain. Energy Rev.* **2019**, *116*, 109465. [CrossRef]
11. Grueger, F.; Möhrke, F.; Robinius, M.; Stolten, D. Early power to gas applications: Reducing wind farm forecast errors and providing secondary control reserve. *Appl. Energy* **2017**, *192*, 551–562. [CrossRef]
12. Cobos, N.G.; Arroyo, J.M.; Alguacil, N.; Street, A. Robust energy and reserve scheduling under wind uncertainty considering fast-acting generators. *IEEE Trans. Sustain. Energy* **2018**, *10*, 2142–2151. [CrossRef]
13. Zhang, Y.; Le, J.; Zheng, F.; Zhang, Y.; Liu, K. Two-stage distributionally robust coordinated scheduling for gas-electricity integrated energy system considering wind power uncertainty and reserve capacity configuration. *Renew. Energy* **2019**, *135*, 122–135. [CrossRef]
14. Shin, J.; Lee, J.H.; Realff, M.J. Operational planning and optimal sizing of microgrid considering multi-scale wind uncertainty. *Appl. Energy* **2017**, *195*, 616–633. [CrossRef]
15. Bludszuweit, H.; Domínguez-Navarro, J.A. A probabilistic method for energy storage sizing based on wind power forecast uncertainty. *IEEE Trans. Power Syst.* **2010**, *26*, 1651–1658. [CrossRef]
16. Liu, Y.; Wu, X.; Du, J.; Song, Z.; Wu, G. Optimal sizing of a wind-energy storage system considering battery life. *Renew. Energy* **2020**, *147*, 2470–2483. [CrossRef]
17. Bui, V.H.; Hussain, A.; Kim, H.M. Double deep q-learning-based distributed operation of battery energy storage system considering uncertainties. *IEEE Trans. Smart Grid* **2019**, *11*, 457–469. [CrossRef]
18. González, J.S.; Payán, M.B.; Santos, J.R.; Rodríguez, Á.G.G. Maximizing the overall production of wind farms by setting the individual operating point of wind turbines. *Renew. Energy* **2015**, *80*, 219–229. [CrossRef]
19. Yang, M.; Zhang, L.; Cui, Y.; Zhou, Y.; Chen, Y.; Yan, G. Investigating the wind power smoothing effect using set pair analysis. *IEEE Trans. Sustain. Energy* **2020**, *11*, 1161–1172. [CrossRef]
20. Bui, V.H.; Hussain, A.; Kim, H.M. Optimal operation of wind farm for reducing power deviation considering grid-code constraints and events. *IEEE Access* **2019**, *7*, 139058–139068. [CrossRef]
21. Ma, X.Y.; Sun, Y.Z.; Fang, H.L. Scenario generation of wind power based on statistical uncertainty and variability. *IEEE Trans. Sustain. Energy* **2013**, *4*, 894–904. [CrossRef]
22. Xu, J.; Yi, X.; Sun, Y.; Lan, T.; Sun, H. Stochastic optimal scheduling based on scenario analysis for wind farms. *IEEE Trans. Sustain. Energy* **2017**, *8*, 1548–1559. [CrossRef]
23. Shoaib, M.; Siddiqui, I.; Rehman, S.; Khan, S.; Alhems, L.M. Assessment of wind energy potential using wind energy conversion system. *J. Clean. Prod.* **2019**, *216*, 346–360. [CrossRef]
24. Bui, V.H.; Hussain, A.; Lee, W.G.; Kim, H.M. Multi-objective optimization for determining trade-off between output power and power fluctuations in wind farm system. *Energies* **2019**, *12*, 4242. [CrossRef]
25. *IBM ILOG CPLEX V12.6 User's Manual for CPLEX 2015, CPLEX Division*; ILOG: Incline Village, NV, USA, 2015.

sustainability

Article

Expert Views on the Future Development of Biogas Business Branch in Germany, The Netherlands, and Finland until 2030

Erika Winquist [1,*], Michiel Van Galen [2], Simon Zielonka [3], Pasi Rikkonen [1], Diti Oudendag [2], Lijun Zhou [3] and Auke Greijdanus [2]

[1] Natural Resources Institute Finland (Luke), P.O. Box 2, 00791 Helsinki, Finland; pasi.rikkonen@luke.fi
[2] Wageningen Economic Research, Wageningen University & Research, Prinses Beatrixlaan 582-528, Postbus 29703, 2502 LS Den Haag, The Netherlands; michiel.vangalen@wur.nl (M.V.G.); diti.oudendag@wur.nl (D.O.); auke.greijdanus@wur.nl (A.G.)
[3] State Institute of Agricultural Engineering and Bioenergy, University of Hohenheim, 70599 Stuttgart, Germany; simon.zielonka@uni-hohenheim.de (S.Z.); Lijun.Zhou@uni-hohenheim.de (L.Z.)
* Correspondence: erika.winquist@luke.fi

Abstract: To be able to meet the European Union's energy and climate targets for 2030, all member states need to rethink their energy production and use. One potential renewable energy source is biogas. Its role has been relatively small compared to other energy sources, but it could have a more central role to solve some specific challenges, e.g., to reduce carbon dioxide (CO_2) emissions from traffic, or to act as a buffer to balance electricity production with consumption. This research analyses how the future of the biogas business in three case study countries is developing until 2030. The study is based on experts' views within the biogas business branch in Germany, The Netherlands, and Finland. Both similarities and differences were found among the experts' answers, which reflected also the current policies in different countries. The role of biogas was seen much wider than just to provide renewable energy, but also to decrease emissions from agriculture and close loops in a circular economy. However, the future of the biogas branch is much dependent on political decisions. To be able to show the full potential of biogas technology for society, stable and predictable energy policy and cross-sector co-operation are needed.

Keywords: expert survey; renewable energy; biogas; biomethane; biogas plant; business model; political support system

Citation: Winquist, E.; Van Galen, M.; Zielonka, S.; Rikkonen, P.; Oudendag, D.; Zhou, L.; Greijdanus, A. Expert Views on the Future Development of Biogas Business Branch in Germany, The Netherlands, and Finland Until 2030. *Sustainability* **2021**, *13*, 1148. https://doi.org/10.3390/su13031148

Academic Editor: Alberto-Jesus Perea-Moreno
Received: 31 December 2020
Accepted: 20 January 2021
Published: 22 January 2021

Publisher's Note: MDPI stays neutral with regard to jurisdictional claims in published maps and institutional affiliations.

1. Introduction

Renewable energy production is growing fast in the European Union (EU) and globally. In the first half of 2020, renewable electricity generation in the EU exceeded fossil fuel generation for the first time ever. This was partly due to the 7% fall in electricity demand because of the coronavirus (COVID-19) pandemic. However, especially the electricity generation using wind and solar energy has grown over a longer time period, from 13% of total electricity generation in 2016 to 21% in the first half of 2020 [1].

The development is most welcomed because the EU aims to be carbon neutral by 2050 [2]. Recently, even China announced to have CO_2 emissions peak before 2030 and achieve carbon neutrality before 2060 [3]. China's commitment is crucial when mitigating climate change as it is responsible for around 28% of global emissions. Finnish emissions might be less important globally, but the goal is even more ambitious; Finland aims to achieve carbon neutrality by 2035 [4].

Individual EU member countries have varying targets for the share of sustainable energy sources and various ways of achieving their renewable energy targets. Replacing fossil fuels with renewable energy sources is, however, much more than just switching the fossil raw materials to renewable ones. Unlike many renewables, fossil fuels are flexible to use in versatile applications and easy to store. Thus, the whole energy system needs to be built again on a renewable basis [5]. Instead of few large energy sources, several energy sources are integrated in the renewable system [6]. Moreover, instead of centralised solutions, the energy is produced locally [7].

The new renewable energy system must tackle several problems—how to balance electricity production with consumption, how to arrange the energy needed for traffic, and how to ensure local energy security and affordability. Biogas could provide solutions to each of these questions, although not alone because biomass resources are limited. In addition to biogas and biomethane obtained by upgrading biogas, corresponding renewable alternatives to natural gas can be produced by power-to-gas from hydrogen produced with renewable electricity and CO_2 captured from industrial processes, and synthetic natural gas (SNG) from biomass gasification [8]. These can both increase the production potential of biomethane and use the same existing infrastructure as natural gas.

The German energy transition (Energiewende) was the first attempt to transform a centralised fossil-based energy system into a local renewable-based system. The generous feed-in tariffs (FiT) enabled renewable electricity production, especially from biogas. Today, Germany is the world leader, with 9527 biogas plants by the end of 2019 [9] and a 13.0% share of biogas/biomethane in renewables-based electricity generation [10]. However, by changing the FiTs to a tendering based system, the current subsidy system favors wind and solar over biogas and large production facilities over small ones [11]. To be able to maintain the production, the biogas plants need to find cost savings, improvements in energy efficiency, and new business models.

This research focusses on the biogas business and its prospects toward 2030. Based on experts' views in the biogas and energy branch in Germany, The Netherlands, and Finland, this research analyses how the future of the biogas business in three case study countries is developing until 2030. By using an expert survey method, expert views of the future are used to map the probable and desirable future views.

The research questions are the following:

- How is the business environment of the renewable energy production evolving until 2030 in the case study countries and the EU?
- How do experts see the probable and desirable future paths of the use of biogas and its role in the energy transition towards renewable energy?
- Which income sources will be more significant in the future for the biogas business branch?

2. Background

Biogas is a mixture of methane (50–70%) and carbon dioxide (30–50%), whereas natural gas is almost pure methane (CH_4). However, also biogas can be upgraded to biomethane (>92% CH_4). Biogas is formed when micro-organisms degrade organic compounds in anaerobic conditions and the process is called anaerobic digestion (AD). Biogas is still collected from old landfill areas, but thanks to the Landfill Directive (1999/31/EC), which obliges the member states to reduce the amount of biodegradable municipal waste that they landfill to 35% of 1995 levels by 2016, the volumes are decreasing. On the contrary, biogas production in reactor plants is increasing. Suitable raw materials are municipal sewage sludge and biowaste, side streams from the food and paper processing industries, as well as from agriculture such as manure, grass, straw, and other crop residues. Biogas can be used for the production of heat, combined heat and power (CHP), and upgraded to biomethane, which can be used as traffic fuel or to replace natural gas in various applications.

2.1. Overview of Current Biogas Production in Europe

The biogas production in the European Union represents roughly half of the global biogas production [12]. The relative importance of biogas in the EU is mainly thanks to Germany which represents half of the EU production. Germany was one of the first European countries to implement a subsidy for renewable electricity and biogas production. Already in 1991, the Electricity Feed-In Law was introduced, and in 2000, the Renewable

Energy Sources Act (EEG, or Erneuerbare Energien Gesetz). The EEG had several updates (2004, 2009, 2012, and 2014) before, in 2017, the basis for its support changed fundamentally to an auction model with lower maximum achievable tariffs. In addition, due to the strong position of Germany in the European biogas production, the political changes in Germany reflects the whole EU level.

The change in subsidy levels can be seen also in the biogas production development in Germany (Figure 1). Until 2015, there was strong growth, but after that, the production has stayed at the same level. The biomethane production data was not easily available. Thus, the biomethane addition to the natural gas network (or use as traffic fuel in the case of Finland) has been used as an indicator of the development of the biomethane market (Figure 1). Biomethane addition to the gas grid decreased somewhat in Germany in 2017. However, according to the EBA report published in 2020 [13], five new biomethane upgrading plants were built in 2018, which would indicate that the interest in biomethane upgrading is still increasing.

In The Netherlands, biogas production has continued the slow growth, whereas biomethane addition to the natural gas network is growing fast (Figure 1). In Finland, there was a stepwise growth in biogas production in 2017, but after that year, the production has stayed at the same level. In 2016, the state-owned Gasum Ltd. entered the market, buying two companies with seven biogas plants and becoming the largest biogas producer in Finland [14]. The same year, Gasum also built one additional larger biogas plant and started biogas upgrading to biomethane, as well as expanding the gas filling station network. The use of biomethane as traffic fuel has continued growing after that (Figure 1).

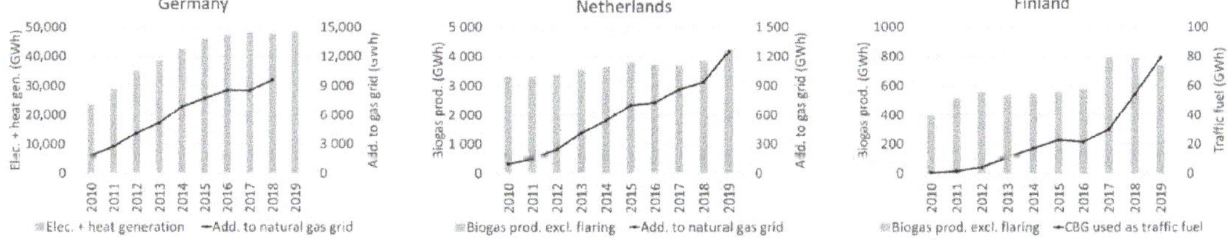

Figure 1. Primary axis: production of electricity and heat from biogas (Germany) [15,16]/biogas production excl. flaring (The Netherlands, Finland) [17,18]; secondary axis: addition of biomethane to the natural gas grid (Germany, The Netherlands) [19–21]/use of biomethane as traffic fuel (Finland) [22].

Biogas consumption and production potential in the case study countries can be estimated both with the natural gas consumption, which describes the existing infrastructure also available for biomethane use, and the availability of agricultural biomasses, which provide the largest raw material reserve for biogas production. In all case study countries, the estimated biogas production represents only a fraction of the natural gas consumption (Table 1). Especially in The Netherlands, natural gas consumption is at a high level compared to the population. However, because of safety reasons for the inhabitants of the province of Groningen, where the production of natural gas has caused earthquakes, the aim is to phase out natural gas until 2050 [23]. The focus in gas transition is on energy savings, replacing natural gas with biomethane or hydrogen gas, producing heat with other renewable energy options, and using natural gas only as feedstock for the chemical industry [24].

Although Finland has nearly the same total area as Germany, the cultivated area in Finland is at the same level as in The Netherlands (Table 1). The cultivated area correlates with the availability of energy crops for biogas production but also with side streams from other crop production such as grass cultivated as green manure, catch and cover crops, straw, and crop residues. Despite the rather limited cultivated area in The Netherlands, the number of livestock is high, and thus the estimated biomethane potential in the collectable manure is almost half of that in Germany (Table 1). If only manure biomethane potential is considered, both The Netherlands and Finland could double their current biogas production. Only in Germany, manure biomethane potential is (22,130 GWh) just half of the current production (48,747 GWh) (Table 1) [25].

Energy crops covered nearly half of the biogas raw material supply in Germany in 2018. Almost the same share is covered by agricultural residues (including manure), and only some biowaste/municipal waste (i.e., organic fraction of municipal solid waste) and industrial side streams (food and drink) are used [13]. Previously, the share of energy crops, particularly maize, has been even larger in Germany, but the utilisation of maize silage and corn has now been limited since the EEG 2017. Initially, from the beginning of 2017, maize was limited to a maximum of 50% for the mass-based substrate input, then later to 47% in 2019–2020 and further to 44% in 2021–2022 [26].

In The Netherlands, when the landfill plants are excluded, the produced biogas originates from biowaste/municipal waste (ca. 40%), co-digestion of agricultural/municipal/industrial side streams (ca. 40%), and sewage sludge (ca. 20%) [27]. In Finland, biogas production relies strongly on biowaste/municipal waste (ca. 90%), the rest being sewage sludge (ca. 5%) and agricultural residues (ca. 5%) [13].

Table 1. Summary table of case study countries (data from 2019 unless reported otherwise).

	Germany	The Netherlands	Finland	Ref.
Population (10^6)	83.0	17.3	5.5	
Biogas production excl. flaring (GWh)	48,747 [a]	4210	740	[10,17,18]
Biogas production per capita (kWh)	587 [a]	243	135	[10,17,18]
Biomethane production (GWh) [b]	10,292	1574	105	[13]
Biomethane production per capita (kWh) [b]	124	91	19	[13]
Natural gas consumption (GWh)	887,000	368,000	20,360	[21,28]
Total area (10^3 km^2)	357	42	338	
Cultivated area (10^3 km^2)	185	18	23	
Collectable manure (10^6 t) [c]	133	52	8	[25]
Realistic manure biomethane potential (GWh) [c]	22,130	9620	1250	[25]

[a] elec. + heat + vehicle fuel, excl. efficiency losses, [b] data from 2018, [c] data from 2013 (1 Nm^3 CH$_4$ = 10 kWh).

2.2. EU Level Directives and Goals

The aim of the original Renewable Energy Directive (2009/28/EC) was to promote renewable energy production in the EU, mitigate climate change, and increase the share of local energy production vs. imported fossil energy sources. The Renewable Energy Directive was revised in December 2018 (RED II) to better meet the emission reduction commitments under the Paris Agreement (December 2015) and to move the legal framework to 2030 [29]. Special emphasis in the revision is on the sustainability criteria for bioenergy, which will also include biomass and biogas for heating, cooling, and electricity generation. This will further steer the development from first-generation biofuels, where raw materials or land use is competing with food production, to second-generation biofuels exploiting various side streams and lignocellulosic raw materials.

Highlights of Revised Renewable Energy Directive (2018/2001/EU) [29] include the following:

- renewables should be 32% of the final energy consumption by 2030;
- national energy and climate plans (NECPs) for 2021–2030 (submitted to the European Commission by the end of 2019);
- renewable sources should account for 14% of transport fuels by 2030;
- strengthened sustainability criteria for bioenergy (including biomass and biogas for heating, cooling, and electricity generation);
- enabling self-production and -consumption of renewable energy;
- the original renewable energy directive will be replaced by 30 June 2021.

To meet the EU's energy and climate targets for 2030, EU member states need to establish a 10-year integrated national energy and climate plan (NECP) for the period from 2021 to 2030. However, member states can decide the structure and scope of their plans individually. Some member states (e.g., Finland and France) brought up the role of biogas and biomethane, whereas others hardly mentioned it at all (e.g., Germany and The Netherlands) [13].

The next step in the EU's climate goals is carbon-neutrality by 2050, which is one of the targets in the European Green Deal initiative. The initiative includes all sectors of the economy and requires even higher greenhouse gas (GHG) emission reductions for 2030 than the RED II. Moreover, a European climate law was proposed in March 2020 to ensure to reach these goals [2]. In addition, to boost renewable energy and energy efficiency, the European Green Deal also contains the 'Circular Economy Action Plan' for sustainable industry and the 'Farm to Fork Strategy' for sustainable agriculture. Both initiatives open new possibilities for biogas, which provides renewable energy and the technology to use various side streams and cut emissions from agriculture.

Also linked to more sustainable agriculture, the European Regulation on Fertilizing Products (FPR) was approved in June 2019 [30]. The FPR recognises that fertilising products can be made from organic materials such as compost and digestate, and it establishes harmonised requirements to make them available on the internal market [13].

Furthermore, the 'Directive on the deployment of alternative fuels infrastructure' (2014/94/EU) was enforced in October 2014 [31]. The aim of this directive was to minimise the oil dependence of transport and reduce the environmental effects of transport throughout the EU. The national policy frameworks had to contain targets for alternative transport fuels and their distribution infrastructure, including pressurised gas fuelling points for 2020 and 2030. Although the requirements may be fulfilled with natural gas alone, biomethane can also be used.

2.3. Governmental Support Systems and Goals

2.3.1. Germany

In Germany, biogas production, as well as wind and solar electricity production, are supported through the latest version of the Renewable Energy Sources Act (EEG 2017), where the support system was switched from the feed-in tariffs (FiT) to the auction model (pay-as-bid). The newly built biogas plants with an installed electrical capacity of more than 150 kW_{el} and already existing biogas plants can participate in auctions [26]. Power generation within the framework of the tender model of the EEG 2017 offers a financing possibility, especially for very cost-effective plants, which typically means very large plants. So far, the amount of electricity put out to tender is far from being used. This shows how tough the conditions of the tender are for plant operators. However, small units (up to a maximum of 100 kW), as well as liquid manure plants (higher FiT) and waste plants (FiT did not change), are still supported by a FiT support scheme [32]. The EEG Amendment 2021, which will come into effect on 1 January 2021, is sticking to the basic principle of making renewable power producers more market-oriented. Currently, the majority of the biogas plants in Germany, depend on the FiT, get lower price for feeding electricity to the grid than in their original EEG contract. The overall number of biogas plants is projected to face decreasing in 2020 for the first time [9].

Germany's Integrated National Energy and Climate Plan (NECP) contains the following goals to contribute to the achievement of the EU energy targets in 2030: (1) increasing energy efficiency by reducing primary energy consumption by 30% by 2030 compared to 2008, and (2) expansion of the share of renewable energies to 30% of gross final energy consumption in 2030. In addition, the NECP confirms the national GHG emission reduction target for 2030 of at least 55% compared to 1990, and the commitment to pursue GHG neutrality as a long-term goal by 2050 [33,34]. Specific biogas-related targets included in NECP are (1) 30% manure digestion by 2025 and (2) gas-tight storage of manure on up to 70% of biogas plants [35].

2.3.2. The Netherlands

The main support instrument for biogas and biomethane in The Netherlands is currently the so-called SDE++ regulation (Stimulering duurzame energieproductie en klimaattransitie), which has replaced a former regulation SDE+ that was active between 2013 and 2020. Both regulations, in addition to the earlier SDE regulation from 2008, intent to stimulate the production of renewable energy from all kinds of sources. The main difference in the new SDE++ regulation is that it is also applicable to projects that aim to reduce CO_2-emissions by, e.g., carbon capture or the use of excess heat from other sources.

The regulation in 2020 opens in four phases with an increased maximum amount of subsidy per reduced ton of CO_2. The subsidy compensates for the difference between the production costs of the renewable energy or CO_2 reduction technique and the market prices of the competing non-renewable energy (FiT of the non-profitable portion of production costs) for 12 to 15 years. The tariffs are guaranteed minimum income, which means that the scheme only pays out if energy prices are lower than the prices in the FiT for a certain category [26].

In The Netherlands, one type of tax allowance is currently relevant for renewable energy production from biomass—the ODE tax (Opslag Duurzame Energie- en Klimaattransitie), i.e., the surcharge for sustainable energy and climate transition paid on electricity and natural gas. Energy Investment Allowance (EIA scheme) is also available for eligible biogas installations in The Netherlands [13]. The SDE++ subsidy, however, cannot be used in combination with investment support due to EU restrictions on state support.

The biogas policies in The Netherlands are relatively stable in the sense that subsidies for biogas plants have continued in the so-called SDE regulation, now called SDE++. But there has been a significant shift in focus on agricultural biogas installations from co-fermentation of manure towards mono-fermentation of manure. The agricultural biogas production has grown considerably until 2011, but since then, the number of co-fermentation plants decreased. In 2019, there were 89 installations left [17]. The decrease in co-fermentation projects was due to high costs of coproducts and low electricity prices resulting from competition from solar and wind energy. In addition, government subsidies for co-fermentation have somewhat decreased, while projects that produce biomethane are stimulated. Mono-fermentation is less demanding in terms of management because no off-farm inputs have to be bought. Furthermore, new mono-fermentation installations have been developed and implemented that better suit the different scales of farms. In The Netherlands, a large dairy cooperative, FrieslandCampina, has been the driving force behind mono-fermentation at farm scale in the so-called Jumpstart-program. In 2020, there was considerable enthusiasm among dairy farmers to participate in the mono-fermentation project. One of the aspects of the approach is for farmers to lease the installation from the cooperative instead of having to entirely buy it themselves. In recent years, therefore, a shift can be observed from co-fermentation to mono-fermentation of manure on farms.

The target for 2030 in the Climate Agreement is to replace 70 PJ (19 TWh) of natural gas used by households and industry with 2 billion m^3 green gas. A roadmap for green gas was initiated in 2019. However, considering the current production, this will only be possible by developing several big gasification and digestion plants in the future [26]. In the roadmap, the Dutch government acknowledges that gas will play a role in the future energy supply. The roadmap predicts that 30–50% of the final demand for energy will consist of gas, either in the form of methane or hydrogen gas. A major driver for developing green gas is that the Dutch government has committed to decreasing the role of natural gas. In the Dutch Climate Agreement, the government, businesses, and societal organizations have agreed to disconnect all houses in The Netherlands from the natural gas grid by 2050 [23].

2.3.3. Finland

As in Germany, the FiTs for renewable energy were replaced with a premium system from the beginning of 2019. The new system is technologically neutral, and those renewable energy plants that offer electricity at the lowest premiums will be accepted into the system.

No biogas projects were proposed to the authorities in the auction in 2018 (1.4 TWh in total), and all projects that were accepted into the premium program use wind power [36]. New auctions have not been announced.

Biogas production is currently supported through two separate investment subsidy programs—one for industrial and another for agricultural plants. The investment support for large-scale industrial plants is paid by the Ministry of Employment and Economy. A maximum of 30% of the acceptable investment costs is covered.

The investment support for agricultural plants is paid from the EU Rural Development Programme 2014–2020 by the Ministry of Agriculture and Forestry. Farm-scale plants, which mainly use the energy themselves and do not sell any energy other than electricity outside the farm, are eligible for an investment subsidy of up to 40% of the acceptable investment costs. It is also possible to get the investment subsidy when most of the energy is sold outside the farm, or the farm sells traffic fuel. However, a separate company must be founded for this purpose. The agricultural company can then get an investment subsidy of up to 30% of the investment costs.

In addition, to support biogas production, the use of biomethane as traffic fuel has been exempted from fuel tax. However, the taxation of biomethane as traffic fuel is currently under discussion. The taxation would allow using the biofuel-blending obligation for biomethane in traffic gas, i.e., the natural gas that is sold as traffic gas would contain a certain amount of biomethane. On the other hand, the farms selling only biomethane could not benefit from this, but instead, the demand could decline as the price increases. Even without tax, biomethane is ca. 20% more expensive than natural gas for the consumer.

Finland's Integrated Energy and Climate Plan [4] has the following main targets by 2030:

- to reduce GHG emissions in the non-emissions trading sector by 39% (compared to 2005);
- renewable energy share of final energy consumption at least 51%;
- renewable energy share of final energy consumption 30% in road transport.

In addition, the plan has the following directly or indirectly biogas-related targets:

- to phase out the use of coal for energy production with minor exceptions;
- to decrease the domestic use of imported oil by 50%;
- to make electricity and heat production nearly emissions-free while also considering the perspectives of security of supply;
- have a minimum of 250,000 electric and 50,000 gas-driven passenger cars on the roads.

Some of these targets were already set by the former government in 2015, such as to have 50,000 gas-driven passenger cars by 2030 [37]. Another biogas-related target was to process 50% of all manure (e.g., in biogas production) by 2025 [38].

Regarding biofuels and biogas, these targets are still far away from the present situation. In 2019, the share of biofuels in road transport was 11% [22], and there were about 9400 gas-driven passenger cars in use, of which 3800 were newly registered (Traficom 2020) [39]. Approximately 6% of manure is currently processed [40]. Both the Ministry of Economic Affairs and Employment and the Ministry of Agriculture and Forestry are working with the implementation of the biogas-specific targets (Ministry of Economic Affairs and Employment 2020) [41]. The use of biomethane as traffic fuel is hoped to be encouraged with blending obligation. On the other hand, the use of manure as raw material for biogas production is planned to have an additional support tool.

3. Materials and Methods

The research data were gathered through three separate surveys in three EU member states, namely, Finland, Germany, and The Netherlands. The questionnaire was first prepared for the Finnish expert community. The survey design and content were planned through a pre-interview round in 2017 [14].

The questionnaire was divided into two parts and several sections. Part I included questions related to the business environment of renewable energy production in general and contained two sections, namely, (A) development of energy policy until 2030, and (B) development of business environment until 2030. Part II included biogas-specific questions and contained four sections, namely, (A) increasing and/or decreasing factors for biogas production and use, (B) income development for centralised biogas plants, (C) income development for farm-scale biogas plants, and (D) significance and roles of biogas technology in the future. The respondents were asked to evaluate statements from the present day to the year 2030 in mind. In Part I, only probable future was considered, whereas in Part II, both probable and desirable futures were included.

The range of responses used a seven-step Likert scale. Answers in Part I varied from totally disagree to totally agree (-3 = totally disagree, -2 = disagree, -1 = slightly disagree, 0 = do not agree or disagree, 1 = slightly agree, 2 = agree, and 3 = totally agree), and answers in Part II varied from decreases significantly to increases significantly (-3 = decreases significantly, -2 = decreases, -1 = decreases slightly, 0 = stays on the current level, 1 = increases slightly, 2 = increases, and 3 = increases significantly). It was also possible to answer "I cannot say" or not to answer at all.

The link to the online questionnaire (Webropol) was sent by email to the chosen respondents (in Finland in November 2018, in Germany in January 2020, and in The Netherlands in June 2020), and the survey was open for about one month. The questionnaire was sent to biogas producers in farm-scale and industrial-scale plants, technology suppliers, consultants, researchers, and policymakers/administration in the biogas field. The idea was to have an extensive and well-balanced expertise coverage within the biogas value chain.

Reminders were sent to the respondents, and a total of 84 responses (Finland 21 responses, Germany 41 responses, and Netherlands 22 responses) were received. Not all respondents answered all questions. Especially in Finland, only 11 respondents out of 21 answered also to biogas-specific questions in Part II. The reason for this was that the original Finnish survey was sent to a wider group of renewable energy experts which were not all specialists in the biogas field. In The Netherlands, 20 respondents out of 22 answered also to Part II, and in Germany, all respondents answered both parts of the survey. Moreover, not all respondents answering Part II also answered the questions related to the desirable future. In Finland, 9 respondents, in The Netherlands, 18 respondents, and in Germany, 40 respondents answered all questions related to both the probable and desirable future in Part II. Even in addition to that, few individual questions were left unanswered by some respondents.

The respondents' expertise was evaluated through background questions. In all countries, the degree of education was asked. Most of the respondents had at least a higher professional education (HBO) or university degree (Germany 93%, The Netherlands 100%, and Finland 86%). In Germany and The Netherlands, the field of education was also asked. Most of the respondents had been studying technology or natural sciences (Germany 80% and The Netherlands 77%) and most of the remainder were studying economics.

In Germany and The Netherlands, experience in the biogas field in years and professional background were also asked. In Germany, 51% of the respondents had less than 10 years' experience in the biogas field and 49% more than 10 years. In The Netherlands, 32% of the respondents had less than 10 years' experience and 68% more than 10 years. For professional background, the respondents could choose farm-scale biogas producer, industrial-scale biogas producer, biogas technology supplier, consultant, researcher, policy-maker/administration, or other. In Germany, many of the respondents were researchers (56%). In The Netherlands, a large group of respondents had identified themselves as 'other' (32%). These are most likely people representing some biogas-related association because the survey was sent to several associations. However, the respondents' expertise covered different parts of the biogas value chain well. Furthermore, the aim of an expert survey was not to have a statistically representative sample, but rather to reach different types of experts through theoretical sampling.

4. Results and Discussion

4.1. Part I: Business Environment of the Renewable Energy Production

A stable and predictable energy policy would encourage companies to do new investments accordingly and thus achieve the policy goals. However, most respondents in all three countries did not see the energy policy to be stable and predictable in the probable future (Figure 2, IA1). German respondents were most pessimistic followed by Finns and Dutch. This reflects the common problem with changing governments and fluctuating political decisions. Especially in Germany, fundamental changes were made to the EEG in 2017, when fixed FiTs were replaced by tenders. Yet, the respondents agreed on stricter climate policies and targets (IA2).

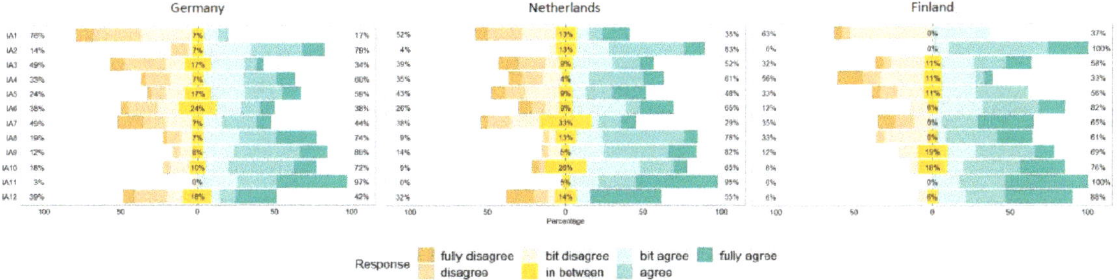

Figure 2. Development of energy policy until 2030 (probable future). IA1. The energy policy will be stable and predictable; IA2. climate policy and targets will become stricter; IA3. the best support form for renewable decentralised production is investment aid; IA4. the best support form for renewable decentralised production is long term production support; IA5. the best support form for renewable decentralised production is a combination of investment aid and production support; IA6. the best support form for renewable decentralised production is tax allowances; IA7. energy subsidies should be neutral in terms of scale and technology; IA8. the basis for energy subsidies should be energy efficiency; IA9. the basis for energy subsidies should be flexible production capacity; IA10. the basis for energy subsidies should be the capability of storing energy; and IA11. the basis for energy subsidies should be a reduction of GHGs; and IA12. any form of energy or fuel should not get permanent subsidies.

The opinion about the best support form for renewable decentralised energy production varied between countries (IA3–IA6). In Germany, 60% of the respondents chose long-term production support followed by a combination of investment aid and production support (59%). Likewise, in Germany, the biogas business branch is used to benefit from the long-term production support. Until the latest version of EEG 2017, the EEG offers 20 years of stability for the individual plant operator. This stability was the reason for the growth of the biogas branch in Germany.

In The Netherlands, most experts agreed with current support for renewable energy, which is mainly consisting of production support and also includes several investment support measures. However, tax allowances were the most popular among the experts (65%). From the question, it is not a priori clear which type of tax allowances are meant. In The Netherlands, the use of electricity that businesses and private households produce themselves from renewable sources (e.g., biogas, landfill gas, sewage gas, and electricity from CHP installations) is exempted from ODE tax. The exemption does not apply to energy consumption; renewable energy and grey energy have the same taxes for consumers.

In Finland, tax allowances were strongly favoured (82%), but investment aid was also seen positively (58%). Currently in Finland, the traffic use of biomethane is exempted from fuel tax. Otherwise, investment support is the leading support measure in Finland, both for the industrial and farm-scale biogas plants.

In all three countries, the reduction of GHGs was considered as the most important basis for energy subsidies (IA8–IA11). Likewise, the respective governments have announced further increases in goals for the reduction of CO_2 and ambitious plans to curb climate change. In Germany and The Netherlands, the flexible production capacity was the second popular option, and in Finland, the capability of storing energy. Especially in Germany, the strong growth in fluctuating wind and solar energy is causing a high demand for balancing energy supply, which will continue to rise sharply in view of the political expansion targets. Therefore, the flexibilization of biogas plants has been promoted since the EEG 2014 and is even a prerequisite for participating in the tenders of the EEG 2017.

Whether the energy subsidies should be neutral in terms of scale and technology (IA7), or any form of energy or fuel should not get permanent subsidies (IA12), divided the opinions both between and within countries. In Germany, neutrality in terms of plant size and technology was never intended under the EEG. The different FiTs for the different forms of energy, such as wind, solar, or biomass, were based on the financial needs of these technologies and their different plant sizes. In addition, bonuses were used to create incentives, e.g., for the use of certain substrates. Both in Germany and The Netherlands, the majority of the respondents were against scale- and technology-neutral subsidies. In Finland, scale- and technology-neutral subsidies were mainly supported (65%), and only Finns agreed strongly against permanent subsidies (88%).

The role of consumers as small producers of both electricity (IB1) and heat (IB2) was seen to become more common in all three countries in the probable future until 2030 (Figure 3).

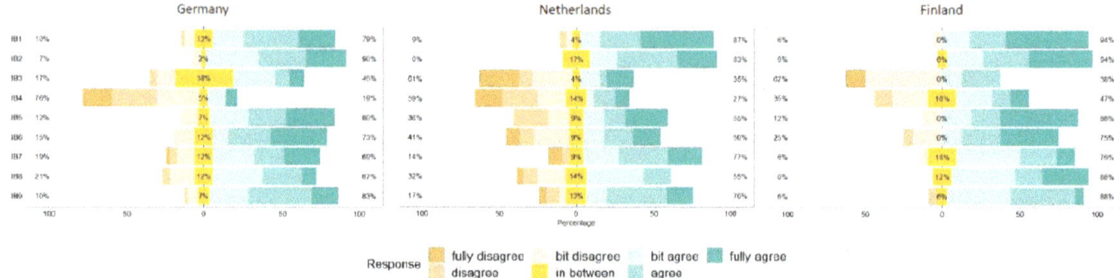

Figure 3. Development of business environment until 2030 (probable future). IB1. New technologies increase smart grids and enhance the role of consumers as small producers; IB2. various decentralised heat production technologies as part of wider heat grid become more common; IB3. energy production with wood chips is economically viable also without support; IB4. energy production with biogas is economically viable also without support; IB5. wind energy is economically viable also without support; IB6. solar energy is economically viable also without support; IB7. local energy companies and farms are co-investing for local renewable energy production; IB8. as the consumer awareness of the energy choices increases, their willingness to pay increases too; and IB9. the amount and quality of energy used in production becomes more relevant for the image of the consumer product (branding).

Renewable energy production without any support was seen most probably economically viable with wind energy (IB5: 55–88%) and second with solar energy (IB6: 50–75%). Wood chips (IB3) were mostly not considered economically viable without support, and biogas (IB4) even less. For wind and solar power, cost reduction has been achieved by technology development, but this has not been the case for biogas because a large part of the costs for biogas plants are the substrates (or their logistical costs), which are not subject to any technological development. Moreover, in Germany, increased administrative requirements and safety regulations included in the EEG 2017 compensated for the cost reduction achieved by technology development. Finns were most positive about the economic viability of biogas without support (47%) followed by Dutch (27%) and Germans (19%). One reason for this might be that, in Finland and The Netherlands, less energy crops are used for biogas production and energy production itself has a smaller role besides waste treatment and nutrient recycling.

The respondents in all three countries believed in co-operation between farms and local energy companies in local renewable energy production (IB7: 69–77%). There also seems to be a great trust in consumers regarding the willingness to pay for environmentally friendly energy and their consumption behaviour. All these possibilities could also benefit the biogas business branch.

4.2. Part II: Biogas Specific Questions

The investment cost for a biogas plant was believed to increase in the probable future in Germany and Finland (56% and 50%) but not so much in The Netherlands (Figure 4: IIA1). The investment costs of biogas plants have increased over time due to the increased safety requirements. However, in the desirable future, most respondents in all countries hoped that the prices would decrease. The state of the biogas technology was believed to increase in all countries in the probable future (70–100%) and even more in the desirable future (IIA2).

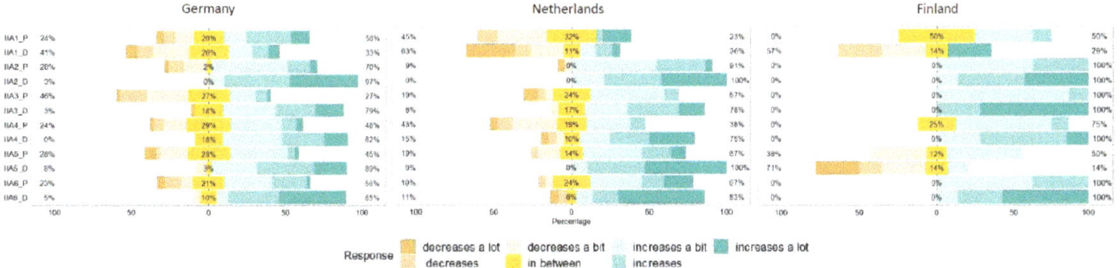

Figure 4. Increasing and/or decreasing factors for biogas production and use (P = probable, D = desirable future). IIA1. Investment cost of biogas plants; IIA2. state of the biogas technology; IIA3. availability of suppliers and technologies of farm-scale biogas plants; IIA4. availability of suitable raw materials for biogas production; IIA5. the price (that a biogas plant can receive) for recycled fertilisers (digestate-based); and IIA6. extension of biogas filling station network.

The Germans were pessimistic about the availability of suppliers and technologies of farm-scale biogas plants in the probable future (IIA3). Of the respondents, 46% believed that the availability would decrease. This might be because the EEG 2017 led to a sharp reduction in plant construction. As a result, some plant manufacturers had to file for insolvency. Suppliers of small liquid manure plants and suppliers with foreign business came through this crisis better. In The Netherlands and Finland, the availability of suppliers and technologies were believed to increase in the probable future, and in all countries, in the desirable future.

In The Netherlands, the availability of suitable raw materials for biogas production shared opinions in the probable future (IIA4). The somewhat unclear answer to this question may be because current production focuses more on waste and residual materials. This increases the number of substrates used, but not their availability. The potential for manure digestion is still very high. In Germany, the respondents favoured an increase in the raw material availability in the probable future (48%), although there are restrictions on the use of maize silage in the EEG 2017 (max. 44% from 2021). Also, in their research, Pehlken et al. concluded that bioenergy supply chains involving alternative biomass and grass from grasslands provide optimisation potentials compared to the current corn-based practice [42]. In Finland, the availability of raw materials was believed to increase strongly.

The question IIA5 was formulated in the first Finnish query as 'The price for recycled fertilisers'. Thus, in Finland, it could have been understood differently as the production cost for recycled fertilisers. In that way, it would be considered positive if the production costs would decrease, and the biogas plant could sell the fertiliser product at a lower price and thus get a larger market share for their products. Otherwise, in Germany and The Netherlands, the price that a biogas plant can receive for recycled fertilisers were believed to increase in the probable future (45–67%). In practice, the achievable prices for digestate-based recycled fertilisers vary greatly from region to region. In areas with a lot of livestock farming and a high density of biogas plants, there is a large surplus of digestate. The processing of the digestate to a commercial fertiliser is cost-intensive and is mainly practiced when the nutrient surplus of the region is so high that a long transport becomes necessary.

The biogas filling station network (IIA6) was believed to extend in all case study countries in the probable future, although in Germany (56%) less than The Netherlands (67%) and Finland (100%). The demand for gas-driven passenger cars is very low in Germany. Only 1% of passenger cars are powered by natural gas or biogas. The utilisation of biogas as a fuel source is almost exclusively realised by feeding it into the natural gas grid and withdrawing it from the balance sheet at natural gas filling stations. The slightly positive tendency regarding the development of natural gas filling stations may be due to the decreasing attractiveness of using biogas for electricity generation. In addition, the fuel sector, through REDII, will have to use more renewable fuels in the future. In Finland, the current and former governments have strongly supported biogas traffic use.

Respondents in all case study countries believed that the price of the competing fuel for biogas (natural gas) would increase in the probable future (Figure 5, IIA7: 57–60%), as well as the availability of suppliers and technologies of small-scale purification units for traffic gas production (IIA8: 62–86%). However, the number of such suppliers is so small that a decrease would not be possible. There is currently no supplier of small turnkey upgrading plants on the German market. However, the components are available, and there are suppliers of natural gas filling stations.

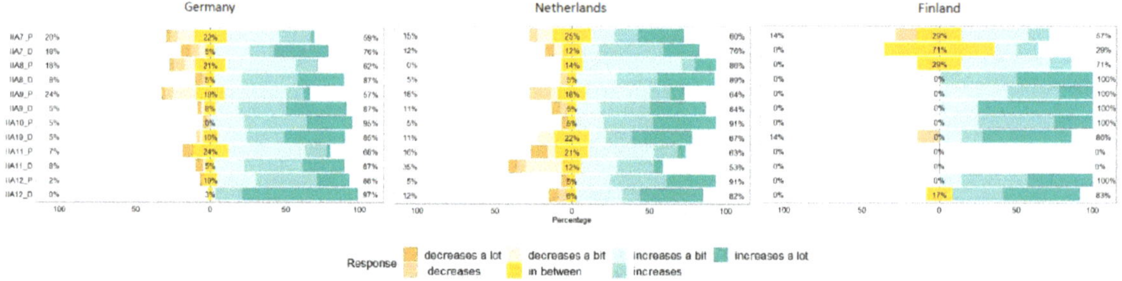

Figure 5. Increasing and/or decreasing factors for biogas production and use (P = probable, D = desirable future). IIA7. Price of the competing fuel for biogas (natural gas); IIA8. availability of suppliers and technologies of small-scale purification units for traffic gas production; IIA9. number of gas cars; IIA10. number of electrical cars; IIA11. number of hydrogen cars (lacking from Finnish survey); and IIA12. favouring low emission cars in the cities (Germany, Finland)/emission limit for city traffic (The Netherlands).

Interestingly, the respondents in all countries wanted to believe that the number of gas cars would increase in the probable future (IIA9: 57–100%), although the trend is seen very differently by the car manufacturers which clearly are gradually backing off from the further development of gas-driven passenger cars. Maybe the reasons for the optimistic rating can be found in heavy-duty vehicles and in RED II. However, in Finland, the number of gas-driven passenger cars is still increasing.

The number of electrical cars was believed to increase even more than gas-driven cars in the probable future (IIA10: 91–100%), which is no surprise. The public focus is strongly on electric cars. The question about the number of hydrogen cars was lacking from the Finnish survey. In Germany and The Netherlands, the respondents believed that the number of hydrogen cars would increase as much in the probable future as natural gas or biogas cars (IIA11: 63–68%). However, in The Netherlands, the experts' opinions about hydrogen cars were divided on what comes to the desirable future. Currently, the availability of filling stations and cars is so low that it is not a real alternative. Nevertheless, both German and Dutch governments have strong hydrogen strategy.

Favouring low emission cars in the cities or emission limits for city traffic will most probably increase in all countries (IIA12: 88–100%). However, the current debate is more about air quality than renewable fuels and GHG emissions.

Centralised plants were referred here as large-scale industrial plants which can use various waste or side streams or agricultural biomasses as raw materials either separately or in co-digestion. Income from production and selling of biomethane for traffic use was seen to be increasing most regarding the centralised biogas plants in all countries, both in the probable and desirable future (Figure 6, IIB5). In Germany and The Netherlands, the two other most increasing income options were selling gas for consumers and industry (IIB4) and heat production (IIB2).

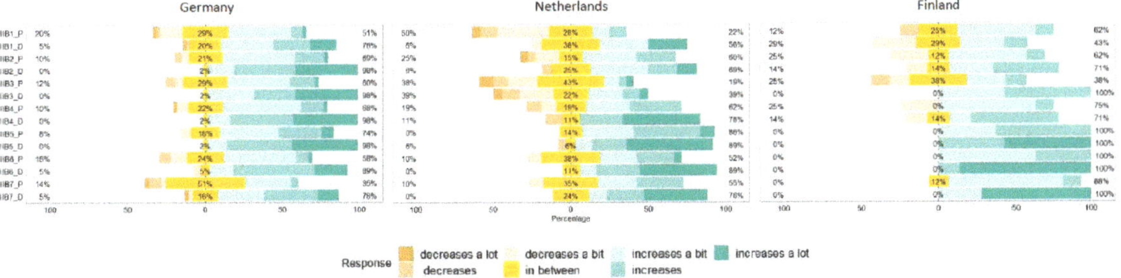

Figure 6. Income development for centralised biogas plants (P = probable, D = desirable future). IIB1. Income from gate fees; IIB2. income from heat production; IIB3. income from combined heat and power production (CHP); IIB4. income from selling gas for consumers and industry (using the gas grid), IIB5. income from production and selling of biomethane for traffic use; IIB6. income from recycled fertilisers; and IIB7. income from biochemicals.

In Finland, recycled fertiliser products and biochemicals were seen to increase most after the biogas traffic use. Biochemicals could have got more votes also in Germany and The Netherlands if the timeframe would have been longer. Maybe many respondents thought that 2030 is too close and biochemicals are still far away from being a common technology.

Despite the preferences, most income options were believed to increase. However, in The Netherlands, most respondents believed that the income from gate fees (IIB1) and CHP production (IIB3) would decrease in the probable future. Already in the current business environment, gate fees are only important for waste plants. The economic feasibility in CHP production is challenging both because of the low electricity price and high maintenance costs of the CHP unit, ca. 0.013 €/kWh$_{el}$ [43]. A further shift towards biomethane and transport fuels, to the disadvantage of the production of electricity and heat from biogas, might be expected.

Farm-scale plants were referred here as single farm plants using mainly agricultural biomasses. Branding agricultural products with carbon-neutral labels (Figure 7, IIC6) was most highly rated in Germany, maybe because it is not common yet. Also, in The Netherlands and Finland, most respondents believed that this will increase strongly. As one example from Finland, Valio Ltd. is aiming for carbon-neutral milk production by 2035, and part of the solution is using cow slurry as biogas raw material and furthermore using biomethane as fuel for the tank trucks collecting milk [44].

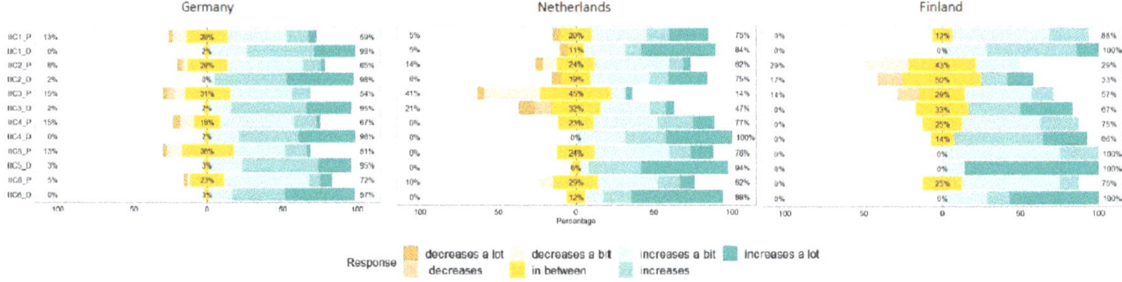

Figure 7. Income development for farm scale biogas plants (P = probable, D = desirable future). IIC1. Improving farm energy self-sufficiency; IIC2. income from selling energy out of the farm—heat production; IIC3. income from selling energy out of the farm—CHP production; IIC4. income from selling energy out of the farm—traffic gas purification; IIC5. improving nutrient self-sufficiency and crop yields; and IIC6. branding agricultural products with a carbon-neutral label (and thus getting higher price).

Both in The Netherlands and Finland, the largest growth potential for income development was believed to be in improving nutrient self-sufficiency and crop yields (IIC5). Biogas plant digestate is a better fertiliser than manure because during AD part of the organic nitrogen is degraded into ammonium nitrogen, which is directly available for the plants. Especially in organic farming, the nutrient use can be improved with biogas plant digestate instead of direct use of animal manure or green manure, i.e., grass or legumes cultivated as part of crop rotation [45].

In Germany, the second-largest income growth potential was seen in selling energy out of the farm in the form of traffic gas (IIC4), which was also highly rated in The Netherlands and Finland. However, Dutch and Finns preferred improving farm energy self-sufficiency as the second option (IIC1). In Finland, there is a clear reason for this. Farms are remotely located and district heating is not possible. Thus, at least some own heat production is needed. The most common heat source is wood chips, but biogas is also a viable option. The only clear decreasing income source was, according to Dutch respondents, selling energy out of the farm in the form of heat and electricity, i.e., CHP production (IIC3). Also, the Germans and Finns saw this among the least promising options. In Finland, the least favourable option was selling energy out of the farm in the form of heat (IIC2), which describes the current situation well. Because of the long distances between farms and other settlements, it is typically not possible to sell heat out of the farm in Finland.

The respondents in all countries were very positive about the many important roles of biogas technology in the future. Treatment of manure and cutting down emissions from agriculture (Figure 8, IID6) got the highest ranking both in the probable and in the desirable future. Also, economically, manure digestion is an extra option with higher support in Germany.

The second most important role was seen in nutrient recycling (IID8). Unlike in composting, where nitrogen is lost through denitrification in the form of nitrogen gas to the atmosphere, nitrogen compounds remain in the digestate, and their fertilisation effect is even improved. Likewise, AD suits well for the treatment of biowaste (IID7). The moisture content does not harm the process as in combustion and the process enables nutrient recycling.

The third-most important role seen by the experts was reducing CO_2 emissions from traffic and improving urban air quality (IID2). These challenges indeed provide great potential for improvement. The transport sector represented 25% of the GHG emissions in 2018 within EU-27 member states [46]. Many European countries are currently promoting electric vehicles (EVs) as a leading GHG mitigation solution for the transportation sector. However, EVs have a high level of production-related emissions. The emissions from their use, on the other hand, depend on the GHG intensity of the electric grid in question. In the worst case, EVs can lead to greater life-cycle GHG emissions than comparable diesel vehicles. The probability that an EV will lead to lower life-cycle GHG emissions than a diesel vehicle is only 75% for Germany and The Netherlands, whereas for Finland it is 99% because of the high share of renewables in electricity production. However, there are several countries in the European Economic Area (EEA), such as Poland, Latvia, and Estonia, where the emissions from EVs most probably exceed those from diesel vehicles [47]. The use of biomethane as a vehicle fuel has one the lowest well-to-wheels GHG emissions, comparable to the use of renewable electricity for EVs, according to the latest JEC Well-To-Wheels report [48]. Unfortunately, the current CO_2 standards for car manufacturers do not recognise biomethane, but the CO_2 emissions for gas-driven cars are calculated based on natural gas. The European Commission will propose a revision of the CO_2 standards for cars and vans by June 2021 and will also review the CO_2 standards for heavy-duty vehicles by 2022 [49]. To be able to meet the aims of the EU's 'Sustainable and Smart Mobility Strategy', i.e., a 90% reduction in transport-related GHG emissions by 2050, the policies should take into account the emissions from the whole life-cycle of a vehicle. In addition, a solution based on several technologies and energy sources would be more resilient and enables selecting the best solution according to the local conditions and needs.

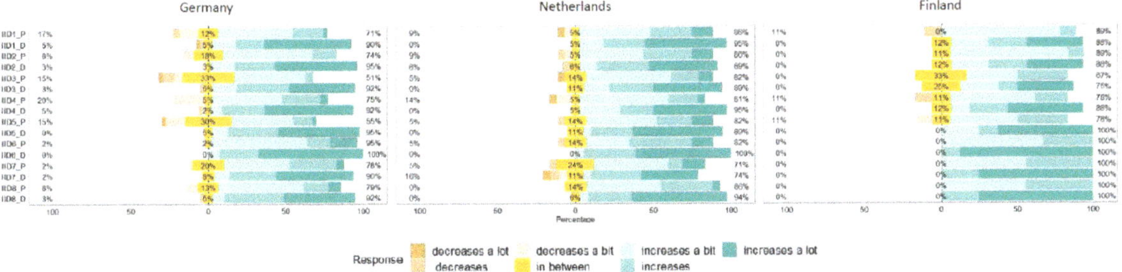

Figure 8. Significance and roles of biogas technology in the future (P = probable, D = desirable future). IID1. Replacing fossil raw materials in the production of energy and fuels; IID2. reducing CO_2 emissions from traffic and improving urban air quality; IID3. replacing fossil raw materials in the production of chemicals and materials; IID4. using biogas for balancing production and storing energy along with other renewable energy sources (e.g., wind, solar); IID5. biogas as part of the decentralised energy production for improving the security of supply; IID6. treatment of manure and cutting down emissions from agriculture; IID7. treatment of other types of biowaste; and IID8. recycling nutrients.

5. Conclusions

The three case study countries had different approaches relating to biogas due to political aims and local conditions. In Germany, biogas has been used for electricity production based on generous FiTs. At present, with the tendering model and lower subsidy levels, research and practice are discussing ways to improve the economic efficiency of biogas plants, such as the use of cost-effective substrates, improved heat utilisation, improved utilisation of digestate as fertiliser, ecosystem services, flexible power production for grid stabilisation or demand-driven production (which is a requirement to enter the tender). However, alternatives without state subsidies are also being sought. One possibility could be biomethane production for gas grid injection or use as traffic fuel. Most of the approaches currently show no economic viability or require the removal of legal hurdles for their implementation. How serious the situation is can be seen from the fact that 2020 is the first year in which the number of biogas plants in Germany decreases.

In The Netherlands, natural gas has been used widely, both in households and by the industry, and the country has a wide gas grid. Due to safety issues, the use of natural gas has been decided to phase down by 2050 and be replaced by biomethane and hydrogen gas. Biomethane could also find more use in traffic, especially heavy-duty vehicles and inland shipping. The current biogas production could be more than doubled only by using manure as a substrate due to the high density of livestock. Biogas production is growing steadily, and biomethane injection to natural gas grid is growing fast.

In Finland, the biogas business branch is challenged with long distances and small markets. Biogas is hoped to give a new income source for agriculture by combining food and energy production, and at the same time, cut down emissions. Both the former and present Finnish governments have supported both biogas production and the use of biomethane as traffic fuel. Biogas production in Finland is still small, measured both in absolute production amount and per capita, but especially, the use of biomethane as traffic fuel is growing fast.

Despite the differences, the biogas business branch in the case study countries also had common drivers and barriers based on expert survey. In the following Table 2, the drivers and barriers are categorised based on a PESTE/PESTLE framework [50,51].

Table 2. The drivers and barriers of biogas production based on expert survey.

PESTE	Drivers	Barriers
Political	- Renewable energy and GHG emission reduction goals - Governmental support mechanisms	- Uncertainty of political decisions - Business models based on subsidies - Biomethane not recognised in CO_2 standards for car manufacturers
Economical	- New income sources from traffic fuel and circular economy solutions (recycled nutrients and biochemicals) - Branding agricultural products	- High logistical costs of raw materials - Low price of end products due to the low price of fossil energy and mineral nutrients - Underdeveloped markets for recycled nutrients
Social	- Significance of 'Green consumers' increases - Co-operation between farms and local energy companies - Decentralised energy production creates jobs for rural communities	- Partly misleading negative images about biogas production ('not in my backyard' phrase) - Too positive images about electrical cars
Technological	- Increase in the number of technological solutions regarding biogas plants and biomethane up-grading	- Lack of interest from car manufacturers to develop new gas-driven vehicles
Environmental	- Emissions' reduction from agriculture - Nutrients recycling - Replacing fossil fuels	- Emissions from biogas plants and digestate storages

Finally, how did the experts see the future role and opportunities for biogas in the three case study countries? The most important role was considered to be the treatment of manure and cutting down emissions from agriculture. The next important role was enabling nutrient recycling, and the third, reducing CO_2 emissions from traffic and improving urban air quality, although they were all highly rated and the differences between these options were small.

To be able to accomplish these tasks, the biogas business must be profitable. But how chould the businesses find new income sources or how could the policymakers create a stable business environment? The promising income sources were different for centralised and farm-scale plants. For centralised plants in all countries, the income from production and selling of biomethane for traffic use was seen to be increasing most. This could indeed be a win-win option for both the environment and the biogas branch. However, the future of this path lies in the policymakers' hands. How biomethane is rated in the future, in terms of CO_2 standards and in comparison to electrical vehicles, will be a central question regarding the biomethane use as traffic fuel.

For farm-scale plants, branding agricultural products with a carbon-neutral label was most highly rated in Germany. Also, in The Netherlands and Finland, there seemed to be a great trust in consumers' willingness to pay for food produced with environmentally friendly energy or with fewer emissions. However, both in The Netherlands and Finland, the largest growth potential for income development was believed to be in improving nutrient self-sufficiency and crop yields.

The expert survey showed the versatile roles and opportunities of biogas technology for societies. While the EU aims to be carbon neutral by 2050, biogas technology could help to achieve this goal by contributing to many sectors of the economy.

Author Contributions: Conceptualization, E.W., P.R., M.V.G., D.O. and S.Z.; Methodology, P.R. and M.V.G.; Formal Analysis, D.O. and A.G.; Investigation, E.W., P.R., M.V.G., D.O. and S.Z.; Writing–Original Draft Preparation, E.W., M.V.G., S.Z., P.R., D.O., L.Z., A.G.; Writing–Review & Editing, E.W.; Visualization, E.W. and D.O. All authors have read and agreed to the published version of the manuscript.

Funding: The Finnish study was part of the FutWend-project: Towards a future-oriented 'Energiewende', funded by the Academy of Finland (grant number 297747).

Institutional Review Board Statement: The survey study follows the regulations of the National Advisory Board on Research Ethics (2009) and the American Psychological Association, including the informed consent, confidentiality and anonymity of the participants.

Informed Consent Statement: Informed consent was obtained from all subjects involved in the study.

Data Availability Statement: Data available on request due to that the surveys were conducted in German, Dutch and Finnish.

Acknowledgments: We gratefully acknowledge the experts in all three case study countries who answered the Webropol survey and gave their time and expertise to our use.

Conflicts of Interest: The authors declare no conflict of interest. The founding sponsors had no role in the design of the study; in the collection, analysis, or interpretation of data; in the writing of the manuscript, and in the decision to publish the results.

Abbreviations

AD	anaerobic digestion
CBG	compressed biogas
CH_4	methane
CHP	combined heat and power
CO_2	carbon dioxide
EEG	Erneuerbare Energien Gesetz
FiT	feed-in tariff
GHG	greenhouse gas
GWh	gigawatt-hour
kWh	kilowatt-hour
NECP	national energy and climate plan
Nm^3	normal cubic meter
RED II	Revised Renewable Energy Directive
SDE	Stimulering duurzame energieproductie en klimaattransitie
SNG	synthetic natural gas

References

1. Jones, D.; Moore, C. Renewables Beat Fossil Fuels-a Half-Yearly Analysis of Europe's Electricity Transition. 2020. Available online: https://ember-climate.org/wp-content/uploads/2020/07/2020-Europe-Half-Year-report.pdf (accessed on 27 December 2020).
2. Proposal for a Regulation of the European Parliament and of the Council Establishing the Framework for Achieving Climate Neutrality and Amending Regulation (EU) 2018/1999 (European Climate Law), COM/2020/80 Final. Available online: https://eur-lex.europa.eu/legal-content/EN/TXT/?qid=1588581905912&uri=CELEX:52020PC0080 (accessed on 22 January 2021).

3. McGrath, M. Climate Change: China Aims for 'Carbon Neutrality by 2060'. *BBC News*. 22 September 2020. Available online: https://www.bbc.com/news/science-environment-54256826 (accessed on 27 December 2020).

4. Ministry of Economic Affairs and Employment. *Finland's Integrated Energy and Climate Plan*; Publications of the Ministry of Economic Affairs and Employment; Ministry of Economic Affairs and Employment: Helsinki, Finland, 2019. Available online: http://urn.fi/URN:ISBN:978-952-327-478-5 (accessed on 22 January 2021).

5. International Renewable Energy Agency. *Transforming the Energy System–and Holding the Line on the Rise of Global Temperatures*; International Renewable Energy Agency (IRENA): Abu Dhabi, UAE, 2019. Available online: https://www.irena.org/publications/2019/Sep/Transforming-the-energy-system (accessed on 27 December 2020).

6. Liebetrau, J.; Kornatz, P.; Baier, U.; Wall, D.; Murphy, J.D. *Integration of Biogas Systems into the Energy System: Technical Aspects of Flexible Plant Operation*. IEA Bioenergy Task 37. 2020. Available online: http://task37.ieabioenergy.com/technical-brochures.html (accessed on 27 December 2020).

7. Ruggiero, S.; Varho, V.; Rikkonen, P. Transition to distributed energy generation in Finland: Prospects and barriers. *Energy Policy* **2015**, *86*, 433–443. [CrossRef]

8. Prussi, M.; Padella, M.; Conton, M.; Postma, E.D.; Lonza, L. Review of technologies for biomethane production and assessment of Eu transport share in 2030. *J. Clean. Prod.* **2019**, *222*, 565–572. [CrossRef] [PubMed]

9. Fachverband Biogas. *Industry Figures 2019 and Forecast of Industry Development 2020*; Fachverband Biogas: Freising, Germany, 2020. Available online: https://www.lee-nds-hb.de/wp-content/uploads/2020/07/03a_neu_20-07-23_Biogas_Branchenzahlen-2019_Prognose-2020.pdf (accessed on 27 December 2020).

10. Federal Ministry for Economic Affairs and Energy (BMWi). *Renewable Energy Sources in Figures–National and International Development 2019*; Federal Ministry for Economic Affairs and Energy (BMWi): Berlin, Germany, 2020. Available online: https://www.bmwi.de/Redaktion/EN/Publikationen/Energie/renewable-energy-sources-in-figures.html (accessed on 17 January 2021).

11. Strippel, F. Economic opportunities for the German biogas market within the framework of the EEG 2017. *Commun. Agric. Appl. Biol. Sci.* **2017**, *82*. Available online: https://www.nijhuisindustries.com/assets/uploads/solutions/Special-Issue-Advance-Trends-in-Biogas-and-Biorefineries-GENIAAL.pdf (accessed on 27 December 2020).

12. Scarlat, N.; Dallemand, J.-F.; Fahl, F. Biogas: Developments and perspectives in Europe. *Renew. Energy* **2018**, *129*, 457–472. [CrossRef]

13. European Biogas Association. *European Biogas Association Statistical Report: 2019 European Overview*; EBA: Brussels, Belgium, January 2020.

14. Winquist, E.; Rikkonen, P.; Pyysiäinen, J.; Varho, V. Is biogas an energy or a sustainability product?—Business opportunities in the Finnish biogas branch. *J. Clean. Prod.* **2019**, *233*, 1344–1354. [CrossRef]

15. Statista. *Bruttostromerzeugung aus Biogas in Deutschland in den Jahren 2000 bis 2019*; Statista: Hamburg, Germany, 2020. Available online: https://de.statista.com/statistik/daten/studie/622560/umfrage/stromerzeugung-aus-biogas-in-deutschland/ (accessed on 28 December 2020).

16. Statista. *Wärmebereitstellung aus Biogas in Deutschland in den Jahren 2003 bis 2019*; Statista: Hamburg, Germany, 2020. Available online: https://de.statista.com/statistik/daten/studie/622766/umfrage/waermebereitstellung-aus-biogas-in-deutschland/ (accessed on 28 December 2020).

17. Statistics Netherlands (CBS). *Biomassa; Verbruik en Energieproductie uit Biomassa per Techniek*; Statistics Netherlands (CBS): The Hague, The Netherlands, 2020. Available online: https://opendata.cbs.nl/statline/#/CBS/nl/dataset/82004NED/table?dl=48664 (accessed on 28 December 2020).

18. Statistics Finland. *127t–Production and Consumption of Biogas by Plant Type, 2017–2019*; Statistics Finland: Helsinki, Finland, 2020. Available online: http://pxnet2.stat.fi/PXWeb/pxweb/en/StatFin/StatFin__ene__ehk/statfin_ehk_pxt_127t.px/ (accessed on 28 December 2020).

19. Statista. *Einspeisevolumen von Biogas in das Gasnetz in Deutschland in den Jahren 2008 bis 2017*; Statista: Hamburg, Germany, 2020. Available online: https://de.statista.com/statistik/daten/studie/269761/umfrage/einspeisevolumen-von-biogas-in-das-gasnetz-in-deutschland/ (accessed on 28 December 2020).

20. Federal Network Agency. *Monitoring Report 2019*; Federal Network Agency: Bonn, Germany, 2020. Available online: https://www.bundesnetzagentur.de/SharedDocs/Downloads/EN/Areas/ElectricityGas/CollectionCompanySpecificData/Monitoring/MonitoringReport2019.pdf;jsessionid=510EA1BB63A93064D88C4B687C7C40F6?__blob=publicationFile&v=2 (accessed on 17 January 2021).

21. BP Statistical Review of World Energy 2020. Available online: https://www.bp.com/content/dam/bp/business-sites/en/global/corporate/pdfs/energy-economics/statistical-review/bp-stats-review-2020-full-report.pdf?utm_source=BP_Global_GroupCommunications_UK_external&utm_medium=email&utm_campaign=11599394_Statistical%20Review%202020%20-%20on%20the%20day%20reminder&dm_i=1PGC%2C6WM5E%2COV0LQ4%2CRQW75%2C1 (accessed on 17 January 2021).

22. Statistic Finland. *12sz–Energy Consumption in Transport, 1990–2019*; Statistic Finland: Helsinki, Finland, 2020. Available online: http://pxnet2.stat.fi/PXWeb/pxweb/en/StatFin/StatFin__ene__ehk/statfin_ehk_pxt_12sz.px/ (accessed on 28 December 2020).

23. NL Ministry of Economic Affairs and Climate Policy. *Energy Transition in The Netherlands–Phasing out of Gas*; NL Ministry of Economic Affairs and Climate Policy: The Hague, The Netherlands, 2018. Available online: https://ec.europa.eu/energy/sites/ener/files/documents/01.b.02_mf31_presentation_nl-fuel_switch-vanthof.pdf (accessed on 29 December 2020).

24. Dumont, M. Circular economy in relation with biogas. In *IEA Task 37 Luke Workshop: Circular Economy in the Food System*; IEA Bioenergy: Jyväskylä, Finland, 8 March 2018. Available online: http://task37.ieabioenergy.com/workshops.html (accessed on 29 December 2020).

25. Scarlat, N.; Fahl, F.; Dallemand, J.-F.; Monforti, F.; Motola, V. A spatial analysis of biogas potential from manure in Europe. *Renew. Sustain. Energy Rev.* **2018**, *94*, 915–930. [CrossRef]

26. Gustafsson, M.; Ammenberg, J.; Murphy, J.D. IEA Bioenergy Task 37 Country Report Summaries 2019. March 2020. Available online: http://task37.ieabioenergy.com/ (accessed on 27 December 2020).

27. Dumont, M.; Siemers, W. Country Update Netherlands. 2019. Available online: https://task37.ieabioenergy.com/country-reports.html (accessed on 29 December 2020).

28. Statistics Finland. *12st–Total Energy Consumption by Energy Source, 2010Q1–2020Q3*; Statistics Finland: Helsinki, Finland, 2020. Available online: http://pxnet2.stat.fi/PXWeb/pxweb/en/StatFin/StatFin__ene__ehk/statfin_ehk_pxt_12st.px/ (accessed on 28 December 2020).

29. Directive (EU) 2018/2001 of the European Parliament and of the Council of 11 December 2018 on the Promotion of the Use of Energy from Renewable Sources. Available online: https://eur-lex.europa.eu/legal-content/EN/TXT/?uri=uriserv:OJ.L_.2018.328.01.0082.01.ENG&toc=OJ:L:2018:328:TOC (accessed on 22 January 2021).

30. Regulation (EU) 2019/1009 of the European Parliament and of the Council of 5 June 2019 Laying down Rules on the Making Available on the Market of EU Fertilising Products and Amending Regulations (EC) No 1069/2009 and (EC) No 1107/2009 and Repealing Regulation (EC) No 2003/2003 (Text with EEA Relevance), PE/76/2018/REV/1. Available online: https://eur-lex.europa.eu/eli/reg/2019/1009/oj (accessed on 22 January 2021).

31. Directive 2014/94/EU of the European Parliament and of the Council of 22 October 2014 on the Deployment of Alternative Fuels Infrastructure. Available online: http://data.europa.eu/eli/dir/2014/94/2020-05-24 (accessed on 22 January 2021).

32. EBA. *Statistical Report of the European Biogas Association 2018*; EBA: Brussels, Belgium, December 2018.

33. Federal Ministry of Environment, Nature Conservation and Nuclear Safety (BMU). *Climate Action Plan 2050–Germany's Long-Term Low Greenhouse Gas Emission Development Strategy*; Federal Ministry of Environment, Nature Conservation and Nuclear Safety (BMU): Bonn, Germany, 2019. Available online: https://www.bmu.de/en/topics/climate-energy/climate/national-climate-policy/greenhouse-gas-neutral-germany-2050/ (accessed on 27 December 2020).

34. Federal Ministry for Economic Affairs and Energy (BMWi). *National Energy and Climate Plan*; Federal Ministry for Economic Affairs and Energy (BMWi): Berlin, Germany, 2020. Available online: https://www.bmwi.de/Redaktion/DE/Downloads/I/integrierter-nationaler-energie-klimaplan.pdf?__blob=publicationFile&v=8 (accessed on 27 December 2020).

35. Liebetrau, J.; Gromke, J.D.; Denysenko, V. IEA Bioenergy Task 37: Country Report Germany 2020. Available online: http://task37.ieabioenergy.com/country-reports.html (accessed on 27 December 2020).

36. Ministry of Economic Affairs and Employment and the Energy Authority. *Support for Seven Projects Awarded through Auction–the Average Price of Accepted Tenders EU2.5 per MWh*; Press Release; Ministry of Economic Affairs and Employment and the Energy Authority: Helsinki, Finland, 1 April 2019. Available online: https://energiavirasto.fi/tiedote/-/asset_publisher/uusiutuvan-energian-tarjouskilpailusta-tukea-seitsemalle-hankkeelle-hyvaksyttyjen-tarjousten-keskihinta-2-5-euroa-mwh?_101_INSTANCE_aRbx5sYgeQOs_languageId=en_US (accessed on 27 December 2020).

37. Ministry of Economic Affairs and Employment. *Government Report on the National Energy and Climate Strategy for 2030*; Publications of the Ministry of Economic Affairs and Employment; Ministry of Economic Affairs and Employment: Helsinki, Finland, December 2017. Available online: http://urn.fi/URN:ISBN:978-952-327-199-9 (accessed on 27 December 2020).

38. Strategic Programme of the Finnish Government. *Finland, a Land of Solutions*; Publications of the Finnish Government; Strategic Programme of the Finnish Government: Helsinki, Finland, October 2015. Available online: https://vm.fi/julkaisu?pubid=6407 (accessed on 27 December 2020).

39. Traficom. *Passenger Cars in Traffic on 31 December by Region, Make, Driving Power and Year*; Traficom: Helsinki, Finland, 2020. Available online: http://trafi2.stat.fi/PXWeb/pxweb/en/TraFi/TraFi__Liikennekaytossa_olevat_ajoneuvot/030_kanta_tau_103.px/table/tableViewLayout1/?rxid=714713ea-4df2-4b82-8e0a-68b51dad9956 (accessed on 10 October 2020).

40. Luostarinen, S.; Tampio, E.; Berlin, T.; Grönroos, J.; Kauppila, J.; Koikkalainen, K.; Niskanen, O.; Rasa, K.; Salo, T.; Turtola, E.; et al. *Means for Advancing the Use of Organic Fertilising Products (Keinoja orgaanisten lannoitevalmisteiden käytön edistämiseen)*; Publications of the Ministry of Agriculture and Forestry; Ministry of Agriculture and Forestry: Helsinki, Finland, 2019. Available online: http://urn.fi/URN:ISBN:978-952-453-941-8 (accessed on 27 December 2020).

41. Ministry of Economic Affairs and Employment. *Biokaasuohjelmaa Valmistelevan Työryhmän Loppuraportti (Final Report of the Biogas Working Group)*; Publications of the Ministry of Economic Affairs and Employment; Ministry of Economic Affairs and Employment: Helsinki, Finland, 2020. Available online: http://urn.fi/URN:ISBN:978-952-327-482-2 (accessed on 27 December 2020).

42. Pehlken, A.; Wulf, K.; Grecksch, K.; Klenke, T.; Tsydenova, N. More Sustainable Bioenergy by Making Use of Regional Alternative Biomass? *Sustainability* **2020**, *12*, 7849. [CrossRef]

43. Hahn, H. *Guideline for Financing Agricultural Biogas Projects-Training Material for Biogas Investors, IEE Project 'BiogasIN' D.3.7, WP 3*; Fraunhofer Institute for Wind Energy and Energy System Technology (IWES): Bremerhaven, Germany, 2011. Available online: http://publica.fraunhofer.de/documents/N-413182.html (accessed on 15 January 2021).

44. Valio Ltd. *Board of Directors' Report and Financial Statements 1 Jan.–31 Dec. 2019*; Valio Ltd.: Helsinki, Finland, 2020. Available online: https://www.valio.fi/vastuullisuus/raportit/ (accessed on 27 December 2020).

45. Koppelmäki, K.; Parviainen, T.; Virkkunen, E.; Winquist, E.; Schulte, P.O.R.; Helenius, J. Ecological intensification by integrating biogas production into nutrient cycling: Modeling the case of Agroecological Symbiosis. *Agric. Syst.* **2019**, *170*, 39–48. [CrossRef]

46. Eurostat. *Greenhouse Gas Emission Statistics–Emission Inventories*; Eurostat, Statistics Explained: Luxembourg, 2020. Available online: https://ec.europa.eu/eurostat/statistics-explained/pdfscache/1180.pdf (accessed on 29 December 2020).

47. Dillman, K.V.; Árnadóttir, A.; Heinonen, J.; Czepkiewicz, M.; Davíðsdóttir, B. Review and Meta-Analysis of EVs: Embodied Emissions and Environmental Breakeven. *Sustainability* **2020**, *12*, 9390. [CrossRef]

48. Prussi, M.; Yugo, M.; De Prada, L.; Padella, M.; Edwards, R. *JEC Well-to-Wheels Report v5. EUR 30284 EN*; Publications Office of the European Union: Luxembourg, 2020. Available online: https://publications.jrc.ec.europa.eu/repository/handle/JRC121213 (accessed on 29 December 2020).

49. European Commission. *Questions and Answers: Sustainable and Smart Mobility Strategy*; European Commission: Brussels, Belgium, 9 December 2020. Available online: https://ec.europa.eu/commission/presscorner/detail/en/qanda_20_2330 (accessed on 29 December 2020).

50. Oraman, Y. An Analytic Study of Organic Food Industry as Part of Healthy Eating Habit in Turkey: Market Growth, Challenges and Prospects. *Procedia Soc. Behav. Sci.* **2014**, *150*, 1030–1039. [CrossRef]

51. Rothaermel, F.T. *Strategic Management: Concepts*, 2nd ed.; McGraw-Hill Education: New York, NY, USA, 2014.

 *sustainability*

Article

Value-Based Building Maintenance Practices for Public Hospitals in Malaysia

Wai Fang Wong [1,*], AbdulLateef Olanrewaju [2] and Poh Im Lim [3]

[1] Lee Kong Chian Faculty of Engineering & Science, Sungai Long Campus, Universiti Tunku Abdul Rahman, Selangor 43000, Malaysia
[2] Department of Construction Management, Universiti Tunku Abdul Rahman, Jalan Universiti, Bandar Barat, Perak 31900, Malaysia; olanrewaju@utar.edu.my
[3] Department of Architecture & Sustainable Design, Lee Kong Chian Faculty of Engineering & Science, Sungai Long Campus, Universiti Tunku Abdul Rahman, Selangor 43000, Malaysia; limpi@utar.edu.my
* Correspondence: Waifang.wong@gmail.com

Abstract: Public hospital buildings in Malaysia have been facing problems and have become subjects of public criticisms due to poor building maintenance practices. A value-based approach which integrates and assimilates the concepts of value can be applied to mitigate maintenance problems in hospital buildings. This study evaluated the causal relationships between value factors and value outcomes of building maintenance in public hospitals in Malaysia. A total of 66 samples were collected via an online questionnaire survey. Analysis was performed using partial least square structural equation modeling (PLS-SEM). Our results reveal that value-adding practices and value co-creation have a positive influence on value outcomes in hospitals. The findings, however, do not support the relationships between factors of user involvement and value outcomes, which merit further investigation. This study concludes that value-adding practice has the strongest impact on value outcomes. Thus, maintenance service providers should assimilate these practices in their services to enhance performance. In addition, the findings also justify the requirement for collaborative working arrangements for value co-creation of building maintenance.

Keywords: building performance; value co-creation; value add; maintenance management; hospital buildings

Citation: Wong, W.F.; Olanrewaju, A.; Lim, P.I. Value-Based Building Maintenance Practices for Public Hospitals in Malaysia. *Sustainability* **2021**, *13*, 6200. https://doi.org/10.3390/su13116200

Academic Editor: Alberto-Jesus Perea-Moreno

Received: 28 April 2021
Accepted: 25 May 2021
Published: 31 May 2021

Publisher's Note: MDPI stays neutral with regard to jurisdictional claims in published maps and institutional affiliations.

1. Introduction

Hospital buildings are primarily designed to provide healthcare functions such as curative nursing and rehabilitation. Hospital infrastructure must be reliable and support the daily functions to ensure continuous operations at all times, including crisis and disaster scenarios [1]. Healthcare premises also need to fulfill safety, comfort, security, and energy efficiency [2]. However, hospital buildings are challenging to maintain, primarily due to the complexity of their infrastructure [3,4]. Poor maintenance services of hospital buildings [5], and construction activities [6], can potentially affect patients' health. Building conditions such as defects were found to affect health problems in buildings [7]. In view of these highly functional requirements, the facilities must always be ready to support medical teams in their operations. Hence, it is important to ensure that hospital buildings are always maintained at an optimal state for excellent healthcare.

Globally, issues of maintenance and poor performance of hospitals have been reported in the extant literature. For instance, assessment of public hospitals in the Gaza Strip, Palestine found that most hospitals adopted corrective maintenance and did not carry out routine inspections for water and plumbing systems [8]. Low performance was reported on vertical transportation, fire protection, telecommunications, and electrical supply in Southwest Nigerian public hospital buildings [3]. An audit in Cuban hospitals also identified mismanagement of maintenance activities by staff in the maintenance department as the root cause of subsequent maintenance problems [9]. More recently, fire safety concerns in hospitals were raised where various fire incidents in Asia and globally were reported and maintenance of firefighting appliances were among the issues mentioned [10,11].

In the context of Malaysia, the healthcare system is categorized into tax-funded public healthcare and private healthcare [12]. Public hospitals contribute 42,424 beds, amounting to 73% of the total hospital beds from 144 public hospitals and medical institutions in the country [13]. Efforts to standardize facilities management (FM) practices in public hospitals were initiated in 1996 through the privatization of hospital support services by only three concession companies (CCs), to improve the overall level of service [14]. The building maintenance component is parked under the facilities engineering maintenance services (FEMS), which is among the six hospital support services that are outsourced. Currently, there are five concession companies nationwide. The concession agreement (CA) was signed between the concession companies and the government, for up to 15 years.

However, despite having a comprehensive contract [15], and a long concession period, weaknesses in public hospital maintenance have persisted unabated. CCs were revealed to be inefficient, lacked competent manpower and training, and heavily rely on their subcontractors [16]. There was also inadequate support from the contractor's top management [16]. Besides that, the contract utilizes a deductive fee system on non-conformance and non-performance by CCs [14], to ensure compliance to contract requirements. This practice requires heavy supervision and monitoring by hospital staff [14]. It was reported that hospital engineers are required to monitor between 2109 and 6625 units of FEMS assets per hospital [17]. Issues of rising operation costs [18], also indicate a lack of value for money. Overall, the maintenance management system is transactional and contractual, and there is a lack of initiatives and collaborative effort among the parties involved, i.e., the contractors, building users and hospital maintenance teams.

In a span of 21 years, healthcare spending as a share of the gross domestic product (GDP) has increased from 3.03% in 1997, to 4.24% in 2017 [19]. Based on 2016 statistics, the spending per GDP for Malaysia is still far behind more developed countries such as France, Germany, Japan, Australia, and the United Kingdom, which spend 9% to 11% of GDP on healthcare [19]. Malaysia is also behind neighboring countries such as Singapore, the Philippines, China, Sri Lanka, and South Korea, but ahead of Thailand, India, Indonesia, and Bangladesh [19].

Despite a constant increase in the allocation of national budget and concession agreements, problems involving poor maintenance are increasing unabated. Some of the issues encountered are experienced in other countries as well [20]. They include budget constraints and high expenditures, customer satisfaction, and complex information and decision-making. Past research works concentrating on hospital building maintenance are fragmented and lean towards the "hard" aspects such as, but not limited to, maintenance strategies, maintenance cost, and overall efficiency. The most researched topics in healthcare facilities management were categorized into IT and decision-making, maintenance costs, sourcing and contracts, and performance measurement [20].

There is a dearth of literature that attempts to explore the potential impact of value concepts as an alternative solution to resolve maintenance issues, particularly in the context of healthcare buildings such as hospitals. Hence, this study assesses the factors contributing to the improvement of maintenance from the aspects of user involvement, value-adding practices, and value co-creation. This research investigates value concepts in a holistic and integrated model for hospital maintenance which have not been explicitly addressed previously. Specifically, the causal relationships between value factors and the value outcomes of hospital maintenance are determined to fill the gap in empirical studies.

2. Literature Review

Internationally, research on hospital maintenance focused on several aspects such as maintenance efficiencies in Israel [21], maintenance manpower in UK hospitals [22], budgeting for university hospitals in Italy [23], building maintenance systems [24], and operational maintenance in Palestine [8]. Recent research trends have dealt with customer satisfaction and service quality [25,26], and performance [27]. Fire safety issues in Asia [10], and automated maintenance systems [28], were investigated. The value-based approach is limited to the aspect of value-adding facilities management in the UK [29].

Similarly, in the Malaysian context, past research on hospital building maintenance is limited and fragmented in various areas, e.g., quality management [30], outsourcing costs [18], audit assessment [16], critical success factors in facilities management [31,32], maintenance strategies [33], and maintenance effectiveness [34]. Recent trends on sustainability and energy saving [35,36], and green hospitals have gained attention [37]. Overall, research on value concepts in building maintenance context is still relatively scarce and lacks theoretical and empirical justifications [4]. A value-based model for building maintenance were developed [38]; however, it is in the context of Malaysian educational buildings and focused on maintenance management functions instead of value-based factors.

The privatization of facilities management services of hospitals in 1996 underscored the Malaysian Government's commitment to combat maintenance issues in public hospitals. However, after two decades of implementation, the performance of hospital buildings is still subject to public criticism. Empirical evidence on building condition assessment revealed that facilities for "Persons with Disabilities" in public hospitals were critical [39], while poor maintenance of fire safety was pointed out as frequent incidences in hospitals [40]. The annual Auditor General's Report has disclosed defect issues, contamination, and failure of facilities [41], planned preventive maintenance and supervision issues in FEMS [17], and high dependency on third-party contractors to rectify defects [42]. Many of the reported problems were preventable and arose explicitly from weaknesses in maintenance management [4,14]. They were likely caused by low-performing contractors, lack of understanding of user expectations [4,43], and lack of emphasis on collaborative working. The maintenance process involves the constant interaction of various systems and parties. Therefore, it is vital to investigate the problems derived from maintenance management from each party individually, as well as from the effect of their collaboration. This encompasses the demand side, supply side, and collaborative working holistically, rather than uni-dimensionally. Value concepts which have gained attention in other industries can potentially help to mitigate problems in maintenance management. However, the value-based approach in hospital maintenance, though promising, remains under-researched. Hence, this study addresses the pertinent research deficiency in the value-based approach for the case of hospital buildings.

The objective of this study is to assess the causal relationships between three value factors (i.e., user involvement, value add, and value co-creation) and value outcomes. The three value factors were investigated from the demand side (user involvement), supply side (value-adding practices) and the collaborative working (value co-creation) perspectives.

2.1. User Involvement

From the demand side, users refer to medical staff, and administrative or support staff who use the hospital facilities to perform day-to-day functions in providing healthcare services to patients. Prior studies provided insight on the needs for the active involvement of users as "partners" rather than as the end recipients of the service [4,43]. This view is in line with the proposition to co-opt customers' competency to increase business competitiveness [44]. The end users are the most unsatisfied stakeholders because their opinions were not heard before decisions were made [45]. The user value system "VALUCRITE" were developed to assess the criteria for academic institutions to enhance user satisfaction toward the maintenance process [46,47]. It is also crucial to measure user satisfaction [26,38,48], to determine whether maintenance services are delivered up to their expectations, and whether there is a gap for any potential improvement. User perceptions should be viewed holistically rather than uni-dimensionally, from their input and expectation, workplace functionality and productivity, and to their satisfaction [49]. However, user involvement needs to be at an appropriate level, to avoid detracting them from their core business roles [49]. Hence, formal user involvement aids the achievement of the objectives of maintenance in hospitals. Thus, in this study, it is hypothesized that user involvement has a positive influence on the value outcomes as follows:

Hypothesis 1 (H1). *User involvement positively influences value outcomes.*

2.2. Value-Adding Practices

The value-adding concept refers to the supply side of the maintenance arrangement. Maintenance service providers are expected to deliver beyond basic transactional maintenance functions. Since building maintenance is a form of service, the role of service providers is crucial to ensure the service is delivered effectively. As the sole service provider, the appointed contractors should be competent, well equipped, and prepared to add value in their services. In the UK, National Healthcare Services (NHS), through partnering arrangements, offer the best value for money, and introduced innovative practices to achieve better service quality and overall corporate image [29]. The reliability of service partners and their ability to solve problems and provide service solutions are identified as factors in value creation for services [50]. Responsiveness to needs is another crucial aspect that customers value in their service partners [38]. Thus, this study explores the proposition that value-adding practices have a positive influence on value outcomes. Hypothesis 2 is presented as follows:

Hypothesis 2 (H2). *Value-adding practices positively influence value outcomes.*

2.3. Value Co-Creation

Value can be co-created through collaborative working. Businesses should shift from traditional company-centric value creation to a new co-creation of unique value with customers [44]. Customers' roles evolved from passive receivers, to be more engaging, active, informed and connected. Value co-creation is a direct result of the interaction among the parties involved [51]. Besides, value co-creation within FM is viewed as a new paradigm of research [52]. In other industries, value creation was explored by investigating collaboration in a different arrangement [53]. A value matrix framework was also proposed in the fashion industry [54]. In the IT industry, value drivers for client and service providers were explored [55], while a strong impact of value co-creation to outsourcing satisfaction was found [56]. Value co-creation can be achieved through the sharing of information [57], intensive cooperation [52], knowledge transfer [55,57], effective communication [52], openness and honesty, and mutual trust and confidence [58], relationship synergies [50], strategic integration [53], strategic alignment [55], and strong governance [58]. Thus, this study explored the proposition that value co-creation has a positive influence on value outcomes as follows:

Hypothesis 3 (H3). *Value co-creation positively influences value outcomes.*

Table 1 presents the three value factors and their sub-factors.

Table 1. Value factors.

Factor	Code	Sub-Factor	Literature Source
User Involvement	USE1	User's Expectation	[38,45,52,59]
	USE2	User's Involvement	[26,38,45,59]
	USE3	User's Satisfaction	[38,48]
Value Adding Practices	VAL1	Integrated Service Solutions	[50]
	VAL2	Innovative Improved Practices	[29]
	VAL3	Value for Money	[29]
	VAL4	Cost Reduction/Saving	[29]
	VAL5	Responsive to Needs	[38,60]
Value Co-creation	JOR1	Sharing of Resources	[52,53,57]
	JOR2	Joint Technology	[57]
	JOR3	Sharing of Information	[57]
	OPE1	Operational Integration	[53]
	OPE2	Intensive Cooperation	[52]
	OPE3	Knowledge Transfer	[55,58]
	COM1	Effective Communication	[52]
	COM2	Transparency of Internal Information	[53,55]
	COM3	Openness and Honesty	[60]
	WWW1	Shared Risks	[53,55]
	WWW2	Mutual Trust and Confidence	[60]
	WWW3	Relationship Synergies	[50]
	STR1	Strategic Integration	[53]
	STR2	Strategic Alignment	[55]
	STR3	Strong Governance	[60]

2.4. Value Outcomes

Value outcomes are what customers perceive, are determined by the beneficiary, and should be viewed holistically [61]. Besides focusing on fulfilling customer satisfaction, the aspects of the organization's wealth generation are important [53]. There are three levels of value outcomes: transactional, business, and strategic outcomes [55]. In the context of public hospitals, business outcomes, or wealth generation are not the objectives. However, the sustainability of hospitals is an important criterion. Hence, in this study, value outcomes for public hospital maintenance are two-fold, emphasizing on daily (1) operational outcomes and (2) strategic outcomes. Operational value outcomes refer to the short-term routine goals to ensure smooth day-to-day operations in the hospital, such as the daily work process. Strategic value outcomes require long-term development as a result of the collaborative effort of parties within the arrangement. Through synergetic actions, the service provider is entrusted with more roles, and treated as a partner in the relationship [55]. Table 2 presents the value outcomes. Figure 1 illustrates the conceptual model for this study.

Table 2. Value outcomes.

Category	Code	Value Outcomes	Literature source
Operational	OVO1	Daily Work Process	[55]
	OVO2	Quality of Output	[53]
	OVO3	Response Time	[53,55]
	OVO4	Reduced Risk	[53]
	OVO5	Health and Safety	[50]
Strategic	SVO1	Skill and Knowledge	[53]
	SVO2	Technology	[53]
	SVO3	Contractor as Partner	[55]
	SVO4	Performance	[47]
	SVO5	User Satisfaction	[29,46,53,59]
	SVO6	Corporate Image	[29]

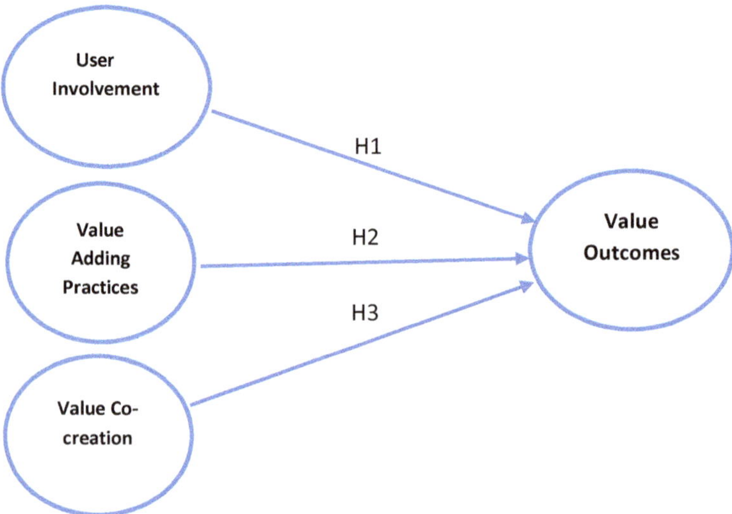

Figure 1. Conceptual model.

3. Methods

This study adopted PLS-SEM using SmartPLS 3.0 software for data analysis [62], for three key reasons. Firstly, the goal of the research is to predict the value factors that influence the value outcomes in building maintenance. PLS-SEM is suitable for a research work that aims to predict key target constructs as compared to other types of SEM techniques, such covariance-based SEM (CB-SEM), which is more appropriate for theory confirmation or rejection [63,64]. Secondly, PLS-SEM is suitable for low sample sizes [63]. In this research, the population is considered low, with only 139 public hospitals in the country. Thirdly, PLS-SEM is also suitable for data that is not normally distributed [63], wherein this study, the data are nonparametric.

Our questionnaire survey collected the following information: (i) respondents' and hospitals' backgrounds; (ii) value factors; (iii) value outcomes of hospital building maintenance. Variables were developed through the adaptation from the literature review and the synthesis of previous studies on value concepts. To present greater discriminant and reliability value, a 6-point Likert scale was used in this questionnaire, ranging from 1 = strongly disagree to 6 = strongly agree [65].

Face-to-face pre-tests were conducted to detect possible problems, bias, ambiguity, or unsuitability in the questionnaire [66,67]. Four respondents were involved, comprising an ex-monitoring consultant of hospital FM, a senior academic, a maintenance manager of a public hospital, and an ex-liaison officer of a public hospital. They were chosen based on experience and expertise either academically or in terms of hospital maintenance. Minor amendments were then made to improve the questions.

Target respondents were engineers from the Engineering Unit in the public hospitals. They were the person-in-charge to monitor, supervise, and control the hospital support services provided by the concession companies [68]. Besides, their roles also include assigning tasks, coordinating, and guiding the appointed users who were involved in the maintenance process. They are the most suitable respondents for the survey in this study.

Our study has attempted to cover all public hospitals in the country with a total of 139 hospitals (excluding medical institutions). The list was identified from the official webpage of the Ministry of Health, Malaysia [13]. The census method was applied, as this was suitable based on our well-defined, accessible, and small population [69]. The questionnaire survey was conducted via an online platform, as it was easier to administer and could reach a wide geographical coverage. Follow-ups via telephone calls were made as a reminder to increase the response rate.

Proposed Model

A value-based building maintenance model consists of four reflective constructs: user involvement, value add, value co-creation, and value outcomes, as shown in Figure 2. Value outcomes is an endogenous construct, while the other three constructs are exogenous. The assessment of the model consists of two steps: assessment of the measurement model (in this study, only reflective models were designed), followed by validation of the structural model. These steps are explained in the subsequent section. The main objective of PLS-SEM is to estimate the path coefficients that maximize the coefficient of determination (R^2) of the endogenous construct. In this study, the endogenous construct is value outcomes. A higher value of R^2 indicates higher predictive accuracy [64]. The value ranges from 0 to 1, with values of 0.26, 0.13 or 0.02 described as substantial, moderate, or weak, respectively, according to Cohen (1989) cited in [63]. The effect size f^2 measures changes in R^2 due to the omission of the exogenous construct from the model, to assess its impact on the endogenous construct. The effect size f^2 value of 0.35, 0.15 or 0.02 indicate substantial, medium, and small effect size, respectively [63].

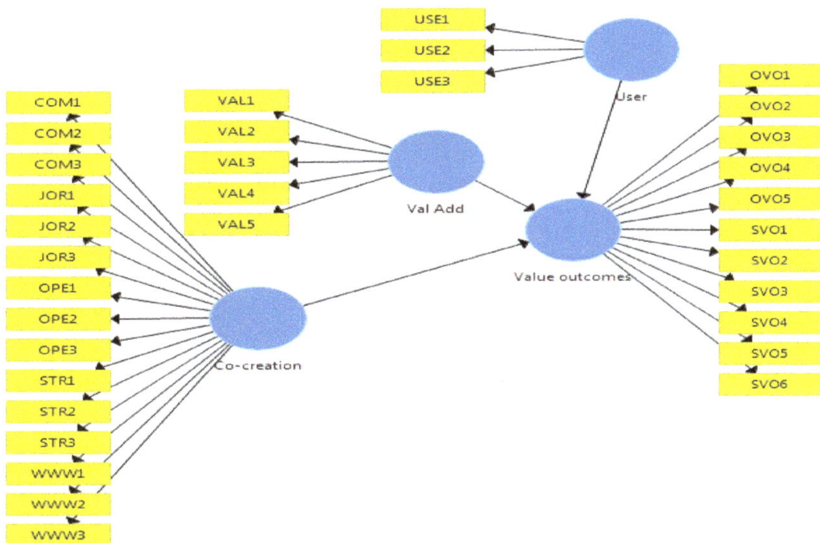

Figure 2. Proposed value-based building maintenance model.

4. Results and Discussions

From the 139 survey forms distributed, a total of 68 responses were collected, amounting to a 48.9% response rate. Out of this, 66 valid responses were accepted. The small sample size is due to the small number of public hospitals in the country, and limited number of target respondents in each hospital. Nonetheless, the 66 usable samples fulfilled the minimum sample criterion of 30 [63], and the 1:10 of structural paths directed to a construct ratio [70], in performing PLS-SEM analysis. Besides, the recommended sample size for R^2 ranges from 0.25 to 0.50, with 5% probability of error, and a model with three arrows pointing to a construct required a sample size between 16 and 37 [64]. This study considered 66 valid samples, with an R^2 value of 0.423.

The profile of respondents is depicted in Table 3. The average age of hospitals is 32 years, ranging from 4 to 133 years. The average number of beds is 274, ranging from

40 to 1600 beds. The average number of staff members for FEMS is 50, ranging from 5 to 414 persons.

Table 3. Profile of respondents.

	Frequency (N = 66)	%
Position		
(i) Engineer	27	40.9
(ii) Assistant engineer	39	59.1
Total:	66	100
Education (Level)		
(i) Diploma	32	48.5
(ii) Bachelor degree	31	47.0
(iii) Master degree	3	4.5
(iv) Others	0	0
Total:	66	100
Academic qualification		
(i) Mechanical engineering	10	15.2
(ii) Electrical engineering	9	13.6
(iii) Civil engineering	42	63.6
(iv) Construction management	1	1.5
(v) Biomedical	1	1.5
(vi) Others	3	4.6
Total:	66	100
Years of Experience		
Mean:	5.19	
Standard deviation:	2.593	
Range:	1–14	

4.1. Reflective Measurement Model

The reflective measurement model was assessed in terms of internal consistency reliability, convergent validity, and discriminant validity. The composite reliability (CR) of this measurement model ranged from 0.756 to 0.927, indicating internal consistency reliability [71]. In terms of indicator reliability/factor loading, the results show that the majority of indicators are above 0.4 of the minimum acceptable factor loadings, ranging from 0.552 to 0.861 [64]. Only 6 out of 34 indicators (17.6%) were discarded due to low loadings; this is within the 20% limit of the overall number of indicators [72]. In terms of convergent validity, the final AVE of all constructs is above the minimum acceptable level of 0.5. Table 4 summarizes the assessment of the measurement model.

Table 4. Measurement model.

Construct	Items	Loadings	AVE	CR
User	USE1	0.685	0.509	0.756
	USE2	0.682		
	USE3	0.769		
Val Add	VAL1	0.757	0.595	0.880
	VAL2	0.857		
	VAL3	0.771		
	VAL4	0.792		
	VAL5	0.668		
Value Co-creation	COM1	0.597	0.502	0.909
	COM3	0.816		
	JOR3	0.741		
	OPE2	0.710		
	OPE3	0.621		

Table 4. *Cont.*

Construct	Items	Loadings	AVE	CR
	STR1	0.738		
	STR2	0.702		
	STR3	0.585		
	WWW2	0.678		
	WWW3	0.850		
Value Outcomes	OVO1	0.633	0.563	0.927
	OVO2	0.815		
	OVO3	0.658		
	OVO5	0.715		
	SVO1	0.728		
	SVO2	0.552		
	SVO3	0.776		
	SVO4	0.861		
	SVO5	0.855		
	SVO6	0.842		

Note: JOR1, COM2, JOR2, OPE1, OVO4, and WWW1 were discarded due to low loadings.

Discriminant validity of this measurement model was established via Fornell and Larcker's criterion [63]. Table 5 indicates that the square root of AVE of the construct is greater than inter-correlation with other constructs, and confirmed discriminant validity. The cross-loadings pattern established that loadings of items are higher with constructs that it is supposed to measure. To further verify this, Table 6 shows that the Heterotrait–Monotrait ratio of correlations (HTMT) values are all below 0.85 which is the stringent criterion by Kline (2011), cite in [63].

Table 5. Discriminant validity using Fornell and Larcker's criterion.

	Co-Creation	User	Val Add	Value Outcomes
Co-creation	0.709			
User	0.272	0.713		
Val Add	0.583	0.387	0.771	
Value Outcomes	0.538	0.299	0.607	0.75

Table 6. Heterotrait–Monotrait ratio of correlations (HTMT) criterion.

	Co-Creation	User	Val Add
Co-creation			
User	0.423		
Val Add	0.654	0.562	
Value Outcomes	0.54	0.393	0.668

4.2. Validation of the Structural Model

Validation of the structural model involves five essential steps to assess lateral collinearity, path coefficient, coefficient of determination, effect size to R^2, and Stone–Geisser Q^2 Predictive Relevance [63]. In terms of lateral collinearity, all three exogenous latent variables have a variance inflator factor (VIF) value below 3.3 suggested by Diamantopoulos and Siguaw (2006) cite in [63], which indicate that there is no collinearity problem. Table 7 shows the results of VIF.

Table 7. Lateral collinearity assessment.

Construct	Value Outcomes (VIF)
Value Outcomes	
Co-creation	1.52
User	1.181
Val Add	1.655

4.3. Hypothesis Testing

Three hypotheses, namely, H1 (User → Value Outcomes), H2 (Val Add → Value Outcomes), and H3 (Co-creation → Value Outcomes) were tested. The significance level was determined by referring to the t-statistics generated using the bootstrapping function in PLS 3.0, and path coefficients were assessed. The 95% confidence interval was adopted.

The results show that for H1, the t-value of 0.569 is lower than the critical value of 1.645, and the p value of 0.285 is over 0.05, indicating that this hypothesis is not supported. H2 has a t-value of 3.476 and a p value of 0.000 ($p < 0.01$, $t > 2.33$), while H3 has a t-value of 2.214 and a p value of 0.014 ($p < 0.05$, $t > 1.645$). These results indicate that both H2 and H3 were supported. Predictors of value co-creation and value add with $\beta = 0.275$ and $\beta = 0.423$, respectively, are positively related to value outcomes.

The coefficient of determination (R^2) for value outcomes is 0.423, which is considered substantial ($R^2 > 0.26$) based on Cohen (1989) as cited in [63]. It explains 42.3% of the variance in value outcomes. In terms of the effect size to R^2, the f^2 value of 0.187 (greater than 0.15) indicates that value add has a medium effect size on value outcomes; while an f^2 value of 0.086 (greater than 0.02) according to Cohen (1988) as cited in [63], indicates that value co-creation has small effect size on value outcomes. Stone–Geisser Q^2 predictive relevance was tested using the blindfolding technique in PLS 3.0. The value of Q^2 of 0.195 is larger than 0 [64], indicating that predictive relevance is established in this model. Table 8 shows the results of hypothesis testing.

Table 8. Hypothesis testing.

H	Relationship	Std Beta	Std Error	t-Value	p Value	Decision	f^2
H1	User → Value Outcomes	0.061	0.112	0.569	0.285	not supported	0.005
H2	Val Add → Value Outcomes	0.423	0.117	3.476	0.000	supported	0.187
H3	Co-creation → Value Outcomes	0.275	0.132	2.214	0.014	supported	0.086

From the analysis, the following causal relationships were established:

1. Value-adding practices positively influence value outcomes.
2. Value co-creation positively influences value outcomes.

A revised model of the value-based building maintenance is shown in Figure 3.

The results from the hypothesis testing found that value-adding practices positively influence value outcomes with a medium effect size. The result is consistent with the case study by Okoroh et al. [29], where value-adding service providers achieved better service quality and enhanced corporate image through FM partnering arrangement in NHS trust. The results provide empirical justification to support their case study's outcome in terms of provision of innovative practices by service providers, value for money, and significant savings in the FM partnership. Besides, the results also support the work by Ali-Marttila et al. [50], that outsourced service providers which provide integrated solutions are more complete in their offering in the industrial maintenance context. Our study confirms the causal relations between value-adding practices which includes integrated service solution, to the value outcomes, in the context of outsourced hospital building maintenance. In hospitals, besides maintenance work of M&E systems and plants, civil engineering works, operation and installation of engineering plants, the outsourced contractors also extend their services to include but not limited to technical advice, equipment loaning, reducing risks and hazards, and prolonging the life of the facilities. In addition, the results also affirm the responsiveness of service providers as one of the crucial aspects [38,58]. Overall, the finding confirms that by providing value-adding practices, contractors are able to improve their achievement of operational or strategic outcomes of hospitals. Ultimately, weaknesses of building maintenance such as lack of planned preventive maintenance [17], and various maintenance issues and user dissatisfaction could gradually be overcome. Besides, heavy reliance on supervision and monitoring by hospitals can be reduced to relieve over-straining of manpower resources [17,73].

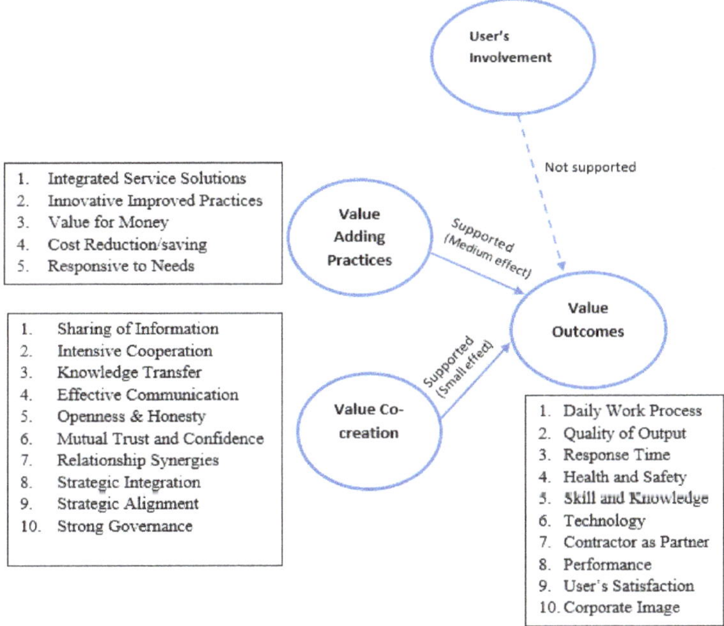

Figure 3. Revised value-based building maintenance model.

Hypothesis testing also found value co-creation positively influences value outcomes, but with a lower effect. It provides empirical evidence that supports the effort of collaborative working among various players to co-create values in the building maintenance context. The result is consistent with outcomes by Sun and Chen [56], where value co-creation was found to have a strong impact on outsourcing satisfaction in the IT field. The finding also supports the notion of sharing of resources of the service-dominant (S-D) logic by Vargo et al. [57], and findings on value co-creation in terms of sharing of information [52,53]. However, sharing of resources in terms of manpower and technology were not justified with the deletion of "Sharing of Manpower Resources" and "Joint Technology" items due to low loadings during PLS-SEM analysis. This can be explained due to the apparent demarcation of provision of manpower and technology by the contractor in the concession contract. However, past research on knowledge transfer notion is supported as a value co-creation practice in hospital maintenance [55,57]. Besides, this study also echoed the findings on the importance of openness and honesty, and mutual trust and confidence in outsourcing strategic partnerships [58]. In Malaysian public hospitals, value co-creation effort is evident in the Sustainable Energy Management Program (SEMP), where both monetary resources and electrical energy saving were achieved. There is a strong commitment from the Ministry of Health in terms of strategic alignment and strong leadership to pilot the projects of government hospitals. Personnel from concession companies and hospitals are trained and certified as Malaysian Certified Healthcare Facility Managers

(CHFMs) [35]. This involves intensive cooperation amongst respective concession companies, engineers, and maintenance teams from hospitals, as well as joint resources and their effective communication.

However, it should be highlighted that the hypothesis "User involvement positively influences value outcomes" was not supported. The findings contradict past studies, which stressed the importance of user involvement [38,48,59]. This could be due to the lack of formal processes used to gather user expectation for decision-making during the early planning stage. Currently, user involvement is confined to implementation and post-maintenance stage such as job requests, complaints, and job completion verification [14], rather than early involvement [45]. User perception needs to be strategically aligned and linked to business processes from the pre-delivery, mid-delivery, and post-delivery stages [49]. Hence, without an understanding of user expectations at the early maintenance planning stage, it is challenging to deliver up to their satisfaction; this explains the reason user complaints are still rampant in public hospitals.

This study refines Value-based Maintenance Management Model by Olanrewaju et al. [38], by adding the value-based practices to their model, which did not explicitly address the relationships of the parties involved in building maintenance. Their model focused on management functions and served primarily as an operational manual for educational buildings. The value factors identified in this study can be assimilated into the management functions of their model, in order to enhance the achievement of value outcomes.

5. Conclusions

At present, the maintenance services of public hospitals in Malaysia are outsourced for better service quality and standardization. The performance of the five concessionaires serving the entire country actually reflects the corporate image of public healthcare facilities in the country. Using an online survey questionnaire, this research investigated the causal relationships between the value factors and value outcomes of building maintenance in public hospitals in Malaysia. The major findings revealed the significant influences which value-adding practice and value co-creation have on the value outcomes in hospitals. However, user involvement was not supported to influence value outcomes, which merits further investigations.

5.1. Theoretical Implications

This research expands the theoretical literature in value concepts by determining the causal relationships between value-adding practices and value co-creation to the value outcomes using PLS-SEM in an integrated model. Previous studies in value concepts were fragmented and investigated value concepts discretely in various fields and objectives. It reveals that value-adding practices by the maintenance service providers has greater effect to the achievement of value outcomes as compared to the effect that value co-creation contributes. This research provides an empirical and statistical assertion that acknowledges concession companies' essential roles in outsourced maintenance in public hospitals.

5.2. Practical Implications

Findings from this study inform maintenance service providers on the significant advantages of being proactive in adding value to their customers. By providing an integrated service solution, innovative improved practices, value for money, cost reduction/saving, and being responsive to users' needs, the contractors would be able to enhance their maintenance service delivery. Findings from this study would also be useful to policymakers in setting benchmarks for the selection of competent contractors, contract renewal, and performance monitoring. It provides justifications for maintenance organizations to shift from traditional, transactional, and contractual relationships, towards a collaborative networking culture. With the concession companies stationed round the clock in the hospital premises, there is a great opportunity for them to interact with the hospital maintenance teams and users. Within this setting, the "joint sphere" allows the providers and customers to interact, or jointly interact, directly or indirectly to co-create new value [51]. Value co-creation can be achieved with joint resources, operations integration, and open communication. In the long concession period, mutual understanding and strategic goals can be achieved through a win-win-win relationship.

5.3. Limitations and Future Research

While this research provides insights on value-based maintenance for hospital buildings, the research has some limitations. For instance, the user involvement factor was not statistically supported to influence the value outcomes. Therefore, further qualitative research such as in-depth interviews could be useful to further explain the quantitative results from this study. In addition, although value-adding practices and value co-creation were found to have a positive influence on value outcomes, the specific importance, and performance of each practice are not known. Hence, it is recommended that further research investigates the critical success factors in implementing the value-based approach in hospital maintenance. Lastly, this research only focused on public hospitals in Malaysia. It is recommended that future research can be designed to carry out comparison or parallel study on private hospitals to gather more insights and explore the best practices in both types of establishments.

Author Contributions: All authors contributed equally to all phases of conceptualization, methodology, investigation, and writing of the manuscript. All authors have read and agreed to the published version of the manuscript.

Funding: This research received no external funding.

Institutional Review Board Statement: Not applicable.

Informed Consent Statement: Informed consent was obtained from all subjects involved in the study.

Data Availability Statement: Further details regarding data can be obtained by contacting the author.

Conflicts of Interest: The authors declare no conflict of interest.

References

1. World Health Organisation. 2020. Available online: https://www.who.int/health-topics/hospitals#tab=tab_2 (accessed on 7 April 2020).
2. Lavy, S.; Shohet, I.M. Integrated maintenance management of hospital buildings: A case study. *Constr. Manag. Econ.* **2004**, *22*, 25–34. [CrossRef]
3. Adenuga, O.; Ibiyemi, A. An Assessment of the State of Maintenance of Public Hospital Buildings in Southwest Nigeria. *Constr. Econ. Build.* **2009**, *9*, 51–60. [CrossRef]
4. Olanrewaju, A.L.; Wong, W.F.; Yahya, N.N.-H.N.; Im, L.P. Proposed research methodology for establishing the critical success factors for maintenance management of hospital buildings. *Int. Symp. Green Sustain. Technol.* **2019**, *2157*, 020036. [CrossRef]
5. Codinhoto, R.; Tzortzopoulos, P.; Kagioglou, M.; Aouad, G.; Cooper, R. The impacts of the built environment on health outcomes. *Facilities* **2009**, *27*, 138–151. [CrossRef]
6. Loo, V.G.; Bertrand, C.; Dixon, C.; Vityé, D.; DeSalis, B.; McLean, A.P.H.; Brox, A.; Robson, H.G. Control of Construction-Associated Nosocomial Aspergillosis in an Antiquated Hematology Unit. *Infect. Control. Hosp. Epidemiology* **1996**, *17*, 360–364. [CrossRef]
7. Abisuga, A.O.; Famakin, I.O.; Oshodi, O.S. Educational building conditions and the health of users. *Constr. Econ. Build.* **2016**, *16*, 19–34. [CrossRef]
8. Enshassi, A.; Farida, E.S.; Suhair, A. Assessment of Operational Maintenance in Public Hospitals Buildings in the Gaza Strip. *Int. J. Sustain. Constr. Eng. Technol.* **2015**, *6*, 2180–3242.
9. Galán, M.H.; Edith, M.D. Management Audit Applied to the Maintenance Department in Hospital Facilities. *Ing. Mec.* **2017**, *20*, 152–159.
10. Salleh, N.M.; Salim, N.A.A.; Jaafar, M.; Sulieman, M.Z.; Ebekozien, A. Fire safety management of public buildings: A systematic review of hospital buildings in Asia. *Prop. Manag.* **2020**, *38*, 497–511. [CrossRef]
11. Shastri, B.A.; Raghav, Y.S.; Sahadev, R.; Yadav, B.P. Analysis of Fire Protection Facilities in Hospital Buildings. In *Advances in Fire and Process. Safety*; Siddiqui, N.A., Tauseef, S.M., Abbasi, S.A., Rangwala, A.S., Eds.; Springer Nature: Singapore, 2018; pp. 183–190. [CrossRef]

12. Malaysia Productivity Corporation. Reducing Unnecessary Regulatory Burdens on Business: Private Hospitals. 2014. Available online: http://www.mpc.gov.my/reducing-unnecessary-regulatory-burdens-rurb-2/ (accessed on 2 January 2021).
13. Ministry of Health Malaysia. Healthcare Facilities (Government). 2019. Available online: http://www.moh.gov.my/ (accessed on 22 February 2020).
14. Fan, H.P. *Privatization of Facility Management in Public Hospitals: A Malaysian Perspective*; Patridge Publishing: Singapore, 2016.
15. Baba, M.; Abdul, H.M. *A Conceptual Contract Framework for Research to Develop a Contract Framework for Outsourcing of Facilities Management in Malaysian Hospitals*; IRERS: Kuala Lumpur, Malaysia, 2008.
16. Ali, M.; Mohamad, W.M.N.B.W. Audit assessment of the facilities maintenance management in a public hospital in Malaysia. *J. Facil. Manag.* **2009**, *7*, 142–158. [CrossRef]
17. National Audit Department Malaysia. *Auditor General's Report 2015 Series 2*; National Audit Department Malaysia: Putrajaya, Malaysia, 2016.
18. Mustapa, F.D.; Muzani, M.; Fuziah, I.; Kherun, N.I. Outsourcing in Malaysian Healthcare Support Services: A Study on the Causes of Increased Operational Costs. In Proceedings of the International Conference on Construction Industry, Sumatra Barat, Indonesia, 21–24 June 2006.
19. Ministry of Health Malaysia. *Malaysia National Health Accounts*; Health Expenditure Report 1997–2017; Planning Division, MOH: Putrajaya, Malaysia, 2019.
20. Yousefli, Z.; Nasiri, F.; Moselhi, O. Healthcare facilities maintenance management: A literature review. *J. Facil. Manag.* **2017**, *15*, 352–375. [CrossRef]
21. Lavy, S.; Shohet, I.M. On the effect of service life conditions on the maintenance costs of healthcare facilities. *Constr. Manag. Econ.* **2007**, *25*, 1087–1098. [CrossRef]
22. Al-Zubaidi, H. Assessing the Demand for Building Maintenance in a Major Hospital Complex. *Prop. Manag.* **1997**, *15*, 173–183. [CrossRef]
23. Vanzanella, C.; Fico, G.; Arredondo, M.T.; Delfino, R.; Viggiani, V.; Triassi, M.; Pecchia, L. Interactive management control via analytic hierarchy process: An empirical study in a public university hospital. *J. Int. Bus. Entrep. Dev.* **2015**, *8*, 144. [CrossRef]
24. Mustapha, Z.; Justice, A. Building Maintenance Systems of Public Health Institutions in Ghana: A Case Study of La General Hospital. *J. Constr. Proj. Manag. Innov.* **2011**, *1*, 155–166.
25. Amankwah, O.; Choong, W.-W.; Mohammed, A.H. Modelling the influence of healthcare facilities management service quality on patients satisfaction. *J. Facil. Manag.* **2019**, *17*, 267–283. [CrossRef]
26. Pheng, L.S.; Rui, Z. *Service Quality for Facilities Management in Hospitals*; Springer: Singapore, 2016. [CrossRef]
27. Jandali, D.; Sweis, R. Factors affecting maintenance management in hospital buildings. *Int. J. Build. Pathol. Adapt.* **2019**, *37*, 6–21. [CrossRef]
28. Yousefli, Z.; Nasiri, F.; Moselhi, O. Maintenance workflow management in hospitals: An automated multi-agent facility management system. *J. Build. Eng.* **2020**, *32*, 101431. [CrossRef]
29. Okoroh, M.; Gombera, P.; John, E.; Wagstaff, M. Adding value to the healthcare sector—A facilities management partnering arrangement case study. *Facilities* **2001**, *19*, 157–164. [CrossRef]
30. Manaf, N.H.A. Quality management in Malaysian public health care. *Int. J. Heal. Care Qual. Assur.* **2005**, *18*, 204–216. [CrossRef]
31. Ghani, A.; Zaid, M.; Abd, Z.; Ibrahim, I.; Musa, Z. Defining the Critical Success Factor in FM Malaysian Healthcare Sector. In Proceedings of the 3rd International Building Control Conference, Kuala Lumpur, Malaysia, 21 November 2013.
32. Ahmad, P.; Nur, A.; Mat, N.A.; Mohd, A.; Janice, L.Y.M.; Mohd, N.J.; Abdul, H.M. Critical Success Factors for Facilities Management Implement in the Healthcare Industry. *Int. J. Real Estate Stud.* **2017**, *11*, 69–83.
33. Rani, N.A.A.; Baharum, M.R.; Akbar, A.R.N.; Nawawi, A.H. Perception of Maintenance Management Strategy on Healthcare Facilities. *Procedia Soc. Behav. Sci.* **2015**, *170*, 272–281. [CrossRef]
34. Omar, M.F.; Ibrahim, F.A.; Omar, W.M.S.W. Key Performance Indicators for Maintenance Management Effectiveness of Public Hospital Building. *MATEC Web Conf.* **2017**, *97*, 01056. [CrossRef]
35. Abdullah, M.S.I.; Noor, M.A.R.; Tauran, Z.A.Z.; Khairul, A.K. Latest Development on Sustainability Programme Initiatives in Malaysian Healthcare Facility Management. In Proceedings of the 37th Conference of the ASEAN Federation of Engineering Organisations, Jakarta, Indonesia, 11–15 September 2019.
36. Kamaluddin, K.A.; Muhammad, S.I.A.; Yang, S.S. Development of Energy Benchmarking of Malaysian Government Hospitals and Analysis of Energy Savings Opportunities. *J. Build. Perform.* **2016**, *7*, 72–87.
37. Sahamir, S.R.; Rozana, Z. Green Assessment Criteria for Public Hospital Building Development in Malaysia. *Procedia Environ. Sci.* **2014**, *20*, 106–115. [CrossRef]
38. Olanrewaju, A.L.; Abdul-Aziz, A.-R. *Building Maintenance Processes and Practices: The Case of a Fast Developing Country*; Springer: Singapore, 2015. [CrossRef]
39. Awang, N.A.; Chua, S.J.L.; Ali, A.S. Building Condition Assessment Focusing on Persons with Disabilities' Facilities at Hospital Buildings. *J. Des. Built Environ.* **2017**, *17*, 73–84. [CrossRef]
40. Ab Ghani, M.Z.; Srazali, A. Comparative Review of Design Requirements for Natural Smoke Ventilation in Hospital Buildings. *J. Malays. Inst. Plan.* **2018**, *16*, 334–344. [CrossRef]

41. National Audit Department Malaysia. *Auditor General's Report 2015 Series 1*; National Audit Department Malaysia: Putrajaya, Malaysia, 2016.
42. Carvalho, M.; Hemananthani, S.; Rahimy, R.; Loshana, K.S. Auditor-General' s Report 2017: Alor Gajah Hospital Comes under Scrutiny. *The Star Online*, 6 August 2018. Available online: https://www.thestar.com.my/news/nation/2018/08/06/ag-report-2017-alor-gajah-hospital-comes-under-scrutiny (accessed on 11 October 2019).
43. Olanrewaju, A.; Fang, W.W.; Tan, S.Y. Hospital Building Maintenance Management Model. *Int. J. Eng. Technol.* **2018**, *7*, 747–753. [CrossRef]
44. Prahalad, C.K.; Ramaswamy, V. Co-creating unique value with customers. *Strat. Leadersh.* **2004**, *32*, 4–9. [CrossRef]
45. Jensen, P.A.; Maslesa, E. Value based building renovation—A tool for decision-making and evaluation. *Build. Environ.* **2015**, *92*, 1–9. [CrossRef]
46. Abdul-Lateef, O.A. Quantitative Analysis of Criteria in University Building Maintenance in Malaysia. *Constr. Econ. Build.* **2010**, *10*, 51–61. [CrossRef]
47. Olanrewaju, A.L.; Mohd, F.K.; Arazi, I. Validation of Building Maintenance Performance Model for Malaysian Universities. *Int. J. Hum. Soc. Sci.* **2011**, *6*, 159–163.
48. Zulkarnain, S.H.; Emma, M.A.Z.; Rahman, M.Y.A.; Nur, K.F.M. A Review of Critical Success Factor in Building Maintenance Management Practice for University Sector. *Int. J. Civ. Environ. Struct. Constr. Archit. Eng.* **2011**, *55*, 215–219. [CrossRef]
49. Tucker, M.; Smith, A. User perceptions in workplace productivity and strategic FM delivery. *Facilities* **2008**, *26*, 196–212. [CrossRef]
50. Ali-Marttila, M.; Marttonen-Arola, S.; Kärri, T.; Pekkarinen, O.; Saunila, M. Understand what your maintenance service partners value. *J. Qual. Maint. Eng.* **2017**, *23*, 144–164. [CrossRef]
51. Grönroos, C.; Voima, P. Critical service logic: making sense of value creation and co-creation. *J. Acad. Mark. Sci.* **2013**, *41*, 133–150. [CrossRef]
52. Coenen, C.; Alexander, K.; Kok, H. Facility management value dimensions from a demand perspective. *J. Facil. Manag.* **2013**, *11*, 339–353. [CrossRef]
53. Bititci, U.S.; Martinez, V.; Albores, P.; Parung, J. Creating and managing value in collaborative networks. *Int. J. Phys. Distrib. Logist. Manag.* **2004**, *34*, 251–268. [CrossRef]
54. Martinez, V.; Bititci, U.S. Aligning value propositions in supply chains. *Int. J. Value Chain Manag.* **2006**, *1*, 6–18. [CrossRef]
55. Joshi, K.P.; Murthy, C. Determining Value Co-Creation Opportunity in B2B Services. In Proceedings of the 2011 Annual SRII Global Conference (SRII 2011), San Jose, CA, USA, 29 March–2 April 2011; pp. 674–684. [CrossRef]
56. Sun, S.Y.; Li, S.C. Achieving Value Co-Creation in IT Outsourcing. *J. Int. Technol. Inf. Manag.* **2016**, *25*, 1–18.
57. Vargo, S.L.; Maglio, P.P.; Akaka, M.A. On value and value co-creation: A service systems and service logic perspective. *Eur. Manag. J.* **2008**, *26*, 145–152. [CrossRef]
58. Dibley, A.; Moira, C. *How to Implement. Best Practice in Strategic Partnerships: An. Outsource Supplier and Client Perspective*; The Henley Centre for Customer Management: London, UK, 2011.
59. Jensen, P.A. The Facilities Management Value Map: a conceptual framework. *Facilities* **2010**, *28*, 175–188. [CrossRef]
60. Dibley, A.; Clark, M. Value Co-Creation in Strategic Partnerships: An Outsourcing Perspective. In *Service-Dominant Logic., Network & Systems Theory and Service Science: Integrating Three Perspectives for a New Service Agenda*; Polese, F., Ed.; Henle Business School: Napoli, UK, 2005.
61. Gummerus, J. Value creation processes and value outcomes in marketing theory. *Mark. Theory* **2013**, *13*, 19–46. [CrossRef]
62. Christian, M.R.; Wende, S.; Becker, J.M. *SmartPLS 3*; SmartPLS GmbH: Bönningstedt, Germany, 2015.
63. Ramayah, T.; Jacky, C.; Francis, C.; Hiram, T.; Mumtaz, A.M. *Partial Least Squares Structural Equation Modeling (PLS-SEM) Using SmartPLS 3.0.*, 2nd ed.; Pearson Malaysia Sdn. Bhd.: Kuala Lumpur, Malaysia, 2018.
64. Hair, J.F.G.; Tomas, M.H.; Christian, R.; Marko, S. *A Primer on Partial Least Squares*, 2nd ed.; SAGE Publications, Inc.: Los Angeles, CA, USA, 2017.
65. Chomeya, R. Quality of Psychology Test between Likert Scale 5 and 6 Points. *J. Soc. Sci.* **2010**, *6*, 399–403.
66. Willis, G.B. Questionnaire Pretesting. In *The SAGE Handbook of Survey Methodology*; Christof, W., Dominique, J., Tom, W., Yang, C.F., Eds.; SAGE Publishing Ltd.: London, UK, 2016; pp. 359–381.
67. Collins, D. Pretesting Survey Instruments: An Overview of Cognitive Methods. *Quality* **2009**, *12*, 229–238. [CrossRef]
68. National Audit Department Malaysia. *Auditor General's Report 2014 Series 3*; National Audit Department Malaysia: Putrajaya, Malaysia, 2015.
69. Mooi, E.; Marko, S. *Concise Guide to Market. Research: The Process, Data and Methods Using IBM SPSS Statistics*; Springer: Berlin/Heidelberg, Germany, 2011.
70. Barclay, D.W.; Christopher, A.H.; Ronald, T. The Partial Least Squares Approach to Causal Modeling: Personal Computer Adoption and Use as Illustration. *Technol. Stud.* **1995**, *2*, 285–309.
71. Hair, J.F.; Risher, J.J.; Sarstedt, M.; Ringle, C.M. When to use and how to report the results of PLS-SEM. *Eur. Bus. Rev.* **2019**, *31*, 2–24. [CrossRef]

72. Hair, J.F.; William, C.B.; Barry, J.B.; Rolph, E.A.; Tatham, R. *Multivariate Data Analysis*, 7th ed.; Pearson Prentice Hall: Hoboken, NJ, USA, 2010.
73. Aliman, K.H. Audit Finds Malaysian Hospitals Understaffed, Underfunded and Overcrowded. The Edge Markets. 15 July 2019. Available online: https://www.theedgemarkets.com/article/audit-finds-malaysian-hospitals-understaffed-underfunded-and-overcrowded# (accessed on 11 October 2019).

sustainability

MDPI

Article

Application of the Multiverse Optimization Method to Solve the Optimal Power Flow Problem in Direct Current Electrical Networks

Andrés Alfonso Rosales-Muñoz [1,†], Luis Fernando Grisales-Noreña [1,†], Jhon Montano [1,†], Oscar Danilo Montoya [2,3,†] and Alberto-Jesus Perea-Moreno [4,*,†]

1   Grupo MATyER, Facultad de Ingeniería, Instituto Tecnológico Metropolitano, Campus Robledo, Medellín 050036, Colombia; andresrosales224822@correo.itm.edu.co (A.A.R.-M.); luisgrisales@itm.edu.co (L.F.G.-N.); jhonrojas7420@correo.itm.edu.co (J.M.)
2   Facultad de Ingeniería, Universidad Distrital Francisco José de Caldas, Bogotá 110231, Colombia; odmontoyag@udistrital.edu.co
3   Laboratorio Inteligente de Energía, Universidad Tecnológica de Bolívar, Cartagena 131001, Colombia
4   Departamento de Física Aplicada, Radiología y Medicina Física, Campus Universitario de Rabanales, Universidad de Córdoba, 14071 Córdoba, Spain
*   Correspondence: g12pemoa@uco.es
†   These authors contributed equally to this work.

Citation: Rosales Muñoz, A.A.; Grisales-Noreña, L.F.; Montano, J.; Montoya, O.D.; Perea-Moreno, A.-J. Application of the Multiverse Optimization Method to Solve the Optimal Power Flow Problem in Direct Current Electrical Networks. *Sustainability* **2021**, *13*, 8703. https://doi.org/10.3390/su13168703

Academic Editor: Michael S. Carolan

Received: 11 July 2021
Accepted: 1 August 2021
Published: 4 August 2021

Publisher's Note: MDPI stays neutral with regard to jurisdictional claims in published maps and institutional affiliations.

Abstract: This paper addresses the optimal power flow problem in direct current (DC) networks employing a master–slave solution methodology that combines an optimization algorithm based on the multiverse theory (master stage) and the numerical method of successive approximation (slave stage). The master stage proposes power levels to be injected by each distributed generator in the DC network, and the slave stage evaluates the impact of each power configuration (proposed by the master stage) on the objective function and the set of constraints that compose the problem. In this study, the objective function is the reduction of electrical power losses associated with energy transmission. In addition, the constraints are the global power balance, nodal voltage limits, current limits, and a maximum level of penetration of distributed generators. In order to validate the robustness and repeatability of the solution, this study used four other optimization methods that have been reported in the specialized literature to solve the problem addressed here: ant lion optimization, particle swarm optimization, continuous genetic algorithm, and black hole optimization algorithm. All of them employed the method based on successive approximation to solve the load flow problem (slave stage). The 21- and 69-node test systems were used for this purpose, enabling the distributed generators to inject 20%, 40%, and 60% of the power provided by the slack node in a scenario without distributed generation. The results revealed that the multiverse optimizer offers the best solution quality and repeatability in networks of different sizes with several penetration levels of distributed power generation.

Keywords: optimal power flow; power flow; optimization algorithms; DC networks; electrical energy; optimization

1. Introduction

1.1. General Context

Electrical energy is essential for modern daily life because people's comfort and socioeconomic conditions depend on it [1]. Such energy can be generated and distributed in the form of Alternating Current (AC) and Direct Current (DC). Although most electrical power and electricity distribution systems work with AC, DC electrical energy is as important as its counterpart. Furthermore, DC offers technical-operational advantages; for instance, as most loads used by people are DC, generating and distributing DC eliminates the need for converting electrical energy from AC to DC. This type of electric current promotes the use of clean and environmentally friendly renewable energy sources because most renewable energy is produced in DC, which eliminates the need for power converters. It also eliminates phasors and reactive elements in the electrical network. This reduces the complexity of the mathematical model and the operation and control of electrical networks. This paper focuses on electrical DC systems due to the advantages mentioned above and because more studies should investigate this type of network [2]. To operate the elements that compose DC electrical networks, i.e., (slack and distributed) loads and power generators, solution methodologies should be used to identify the operating points of these devices, mainly the generators, in order to guarantee the technical conditions of the network (power balance, power and voltage limits, penetration levels of the distributed generators, etc.). In addition, such methodologies should enable network operators or owners to meet the objective functions they have established in order to improve the technical conditions of the system (power losses, voltage stability, loadability of the lines, etc.), as well as its environmental characteristics, or to reduce operating costs (buying electricity from the grid) [3]. This problem, known as Optimal Power Flow (OPF), is about finding the reference points for the power of the Distributed Generators (DGs) in a network in order to meet the objective functions that have been defined.

1.2. State-of-the-Art

To solve the OPF problem in DC networks, the first step is interpreting and solving the load flow or power flow problem. As a result, we can determine the existing voltage in each one of the nodes in the network and the current circulating through the transmission lines thanks to the parameters of the electrical power system, e.g., conductance of transmission lines, power demanded by the loads, and electrical topology [4]. To solve the load flow problem, numerical methods should be employed to solve the nonlinear nonconvex problem that it represents [5]. In recent years, different types of methods have been proposed for this purpose. Some of the classical options include Newton–Raphson [6], Gauss–Seidel, and Gauss–Jacobi [7]. More current alternatives are Taylor series approximation [8], successive approximation [9], triangular matrix formulation [8], sweeping based on graph theory [10], and backward/forward sweeping [11]. To solve the load flow and calculate each one of the previously mentioned variables, this study used the Successive Approximation (SA) method [9]. The latter was selected due to its excellent performance in terms of convergence and processing times for radial as well as meshed networks [11].

The load flow method can describe the electrical behavior of a DC network based on known data of power demanded and generated by the DGs. However, in case of any variation in said power, the power flow should be calculated again to establish the effect of the generation and demand inside the network. To find the most adequate power level that the DGs should inject, the OPF problem should be solved by means of optimization techniques in order to meet the objective function proposed by the DC network operator and/or owner [12]. This guarantees that the electrical constraints that compose DC networks in an environment of distributed generation are always respected [12].

Engineering often faces highly difficult problems, which cannot be solved directly; they are called nonlinear nonconvex problems [13]. For this reason, multiple techniques have been developed to offer a good quality solution to these problems by applying intelligent optimization methods. Some of them are the so-called optimization techniques, which are inspired by the behavior of nature. Optimization methods apply mathematical models and strategies to move within a solution space and find good quality solutions in relatively short times. Every optimization method should be adapted to the problem under analysis, and the impact of the solution produced by each type of method will depend on the difficulty of said problem.

Due to the high efficiency and excellent performance of metaheuristic optimization techniques in solving engineering problems, they are widely applied to solve nonlinear nonconvex problems with continuous variables. Optimization techniques are employed in this study to solve the OPF problem because the mathematical model used here requires nonlinear methods to produce a solution [14]. Furthermore, the mathematical modeling of the problem includes continuous variables. To solve the OPF problem by means of optimization techniques, the problem is generally divided into two stages using a master–slave methodology [14]. The master stage determines the optimal power level that the DGs should inject into the DC network, and the slave stage solves the power flow problem. The objective is to calculate the electrical variables that can be used to establish the impact of the objective function and the problem constraints of each one of the solutions proposed by the master stage. This master–slave methodology enables us to find a good quality solution based on an adequate selection of the optimization techniques to be implemented. Importantly, the solution quality greatly depends on the proposed solution method and the parameters assigned to it.

In the last decade, the specialized literature has proposed different solutions to solve the OPF problem in DC networks. Usually, their objective function is the minimization of power losses in the network [15,16]. In [15], the authors proposed a convex relaxation method based on second-order cone programming. In turn, in [16], the authors employed sequential quadratic convex programming to solve the OPF problem. The studies mentioned above used commercial optimization software to solve the OPF problem, which increases the complexity of the problem and the cost of the solution. Additionally, in such studies, the OPF problem was evaluated in small networks only (10, 16, and 21 nodes), thus ignoring the performance of the methods in bigger networks. Likewise, in [16], the standard deviation of the techniques was not taken into account, which is important to establish how dispersed the results obtained are with respect to the average result. In order to propose methods based on sequential programming that can be developed in any programming language and eliminate the need to acquire specialized (commercial) software, different methodologies have been studied and proposed in recent years to solve the problem addressed here. For example, in [17], the authors employed the black hole (BH) algorithm and the Gauss–Seidel numerical method to generate a master–slave solution strategy for the OPF problem. Remarkably, they did not take into consideration the standard deviation of the techniques they used. In [5], the authors employed the vortex search algorithm and the numerical method by SA to solve the OPF problem. In their study, the solution methodology was evaluated in small DC networks (10 and 21 nodes). By contrast, in [18], the authors used the sine cosine optimization algorithm and the SA numerical method as the solution methodology. Nevertheless, they only analyzed the effect of the methodology in the 21-node system and did not consider the standard deviation of the optimization techniques to solve the OPF problem. This analysis of the state-of-the-art of methodologies that have been applied to solve the OPF problem shows that strategies based on sequential programming should be promoted in order to eliminate the need for commercial software and improve the solution quality. Nevertheless, such methodologies should be validated in test systems of different sizes so that the effectiveness of the proposed solution methodology can be guaranteed in a network of any size. For example, in [14,19], the authors employed the master–slave methodology based on sequential programming to solve the OPF problem. Particularly in [19], three metaheuristic optimization algorithms were implemented as strategies to solve the OPF problem. First, the authors used a continuous genetic algorithm (CGA), which had been presented in [20]. The CGA is inspired by the genetic process of living beings, employing the selection, combination, and mutation process in order to transmit information collected by parents to their descendants. Second, they used the black hole (BH) algorithm reported in [17], which is based on the dynamic interaction of stars and black holes and is used to solve optimization problems with continuous variables, such as the OPF problem. Third, they employed the particle swarm optimization (PSO) metaheuristic, an algorithm inspired by the behavior of schools of fish and bird flocks, to explore the solution space and thus solve the OPF problem [21]. They implemented these three methodologies and evaluated different test systems, demonstrating the effectiveness and robustness of each methodology to solve the problem analyzed in this document. In [14], the authors used the techniques described above plus the ant lion optimizer (ALO), which was recently proposed. The ALO imitates the hunting technique of ant lions, which dig a cone-shaped hole in the ground to catch their prey. Importantly, the techniques mentioned above adopted a master–slave methodology to solve the OPF problem. In the master stage, they used optimization algorithms (CGA, BH, PSO, and ALO) to solve the DG sizing problem; in the slave stage, they implemented the SA numerical method to solve the load flow. The approach called "optimal flow" is a cutting-edge topic that requires more analysis and development in order to improve the operating conditions of the system. In the studies described above, optimization techniques were used as solution methods in order to obtain the best solution to the problem, while considering different energy projects to which it can be applied.

1.3. Proposed Solution Methodology and Main Contributions of This Study

Due to the importance of the optimal power flow problem in DC networks and the need to propose new solution methodologies that are more efficient in terms of solution quality and repeatability, this study proposes a new methodology based on a master–slave strategy to solve the OPF problem in DC networks. In the master stage, the proposed methodology applies the technique called multiverse optimizer (MVO) to find the optimal power level to be injected by the DGs in order to minimize electrical power losses in DC networks. In the slave stage, the numerical SA method is employed to solve the load flow problem and evaluate the objective function and the set of constraints related to every possible solution proposed by the master stage [9]. This study used the 21- and 69-node test systems reported in the specialized literature. In addition, it includes three levels of maximum penetration of distributed generation, i.e., 20%, 40%, and 60% of the power supplied by the slack generator in an environment without DGs. The results of the proposed methodology were compared with those of more popular methodologies reported in the literature: CGA [20], PSO [21], BH [17], and ALO [14]. This demonstrated the effectiveness of the MVO and its behavior regarding solution quality. Additionally, in order to evaluate the repeatability of the algorithms in terms of the solution, each simulation was executed 100 times, and the standard deviation obtained was analyzed. This study makes three contributions:

- A new solution methodology based on a master–slave strategy that combines MVO and SA to solve the optimal power flow problem in DC networks of any size
- A methodology based on average solutions and standard deviation to evaluate the effectiveness and repeatability of the solution methods proposed to solve the OPF problem
- Better results in the solution to the OPF problem in DC networks than those reported in the specialized literature.

1.4. Structure of the Paper

This article is organized as follows: Section 2 proposes a mathematical formulation to solve the optimal power flow problem in DC networks, whose objective function is the reduction of power losses associated with energy transport. Section 3 describes the proposed solution methodology, which is composed of a master–slave strategy that combines the multiverse optimizer and the numerical method of successive approximation. Section 4 details the 21- and 69-node systems used for the tests in this study, in addition to other methods used here to compare and validate the proposed solution methodology. Section 5 reports the simulation results obtained by each one of the proposed solution methodologies. Section 6 presents the analysis of the processing times required by the different solution methods used in this paper. Finally, Section 7 draws the conclusions and proposes future research in this field.

2. Mathematical Formulation

Several objective functions to be minimized can be used to mathematically model the OPF problem. Some of them are the minimization of the operating costs of the network, minimization of emissions of polluting substances into the environment, improvement of voltage profiles, improvement of the loadability in distribution lines, and minimization of electrical power losses in the DC network. The selection of the objective function depends on the system owner or the operator of the DC network, and it is subject to the set of constraints that compose DC networks in an environment of distributed generation. The objective function selected in this study is the minimization of the power losses in the network [19], which is presented below along with its set of constraints.

2.1. Objective Function

The objective function is defined as the value to be minimized depending on the constraints that have been determined, in this case, the minimization of power losses. For that purpose, this study used Equation (1):

$$minP_{loss} = v^T G_L v \tag{1}$$

where P_{loss} denotes the power losses in the electrical network, which directly depend on v and G_L that represent the voltage profiles in each one of the nodes in the DC network calculated based on the load flow and the conductance matrix of the distribution lines, respectively.

2.2. Set of Constraints

The set of equations that compose the constraints of the problem addressed in this paper refer to the technical limits of the pieces of equipment and the operating parameters of the devices that constitute the DC network in an environment of distributed generation. Such equations are detailed below:

$$P_g + P_{DG} - P_d = D(v)[G_L + G_N]v \tag{2}$$

$$P_{DG}^{min} \leq P_{DG} \leq P_{DG}^{max} \tag{3}$$

$$v^{min} \leq v \leq v^{max} \tag{4}$$

$$1^T(P_{DG} - \alpha P_g) < 0 \tag{5}$$

The mathematical interpretation of Equations (2) to (5) is the following: Equation (2) represents the power balance in the DC network, where P_g is the power generated by the slack node; P_{DG}, the power supplied by each one of the DGs installed in the DC network; and P_d, the power demanded by the nodes in the network. Additionally, in this equation, $D(v)$ denotes a symmetrical positive matrix that contains the nodal voltages of the system in its diagonal, and G_N and G_L are the conductances of the transmission lines and the resistive loads connected to the network, respectively. Equation (3) expresses the minimum and maximum power that each DG installed in the network can supply, where P_{DG}^{min} and P_{DG}^{max} are the minimum and maximum powers allowed for the DGs, respectively. Equation (4) presents the voltage regulation limits, where v^{min} and v^{max} are the minimum and maximum load allowed in each node in the system, respectively. Finally, Equation (5) defines the maximum penetration level allowed for the DGs, where α represents the allowable penetration percentage with respect to the power generated by the slack node.

These conditions should be met at all times; for that purpose, the system should be penalized if the aforementioned limits are violated. Equation (6) defines the penalty of the objective function if a constraint is violated.

$$min\ z = \begin{pmatrix} P_{loss} + \beta_1 Ones^T max\{0, v - V^{max}\} \\ + \beta_2 Ones^T min\{0, v - V^{min}\} \\ + \beta_3 Ones^T min\{0, P_g - P_g^{min}\} \\ + \beta_4 Ones^T max\{0, P_{DG} - P_{DG}^{max}\} \\ + \beta_5 Ones^T min\{0, P_{DG} - P_{DG}^{min}\} \\ + \beta_6 max\{0, Ones^T P_{DG} - MDG\} \end{pmatrix} \tag{6}$$

where β_1 to β_6 are the penalty factors (which are usually higher than zero). In this study, each one of them is equal to 1000 to force the optimization methods to respect the limits imposed in Equations (2) to (5). This value was obtained heuristically. Moreover, the $Ones^T$ term is a transposed vector filled with ones that can be used to add together different penalties in the adaptation function. When all these constraints are respected, all the penalty values are null using the $min\{.\}$ and $max\{.\}$ functions, which, in this case, causes z to be equal to P_{loss}.

3. Proposed Solution Methodology

The equations presented in the mathematical formulation above describe a nonlinear optimization problem and represent the OPF problem in DC networks. These equations should be solved using numerical nonlinear methods. Hence, this study proposes dividing the problem into two stages: First, the master stage determines the optimal power level to be injected by the distributed generators located in the DC network. To solve this stage of the problem, this study uses the multiverse optimizer (MVO) [22]. Second, the slave stage calculates the electrical variables that can be used to determine the impact on the objective function, as well as the constraints of the problem of each one of the solutions proposed by the master stage (solves the power flow problem); for this purpose, this methodology implements the SA numerical method. The next subsection describes the master–slave (MVO-SA) methodology proposed in this study:

3.1. Master Stage: Multiverse Optimizer (MVO)

The Big Bang Theory posits that the universe we live in began as a result of a massive explosion [23]. There was nothing before the Big Bang, which was the start of our universe. In turn, the multiverse theory, which is more recent and well known among scientists [24], claims there is more than one Big Bang and that each Big Bang constitutes the beginning of a new universe. In addition, it holds that there are multiple universes that interact and can even clash with each other, and every universe can have different physical laws. The MVO is inspired by this theory, and it uses three of its main concepts: white hole, black hole, and wormhole [22]. A white hole has never been observed in the known universe, but physicists believe that the Big Bang can be considered one of them and the main component for the start of a new universe [25]. White holes expel matter and energy because no object can remain inside them. In turn, black holes, whose behavior is the opposite to that of white holes, have been observed throughout the known universe. They absorb everything that comes close to them: matter and even light. Nothing can escape black holes because their gravitational force is extremely high [26]. Finally, in the multiverse theory, wormholes connect two distant points, i.e., space/time, enabling objects that go through them to travel instantly to any place in the universe and even from one universe to another [27]. Each universe has an inflation rate (cosmic inflation) that causes it to expand in space [28]. One of the cyclic multiverse models [29] holds that multiple universes interact through white holes, black holes, and wormholes to reach a stable situation. This is exactly the inspiration for the MVO algorithm, which is conceptually and mathematically modeled as follows.

The MVO algorithm is based on three concepts of cosmology, i.e., white holes, black holes, and wormholes, which were mentioned above and are mathematically developed to construct the MVO. The search process of the latter can be divided into two stages: exploration and exploitation. White and black holes are used for exploring the search space and discovering regions where the best solutions can potentially be found. In turn, wormholes are used for exploiting the solution space, i.e., a local search around the promising solutions found in the previous stage [22]. The following subsection presents the stages followed in this study for the computational development and to find a solution to the OPF problem using the MVO.

3.1.1. Generation of the Initial Population

Equation (7) is used to generate each one of the objects that compose the initial population of $(U_{i,j})$ universes, where each universe in the population is a possible solution to the problem (power level to be injected by the DGs). In said equation, the i subindex denotes the i-th universe, and j is the j-th object that composes the i element. In this problem, each one of the objects generated in the different universes is a power level to be injected by each one of the DGs installed in the DC network. The values generated for each one of the objects that make up the universe are located within the solution space, which is limited by the technical constraints of the problem (in this case, the maximum and minimum allowable limits of the power to be injected by the DGs). This is possible by implementing upper (ub) and lower bounds (lb), which are assigned to each element in the universe and represent the maximum and minimum power level allowed for each generator in the problem addressed here. In order to explore larger regions of the search space, the first population of universes is generated from $(rand)$ random values in the $[0 - 1]$ range for each object that composes the different universes. These random numbers multiplied by the difference between the limits can be used to generate a larger distribution of particles over the search space.

$$U_{i,j} = ((ub - lb) \cdot rand) + lb \tag{7}$$

The previous equation can be used to find the value of the j object in the i universe. To generate all the universes, this study proposes a matrix of size nxd, where n denotes the number of universes as possible solutions to the problem, and d represents the number of variables in the problem (number of objects that belong to each universe). In (8), U_n is the n-th universe in the $M_{Universes}$ matrix. In the specific case of the OPF problem, the number of columns (d) represents the number of DGs that generate the electrical power inside the DC network (different from the slack node), and its value represents the power to be injected by each DG. To obtain the initial population, the values of each U_n were calculated randomly, respecting the limits established in the equations that rule the OPF problem and employing Equation (7) to generate all the objects in the different universes described in the Mathematical Model section.

$$M_{Universes} = \begin{bmatrix} U_1 \\ U_2 \\ \vdots \\ \vdots \\ U_n \end{bmatrix} = \begin{bmatrix} U_{1,1} & U_{1,2} & \cdot\cdot & \cdots & U_{1,d} \\ U_{2,1} & U_{2,2} & \cdots & \cdots & U_{2,d} \\ \vdots & \vdots & \vdots & \vdots & \vdots \\ \vdots & \vdots & \vdots & \vdots & \vdots \\ U_{n,1} & U_{n,2} & \cdots & \cdots & U_{n,d} \end{bmatrix} \tag{8}$$

3.1.2. Calculating the Objective Function

To evaluate the impact of each one of the universes proposed in $M_{Universes}$ on the objective function of the problem, the objective function is evaluated for every U_n universe (adaptation function) employing the slave stage (which is described in the following subsection). Subsequently, the values obtained for every solution are stored in the $nx1$ vector called $MO_{Universes}$. Thus, we calculate the electrical power losses in every solution for every DG, as well as the penalties in case the constraints established beforehand are not respected.

$$MO_{Universes} = \begin{bmatrix} f([U_{1,2} & U_{1,2} & \cdots & \cdots & U_{1,d}]) \\ f([U_{2,1} & U_{2,2} & \cdots & \cdots & U_{2,d}]) \\ \vdots & \vdots & \vdots & \vdots & \vdots \\ f([U_{n,1} & U_{n,2} & \cdots & \cdots & U_{n,d}]) \end{bmatrix} \tag{9}$$

Finally, the universe that presents the best solution in the $MO_{Universes}$ matrix (in this case, the lowest power losses) is selected as the incumbent solution, storing the objective function in Equation (10) and the configuration of the variables that compose it during the iterative cycle in Equation (11). To carry out this task, this study proposes to sort the vector of objective functions in ascending order, where the first position represents the best solution that has been found in the current iteration. If during an iteration of the algorithm the incumbent solution is improved, the values of $Best\_Fob$ and $Best\_U$

are updated. In Equation (11), $Best\_U$ is a vector of $1xd$, and a denotes the universe that presents the best objective function in each iteration.

$$Best\_Fob = [f(U_{1,d})] \tag{10}$$

$$Best\_U = [U_a] \tag{11}$$

3.1.3. Existence of Wormholes

As previously mentioned, the MVO algorithm divides the search process into two stages: exploration and exploitation. This technique uses white and black holes to explore the search space. In this stage, the algorithm tries to discover the most promising regions where the global optimum can be potentially found. In turn, wormholes are used to exploit the promising search spaces obtained in the exploration stage. This stage presents the convergence of the algorithm toward the global optimum (advances toward the best solution). In the MVO method, each solution is assigned an inflation rate (IR_i) that corresponds to the best P_{loss} obtained thus far by each universe, which is proportional to the aptitude function retrieved by the slave stage when the objects proposed by each one of the universes are evaluated (power levels to be injected by the DGs). Additionally, the following rules are applied to the universes of the MVO:

- The higher the IR, the greater the probability of having a white hole.
- The lower the IR, the greater the possibility of having a black hole.
- Universes with high IR tend to send objects through white holes.
- Universes with low IR tend to receive more objects through black holes.
- Any object in any universe can randomly move toward the best universe through wormholes, regardless of the IR.

Depending on the specific need, white or black holes can change their behavior if the problem to solve is the maximization or minimization of the objective function. When the movement is toward a higher IR, it is a maximization problem, where mainly white holes operate. In turn, when it is toward a lower IR, it is a minimization problem, where mainly black holes operate.

Objects can move between universes through white/black holes (objects can be exchanged inside universes). When a connection is established with a white/black hole between two universes, the universe with the higher IR is considered to have a white hole, and that with the lower IR is considered to have black holes. Objects are then transferred from the white holes in the universe of origin to the black holes in the target universe. To exchange objects through universes and carry out the exploitation, all the universes are assumed to have wormholes, regardless of their IR, in order to randomly transport objects. To provide local changes for each universe and have a high probability of improving the IR using wormholes, it is assumed that wormhole tunnels are always established between a universe and the best universe formed so far. Two coefficients are fundamental for the optimal operation of the MVO: wormhole existence probability (WEP) and travel distance rate (TDR).

Equation (12) below defines WEP as the probability of the existence of wormholes in the universes, which increases linearly over the iterations in order to advance in the exploitation process.

$$WEP = Min + l \cdot \left(\frac{Max - Min}{L} \right) \tag{12}$$

In other words, this equation can generate a rate of change based on the sum of a minimum value $(Min = 0.2)$ and a maximum value $(Max = 1)$, which depend on the algorithm and the iterative process, where l is the current iteration and L is the maximum number of iterations. This enables changes in the exploration and exploitation levels of the optimization technique.

Additionally, in Equation (13), TDR is the factor that defines the rate of variation or distance over which an object can be transported through a wormhole around the best universe obtained thus far. Unlike WEP, TDR increases over the iterations to have a more precise local exploitation/search around the best universe obtained.

$$TDR = 1 - \left(\frac{l^{1/P}}{L^{1/P}} \right) \tag{13}$$

where P denotes the level of precision of the exploitation of the algorithm, which aims to guarantee that, the higher its value, the faster and more precise the search (exploitation) of the algorithm. Depending on the problem addressed, WEP and TDR can be considered constant values, and they change according to the needs of each problem. In this study, as indicated at the start of the problem, P must be a variable due to the nonlinear parameters that the equations use.

Figure 1 shows the existing relationship between WEP and TDR. It indicates that WEP grows in a linear fashion from its initial value (0.2 in this case) up to 1 as the iterations increase. TDR starts from an initial value of 0.6 and falls to its minimum value $(TDR = 0)$ in an exponential manner (e^{-x}). Thus, these factors can be used to control the convergence of the algorithm by controlling the step forward of the universes inside the solution space.

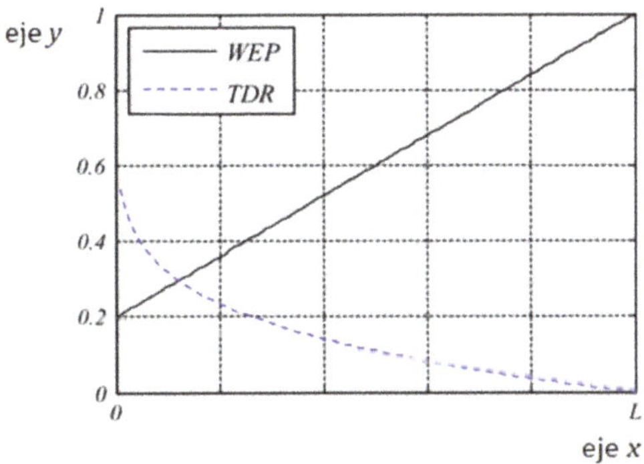

Figure 1. WEP vs. TDR, taken from [22].

3.1.4. Evolution of the Universes in the Iterative Process

For the evolution of the universes contained in the population, the MVO generates two strategies to alter its universes from one iteration to the next so that they finally converge toward the universe with the best characteristics. The first change made to the universes refers to the interaction between white and black holes; the second one is the generation of wormholes. Each iteration of the algorithm enables each one of the universes contained in the population to apply or not both movement strategies based on the evaluation of the local incumbent (of each universe) and the global incumbent of the set of universes, as well as the generation of random decision numbers. Thus, the objects inside each universe can be changed or not depending on a decision that contains a greedy random factor that boosts the exploration of the algorithm [13].

3.1.5. Updating the Universes Using the Interaction between White and Black Holes

The MVO presents the relationship between white and black holes, which enables the exchange of objects between universes through white/black holes and where a universe can update the value of an object based on the information of the other object from another universe. The interaction between them occurs as a function of IR. Objects travel from a white hole of origin to a target black hole. Thus, objects in a universe disappear through a black hole and are replaced with others coming from white holes. Equation (14) shows how the normalized inflation rate of the i-th universe is obtained, where the values of IR calculated for each universe in the population are normalized. When this equation is applied to the set of universes, we obtain the vector of $nx1$ (where n is the number of solutions or universes) with normalized values in a $[0-1]$ range.

$$NI(IR_i) = \left(\frac{IR_i}{max(IR_i)} \right) \tag{14}$$

The white/black tunnel is produced by Equation (15), where each j object in the i universe can be replaced with the j object of the k universe, and the position of k is selected by means of a roulette wheel that offers the option of randomly selecting any universe in the solution space. The exchange of objects between universes takes place for the same element, and it is performed by comparing a random number $r1$ with the normalized inflation rate $(NI(IR_i))$.

$$U_{i,j} = \begin{cases} U_{k,j} & r1 < NI(IR_i) \\ U_{i,j} & r1 \geq NI(IR_i) \end{cases} \quad \forall_i = 1,2,\cdots,n; \quad \forall_j = 1,2,\cdots,d \tag{15}$$

where $U_{k,j}$ is the j object of the k universe selected by the roulette wheel, and it is transported through the white/black tunnel to the i universe. An object is transported to the i universe if and only if the $r1$ random value in the $[0-1]$ range is lower than the $NI(IR_i)$. Otherwise, there is no exchange of objects through white/black tunnels; that is, the object of the $U_{i,j}$ universe will not be updated and it will keep its current value. To determine the k-th universe to which the j object will be moved, Equation (16) is implemented. In said equation, k is determined using the roulette wheel function [30], which selects a random number in the $[1-n]$ range, where n is the number of solutions proposed for the problem.

$$k = RouletteWheelSelection(-NI) \tag{16}$$

Algorithm 1 presents the pseudocode to select the k universe by means of Equation (16).

Algorithm 1 Proposed pseudocode of the roulette wheel to select the k universe to transport the j element to the i universe.

```
1:  function choice = RouletteWheelSelection(weights)
2:  accumulation = cumsum(weights);
3:  p = rand() * accumulation(end);
4:  chosen_index = -1;
5:  for index = 1 : length(accumlation) do
6:      if (accumlation(index) > p) then
7:          chosen_index = index;
8:          break
9:      end if
10:     chose = chosen_index;
11: end for
```

Importantly, as this is a minimization problem, the selection using the roulette wheel should be performed with $-NI$. If the problem to be solved is a maximization one, $-NI$ should be changed for the positive NI.

3.1.6. Updating Universes Based on Wormholes

In the MVO algorithm, the optimization process starts with the creation of a set of random universes. When the objects in each one of the universes are updated through the interaction of white and black holes at each iteration, the objects in universes with high IR tend to move to the universes with low inflation rates through multiple white/black tunnels. Importantly, the IRs represent the evaluation of the objective function, in this case, power losses. As this is a minimization problem, the objects tend to move to universes with a lower IR (fewer power losses). Meanwhile, each universe experiences random teleportation of its objects through wormholes to the best universe $(Best\_U(1,j))$. This process is repeated until a stopping criterion is met (for instance, a maximum number of iterations). To guarantee the changes in each universe, there should be a high probability of improvement of the IR by means of wormholes, assuming that the wormhole tunnels are always established between new universes randomly generated based on $r4$ and $Best\_U(1,j)$, i.e., the best universe, which are represented in Equation (17).

$$U_{i,j} = \begin{cases} \begin{cases} \left[Best\_U_{(1,j)} + TDR \cdot ((ub - lb) \cdot r4 + lb) \right] & r3 < 0.5 \\ \left[Best\_U_{(1,j)} - TDR \cdot ((ub - lb) \cdot r4 + lb) \right] & r3 \geq 0.5 \end{cases} & r2 < WEP \\ U_{i,j} & r2 \geq WEP \end{cases} \tag{17}$$

$$\forall_i = 1, 2, \cdots, n; \ \forall_j = 1, 2, \cdots, d$$

This equation can be used to generate new j objects for the i $(U_{i,j})$ universe at each iteration, which depends on WEP and $r2$, i.e., a random number in the $[0 - 1]$ range. If $r2$ is lower than WEP, a new j element will be generated for the i universe as a function of the value that $r3$ takes, i.e., a random number in the $[0 - 1]$ range. If $r3$ is lower than 0.5, the lower bound (lb) is subtracted from the upper bound (ub). The result is multiplied by a random number $(r4)$ in the $[0 - 1]$ range. The value thus obtained is added lb and then multiplied by the TDR parameter. Finally, the outcome of this operation is added together with $Best\_U(1,j)$, which is the j object of the best universe obtained thus far (incumbent solution). If $r3$ is higher than or equal to 0.5, the lower bound (lb) is subtracted from the upper bound (ub), and the result is multiplied by $r4$. The resulting value is added to the lb bound and then multiplied by the TDR parameter. Finally, the value obtained with this operation is subtracted with $Best\_U(1,j)$. However, if $r2$ is greater than or equal to WEP, a new j object will not be created in the i universe; instead, the current value of the j object will be maintained until $r2$ is lower than WEP.

Black holes, white holes, and wormholes are updated at each iteration until the algorithm meets the stopping criteria described below.

3.1.7. Stopping Criteria

The master stage uses two stopping criteria: (1) the counter reaches the maximum number of iterations or (2) the algorithm reaches the limit of non-improvement iterations. These stopping criteria are detailed as follows:

- The master stage will finish the iterative process when the incumbent solution (the best obtained thus far) is not updated after an x number of consecutive iterations. This is possible by adding a non-improvement counter to the code $(Iter\_NM)$.
- The computational analysis will finish the iterative process when the algorithm reaches a maximum number of allowable iterations (L).

3.2. Slave Stage

The slave stage calculates the adaptation function associated with each solution (universe) provided by the master stage. That is, the slave stage enables us to determine the electrical variables of the system in each possible scenario of power injection by the DGs, which is used to establish the electrical power losses in the DC network (i.e., the objective function to be solved in this study), as well as the set of constraints that composes the problem. For this purpose, it is necessary to solve the load flow in the DC network by means of an iterative method and thus determine the impact of the proposed solution and its compliance with the constraints of the OPF problem. This study aims to solve the load flow using the SA method presented in [4], which was selected thanks to its excellent performance in terms of solution and convergence time. In order to implement said method, the following equation can be used to find the solution to the load flow:

$$G_{dd} \cdot v_d = -D_d^{-1}(v_d)P_d - G_{dg} \cdot v_g \tag{18}$$

where G_{dd} is a positive symmetrical matrix that contains the conductances of the conducting lines and the resistive loads of the DC network, except for the slack node; v_g is the voltage profile of the slack generator; and v_d is the voltage in the demand nodes of the system.

By means of a mathematical development applied to (18), we can obtain the equation that can be used to calculate the nodal voltages in the demand nodes:

$$v_d = -G_{dd}^{-1}[D_d^{-1}(v_d)P_d + G_{dg} \cdot v_g] \tag{19}$$

In order to calculate the v_d voltage profiles, an iterative process should be implemented to find such values with an almost-null convergence error. For that purpose, a t counter should be added to Equation (19). As a result, the equation is:

$$v_d^{t+1} = -G_{dd}^{-1}D_d^{-1}(v_d^t)P_d + G_{dg} \cdot v_g \tag{20}$$

Finally, Algorithm 2 presents the pseudocode of the master–slave methodology (MVO-SA) that solves the OPF problem.

Algorithm 2 Hybrid MVO-SA optimization algorithm.

1: Load system data
2: Initialize the parameters of the algorithms
3: Generate initial population of universes ($M_{Universes}$)
4: Calculate adaptation function employing slave stage ($MO_{Universes}$)
5: Select the incumbent solution ($Best\_U$)
6: Initialize P, Max, and Min parameters
7: **while** $l \leq L$ **do**
8:     Normalize inflation rate;
9:     Initialize WEP and TDR parameters
10:    **for** *Each universe*(i) **do**
11:      **for** *Each object*(j) **do**
12:       $r1 = random[0,1]$;
13:       **if** $r1 < NI(IR_i)$ **then**
14:        $k = RouletteWheelSelection(-NI)$;
15:        $U(i,j) = U(k,j)$;
16:       **end if**
17:       $r2 = random[0,1]$;
18:       **if** $r2 < WEP$ **then**
19:        $r3 = random[0,1]$;
20:        $r4 = random[0,1]$;
21:        **if** $r3 < 0.5$ **then**
22:         $U(i,j) = Best\_U(1,j) + TDR * (ub - lb) * r4 + lb$;
23:        **else**
24:         $U(i,j) = Best\_U(1,j) - TDR * (ub - lb) * r4 + lb$;
25:        **end if**
26:       **end if**
27:      **end for**
28:    **end for**
29:     Calculate adaptation function by means of SA
30:     Update incumbent solution
31:     $l = l + 1$;
32: **end while**

3.3. Comparison of Methods

To validate the solution method proposed in this paper, it was compared with four other solution methodologies reported in the literature: PSO, BH, CGA, and ALO. They were selected due to their excellent performance and convergence in the solution to the OPF problem in DC networks [14,19]. All these methods employ a master–slave methodology. The master stage employs the respective optimization algorithm, and the slave stage uses the SA method proposed in [4] to solve the load flow. To conduct the tests and improve the impact on the electrical systems, the 21- and 69-node systems presented in [2,21,31] were adapted here. Such impact was evaluated using three percentages of penetration of the electrical power supplied by the slack generator: 20%, 40%, and 60%.

To guarantee the same conditions, each one of the solution methodologies used in this study was tuned in each one of the test systems (21 and 69 nodes). The objective of this step is that each one of the techniques achieves the best possible solution in terms of electrical power losses. To carry out said tuning, this study implemented the PSO optimization algorithm used in [21]. The tuning parameters used in this case were a population size in a [2–100] range, a maximum number of iterations in a [1–1000] range, and a number of non-improvement iterations in a [1–1000] range. In this tuner, the number of particles was 10 and the number of iterations was 300. Additionally, for the MVO, the P parameter was optimized employing a [1–10] range, which represents the level of precision of the algorithm's exploitation. Tables 1 and 2 report the results obtained after the tuning of each one of the techniques for the 21- and the 69-node system, respectively.

Table 1. Parameters of the continuous methods employed here in the master stage for the 21-node system.

	21-Node System				
Method	**MVO**	**ALO**	**BH**	**CGA**	**PSO**
Number of particles	71	79	67	52	49
Maximum iterations	613	769	317	592	679
Non-improvement iterations	504	441	317	346	263
P parameter	8	—	—	—	—

Table 2. Parameters of the continuous methods employed here in the master stage for the 69-node system.

	69-Node System				
Method	**MVO**	**ALO**	**BH**	**CGA**	**PSO**
Number of particles	86	77	35	40	58
Maximum iterations	656	182	566	622	723
Non-improvement iterations	584	182	566	443	252
P parameter	7	—	—	—	—

4. Test Scenarios and Considerations

In order to evaluate the effectiveness and robustness of the methodology proposed in this study, two test systems were implemented: a 21-node system [31] and a 69-node system [21]. These systems are widely used in the specialized literature to validate the impact of optimization techniques implemented to solve the OPF problem. Such test systems are composed of constant power loads and a single slack generator in a scenario without DGs, which constitutes the "base case".

4.1. 21-Node System

Figure 2 presents the 21-node system. The base case of this system employs a base voltage of 1 kV and a base power of 100 kW. In this system, the slack generator produces 581.6 kW, and the power losses amount to 27.603 kW. With respect to the current limits of the branches, homogeneous networks were considered for both systems. A 900-kcmil conductor ($I_{ij}^{max} = 520$ A) was selected for the 21-bus test system based on the operation of the base case (without DGs installed). The voltage bounds of both test systems were set at $+/-$ 10% of the nominal voltage [32]. Because this study addresses the OPF problem, the location of the DGs here is the same as that reported in [17]; therefore, they are located at nodes 9, 12, and 16 in the network. Importantly, the evaluation of the base case does not consider power injection or the installation of such devices in the network. In this test system, the minimum power was 0 kW for all the DGs at all levels of penetration.

In addition, the maximum powers of distributed generation were 116.3207, 232.6414, and 348.9620 kW, which correspond to the 20%, 40%, and 60% of the power provided by the slack generator in the base case, respectively.

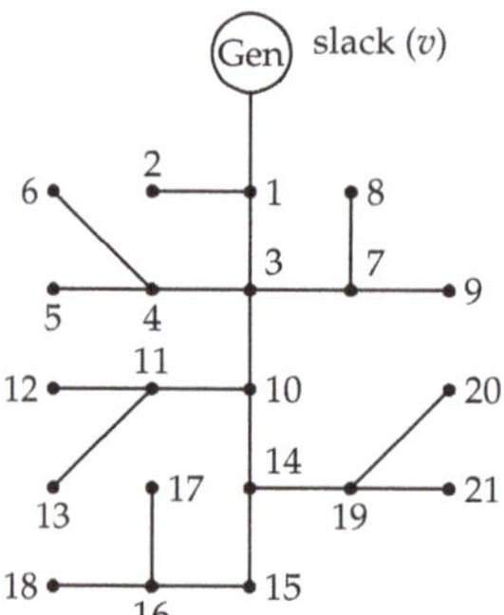

Figure 2. The 21-node system.

Table 3. Electrical parameters for the 21-node system.

Output Node	Input Node	R (pu)	P(pu)	Output Node	Input Node	R (pu)	P(pu)
1 (Slack)	2	0.0053	−0.70	11	12	0.0079	−0.68
1	3	0.0054	0	11	13	0.0078	−0.10
3	4	0.0054	−0.36	10	14	0.0083	0
4	5	0.0063	−0.04	14	15	0.0065	−0.22
4	6	0.0051	−0.36	15	16	0.0064	−0.23
3	7	0.0037	0	16	17	0.0074	−0.43
7	8	0.0079	−0.32	16	18	0.0081	−0.34
7	9	0.0072	−0.80	14	19	0.0078	−0.09
3	10	0.0053	0	19	20	0.0084	−0.21
10	11	0.0038	−0.45	19	21	0.0082	−0.21

Table 3 shows the connections between the nodes in the 21-node network, specifying the output and input nodes, the resistance of each conduction line, and the powers demanded by each one of the nodes in the system in *p.u.*

4.2. The 69-Node System

Figure 3 shows the electrical diagram of the 69-node system, which is composed of 68 lines and 69 nodes. Originally, this test system operates in DC [21], but thanks to the adaptation in the previously cited document, it was possible to implement it for DC networks. For such adaptation, the base voltage was 12.66 kV, the power base was 100 kV, and the active elements were disregarded.

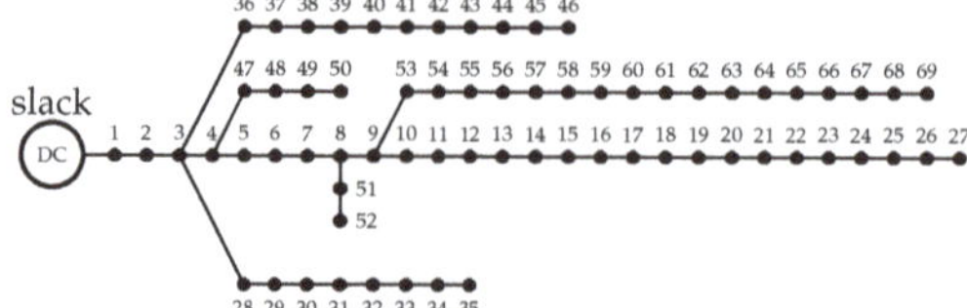

Figure 3. The 69-node system.

In this test system, the slack generator produces 4043.1 kW, and the power losses equal 153.85 kW. With respect to the current limits, a 400-kcmil electrical conductor (I_{ij}^{max} = 335 A) was selected for this test system considering a non-telescopic configuration, as well as the same voltage limits used in the 21-bus test system [32]. Finally, the location of the DGs was defined as in [21], i.e., at nodes 26, 61, and 66. In this test system, the minimum power was 0 kW for all the DGs at all levels of penetration. In addition, the maximum powers of distributed generation were 808.6195, 1617.2390, and 2425.8585 kW, which correspond to 20%, 40%, and 60% of the power provided by the slack generator in the base case, respectively.

Table 4 details the connection between the nodes in the network of the 69-node system, the output and input nodes, the resistance of each conduction line, and the power demanded by each one of the nodes in the system.

Table 4. Electrical parameters for the 69-node system.

Output Node	Input Node	$R\ (\Omega)$	P (kW)	Output Node	Input Node	$R\ (\Omega)$	P (kW)
1	2	0.0005	0	3	36	0.0044	−26
2	3	0.0005	0	36	37	0.0640	−26
3	4	0.0015	0	37	38	0.1053	0
4	5	0.0215	0	38	39	0.0304	−24
5	6	0.3660	−2.6	39	40	0.0018	−24
6	7	0.3810	−40.4	40	41	0.7283	−102
7	8	0.0922	−75	41	42	0.3100	0
8	9	0.0493	−30	42	43	0.0410	−6
9	10	0.8190	−28	43	44	0.0092	0
10	11	0.1872	−145	44	45	0.1089	−39.2
11	12	0.7114	−145	45	46	0.0009	−39.2
12	13	10.300	−8	4	47	0.0034	0
13	14	10.440	−8	47	48	0.0851	−79
14	15	10.580	0	48	49	0.2898	−384
15	16	0.1966	−45	49	50	0.0822	−384
16	17	0.3744	−60	8	51	0.0928	−40.5
17	18	0.0047	−60	51	52	0.3319	−3.6
18	19	0.3276	0	9	53	0.1740	−4.35
19	20	0.2106	−1	53	54	0.2030	−26.4
20	21	0.3416	−144	54	55	0.2842	−24
21	22	0.0140	−5	55	56	0.2813	0
22	23	0.1591	0	56	57	15.900	0
23	24	0.3463	−28	57	58	0.7837	0
24	25	0.7488	0	58	59	0.3042	−100
25	26	0.3089	−14	59	60	0.3861	0
26	27	0.1732	−14	60	61	0.5075	−1244
3	28	0.0044	−26	61	62	0.0974	−32
28	29	0.0640	−26	62	63	0.1450	0
29	30	0.3978	0	63	64	0.7105	−227
30	31	0.0702	0	64	65	10.410	−59
31	32	0.3510	0	65	66	0.2012	−18
32	33	0.8390	−10	66	67	0.0047	−18
33	34	17.080	−14	67	68	0.7394	−28
34	35	14.740	−4	68	69	0.0047	−28

5. Simulations and Results

This section analyzes the results obtained from the implementation of the solution methods in the two test systems. All the simulations were conducted in Matlab (version 2018b), a numerical computing system, running on a desktop computer with 4GB of RAM, an Intel® Core™ i5-8250U CPU @1.60GHz 1.80GHz processor, a 225-GB solid state drive, and Windows 10 PRO. All the methodologies were executed 100 times in order to evaluate the repeatability of the solutions and the standard deviation of each one of the techniques. The following subsection presents their results in the two test systems.

5.1. 21-Node System

Table 5 reports the results of each one of the methods for the OPF problem in the 21-node system at three maximum levels of penetration of distributed generation: 20%, 40%, and 60% of the power provided by the slack node in the base case. From left to right, said table specifies the proposed solution method; the nodes where the DGs are located and the power each one of them injects into the network in (kW); the minimum power losses (P_{loss}) in (kW) and the percentage of reduction compared to the base case (%); the average value of P_{loss} in (kW) and the average reduction with respect to the base case (%); the standard deviation in percentage, which was obtained after each solution method was executed 100 times; the worst voltage; and the maximum current in the DC grid employing the distributed power injection proposed by each solution methodology. Figures 4 and 5 are bar charts that show the differences, in percentage, between the algorithms and the new proposed methodology regarding minimum P_{loss} and average P_{loss}, respectively.

Table 5 can be used to compare the results obtained by different solution methodologies proposed for the 21-node system. Columns 6 and 7 in this table indicate that all the solution methods satisfied the voltage and current bounds established here for the 21-bus test system. Figures 4 and 5 were created to analyze the minimum and average reduction of power losses in the system. Figure 4 details the minimum loss reductions obtained by each methodology at different DG penetration levels: 20%, 40%, and 60%. In the first case, i.e., a 20% penetration of DG, PSO reduced the minimum P_{loss} by 52.2382%, outperforming the MVO by only 5×10^{-5}%. The MVO took the second place in this regard (52.2381%) but outperformed the ALO by 0.0039%. The CGA and BH are in fourth and fifth position, respectively, outperformed by the MVO by 0.0485% and 0.4256%, respectively. The proposed methodology achieved an average reduction in minimum P_{loss} of 0.1195%. In the scenario that allows a 40% power injection, the MVO presented the greatest reduction in minimum P_{loss} with respect to the base case: 77.8233%. It outperformed PSO, ALO, CGA, and BH by 5×10^{-5}%, 0.0013%, 0.0058%, and 0.1953%, respectively. In the case of 60% penetration, the MVO and PSO exhibited the same reduction in minimum P_{loss}, i.e., 89.9083%, thus outperforming the ALO by 0.0012%, the CGA by 0.0176%, and the BH by 0.1091%. In terms of average P_{loss} in the 21-node system, the MVO outperformed all the other solution techniques at the three penetration percentages (see Figure 5). At a 20% penetration, the MVO presented an average reduction in P_{loss} of 52.2363%, thus outperforming PSO, ALO, CGA, and BH by 0.1168%, 0.3007%, 0.3635%, and 3.2155%, respectively. At a 40% penetration of distributed generation, the MVO presented a reduction in average P_{loss} of 77.8229%, thus outperforming ALO by 0.0272%, PSO by 0.0950%, CGA by 0.0958%, and BH by 1.5013%. Finally, at a 60% penetration, the MVO exhibited a reduction in mean P_{loss} of 89.9080%, thus outperforming ALO by 0.0082%, PSO by 0.0590%, CGA by 0.0954%, and BH by 0.9285%. This demonstrates that the MVO is superior to the other techniques in terms of precision and repeatability.

Table 5. Results of the simulations of the 21-node system.

Method	Node /Power (kW)	Minimum (kW) /Reduction (%)	Average (kW) /Reduction (%)	STD (%)	Vworst (p.u)	Imax (A)
21-Node System						
		Power Losses				
Without DGs	- - -	27.603	- - -	- - -	(0.9–1.1)	520
20% Penetration						
MVO	9/0.0004 12/17.96 16/98.36	13.1822/52.24	13.1828/52.24	0.003	0.96/20	380.60
ALO	9/0.03 12/16.85 16/99.44	13.1833/52.23	13.2658/51.94	1.14	0.96/20	380.60
BH	9/1.15 12/32.73 16/82.44	13.2997/51.81	14.0703/49.02	2.418	0.95/17	380.72

Table 5. *Cont.*

Method	Node /Power (kW)	Power Losses		STD [%]	Vworst (p.u)	Imax (A)
		Minimum (kW) /Reduction (%)	Average (kW) /Reduction (%)			
21-Node System						
Without DGs	- - -	27.603	- - -	- - -	(0.9–1.1)	520
CGA	9/0.07	13.1957/52.19	13.2831/51.87	0.279	0.96/20	380.76
	12/17.59					
	16/98.52					
PSO	9/0	13.1823/52.24	13.2150/52.12	0.783	0.96/20	380.60
	12/17.73					
	16/98.59					
40% Penetration						
MVO	9/30.60	6.1208/77.82	6.1209/77.82	0.002	0.97/20	257.22
	12/72.97					
	16/129.06					
ALO	9/30.50	6.1211/77.82	6.1284/77.80	0.8	0.97/20	257.22
	12/72.56					
	16/129.58					
BH	9/41.01	6.1747/77.63	6.5352/76.32	3.238	0.97/20	257.81
	12/67.44					
	16/123.66					
CGA	9/32.71	6.1224/77.82	6.1473/77.73	0.236	0.97/20	257.23
	12/72.06					
	16/127.87					
PSO	9/30.43	6.1208/77.82	6.1471/77.73	1.243	0.97/20	257.22
	12/73.22					
	16/128.99					
60% Penetration						
MVO	9/93.33	2.7853/89.91	2.7854/89.91	0.002	0.98/20	137.56
	12/107.48					
	16/148.16					
ALO	9/93.09	2.7856/89.91	2.7876/89.90	0.044	0.98/20	137.57
	12/108.49					
	16/147.38					
BH	9/91.48	2.8154/89.80	3.0416/88.98	4.76	0.98/20	139.38
	12/110.59					
	16/145.10					
CGA	9/92.13	2.7902/89.89	2.8117/89.81	0.449	0.98/20	137.66
	12/105.11					
	16/151.64					
PSO	9/93.34	2.7853/89.91	2.8017/89.85	2.007	0.98/20	137.56
	12/107.45					
	16/148.18					

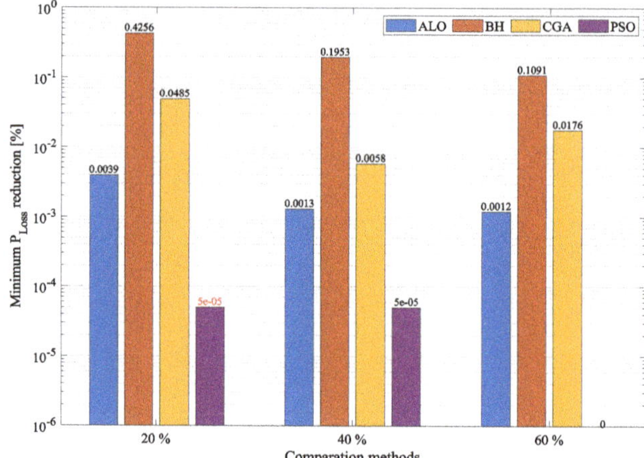

Figure 4. Percentage of reduction in minimum losses obtained by the MVO in the 21-node system compared to other methodologies.

To complete the analysis of the 21-node system, the repeatability of the solutions was examined. For that purpose, Figure 6 presents the reduction in standard deviation obtained by the proposed solution methodology compared to that of its counterparts. This figure indicates that the MVO outperforms all the other optimization techniques at all the penetration levels allowed in the 21-node system. At 20% penetration, the MVO outperforms the CGA, PSO, ALO, and BH by 0.27560%, 0.78019%, 1.13683%, and 2.41501%, respectively. At 40% DG penetration, the MVO outperforms the CGA (in second place) by 0.23380%, which is followed by ALO (in third place) with a difference of 0.79843%. PSO and BH are in fourth and fifth place, outperformed by 1.24157% and 3.23596%, respectively. Finally, at a 60% penetration, the MVO presents an average reduction of 1.1834% with respect to the other techniques. Based on these data, it can be concluded that the solution is highly reproducible in this system, which guarantees that a good solution is obtained every time the algorithm is executed. By contrast, all the other methods exhibit a standard deviation higher than that of the technique proposed in this paper.

Figures 4–6 use a logarithm with base 10 to better visualize the differences between the techniques. In Figure 4, the number in red indicates that the MVO was outperformed by PSO.

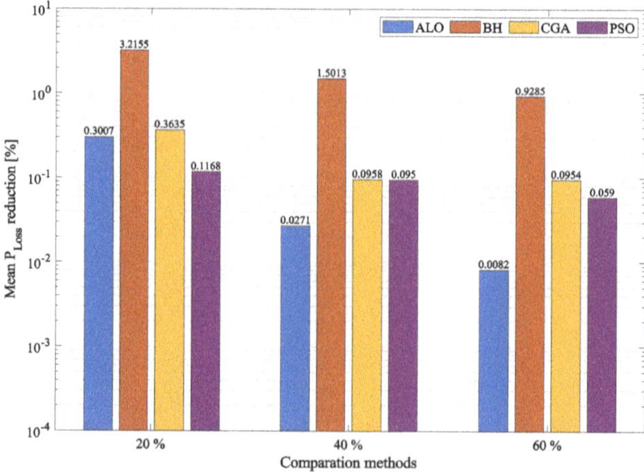

Figure 5. Percentage of reduction in average losses obtained by the MVO in the 21-node system compared to other methodologies.

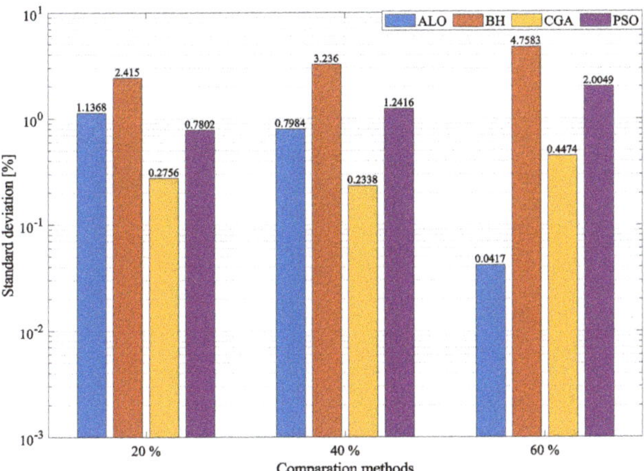

Figure 6. Percentage of standard deviation obtained by the MVO in the 21-node system compared to other methodologies.

5.2. The 69-Node System

Table 6 presents the results of the techniques used to solve the OPF problem in the simulation of the 69-node system. Columns 6 and 7 in this table indicate that all the solution methods satisfied the voltage and current limits established here for the 69-bus test system.

Table 6, which was organized the same way as Table 5, presents the results of the solution methodologies proposed for the 69-node system. As in the case of the 21-node system, Figures 7 and 8 can be used to compare the percentages of minimum P_{loss} and average P_{loss}, respectively. Regarding minimum P_{loss}, Figure 7 reports the reduction in minimum power losses obtained by the MVO compared to the other methods. At a 20% penetration, PSO achieved a reduction percentage of 63.2853%, outperforming the MVO (in second place) by an almost negligible percentage of $7x10^{-6}$%. In this scenario, the MVO outperformed the ALO, CGA, and BH in terms of minimum losses by 0.0479%, 0.0646%, and 0.8576%, respectively. At a 40% power injection, the MVO achieved the greatest reduction in minimum P_{loss} with respect to the base case: 90.9052%. Hence, it outperformed PSO, ALO, CGA, and BH by 0.0004%, 0.0104%, 0.0116%, and 0.4025%, respectively. In the case of 60% penetration, the MVO and PSO produced the same reduction in minimum P_{loss} (96.3888%), thus outperforming ALO by 0.0013%, CGA by 0.0004%, and BH by 0.2134%. Figure 8, which reports the reduction in average P_{loss} in the 69-node system, indicates that the MVO outperforms all the other techniques at the three penetration percentages, except for PSO at a 60% penetration, where they produced the same result. At a 20% penetration, the MVO achieved a reduction in average P_{loss} of 63.2828%, thus outperforming PSO, ALO, CGA, and BH by 0.1380%, 0.5912%, 0.3855%, and 3.7973%, respectively. At a 40% penetration, the MVO produced a reduction in average P_{loss} of 90.9049%, thus outperforming PSO by 0.2080%, ALO by 0.2822%, CGA by 0.1043%, and BH by 3.0280%. Finally, at a 60% penetration level, the MVO and PSO presented the same reduction in mean P_{loss}, i.e., 96.3888%, thus outperforming ALO by 0.1142%, CGA by 0.0156%, and BH by 1.7997%. Although the results in the 69-node system are similar, in most cases, the MVO outperforms the other techniques employed here for comparison, especially in terms of average P_{loss}.

To complete the analysis of the 69-node system, the repeatability of the solutions proposed here for the OPF problem was examined using the standard deviation. In Figure 9, the MVO outperforms most of the other optimization techniques in terms of each one of their average standard deviations, except for PSO at a 60% penetration, which produced a result 6×10^{-6}% better, a difference that is almost negligible. At a 20% penetration, the MVO outperforms all the other techniques, with an average standard deviation of 1.6131%; likewise, at a 40% penetration, the MVO outperforms all the other algorithms, with an average standard deviation of 5.3906%. Finally, although PSO outperformed the MVO at a 60% penetration, the latter presents an average standard deviation of 7.0432% with respect to all the methods used for comparison. These data show that the MVO is the most precise and repeatable technique in different test systems at multiple penetration percentages.

Table 6. Results of the simulations of the 69-node system.

Method	Node /Power (kW)	Power Losses		STD (%)	Vworst (p.u)	Imax (A)
		Minimum (kW) /Reduction (%)	Average (kW) /Reduction (%)			
Without DGs	- - -	153.85	- - -	- - -	(0.9–1.1)	335
20% Penetration						
MVO	$26/5.74\times 10^{-5}$ 61/564.21 66/244.40	56.4856/63.29	56.4903/63.28	0.011	0.961/64	247.80
ALO	26/0 61/616.06 66/192.22	56.5594/63.24	57.3990/62.69	1.159	0.961/64	247.83
BH	26/0.29 61/330.66 66/471.60	57.8050/62.43	62.3316/59.49	4.494	0.962/61	248.38
CGA	26/1.60 61/560.06 66/246.25	56.5850/63.22	57.0826/62.90	0.418	0.961/64	247.86
PSO	$26/8.15\times 10^{-8}$ 61/566.67 66/241.94	56.4856/63.29	56.7017/63.15	0.407	0.961/64	247.80
40% Penetration						
MVO	26/157.93 61/1214.34 66/244.97	13.9923/90.91	13.9929/90.90	0.005	0.985/21	180.57
ALO	26/156.07 61/1234.42 66/226.40	14.0084/90.89	14.4271/90.62	2.378	0.985/21	180.60
BH	26/141.05 61/1093.53 66/369.47	14.6116/90.50	18.6515/87.88	12.623	0.984/21	181.66
CGA	26/156.56 61/1189.74 66/270.64	14.0101/90.89	14.1533/90.80	0.644	0.985/21	180.59
PSO	26/158 61/1211.24 66/247.99	13.9929/90.90	14.3129/90.70	5.936	0.985/21	180.57

Table 6. Cont.

Method	Node /Power (kW)	Power Losses		STD (%)	Vworst (p.u)	Imax (A)
		Minimum (kW) /Reduction (%)	Average (kW) /Reduction (%)			
69-Node System						
Without DGs	- - -	153.85	- - -	- - -	(0.9–1.1)	335
60% Penetration						
MVO	26/375.11 61/1588.50 66/245.73	5.5558/96.39	5.5558/96.39	0.006	0.995/12	133.13
ALO	26/380.38 61/1584.26 66/250.89	5.5577/96.39	5.7315/96.27	6.332	0.995/12	132.64
BH	26/401.27 61/1417.44 66/343.43	5.8840/96.18	8.3247/94.59	21.543	0.995/12	136.89
CGA	26/373.60 61/1589.01 66/242.18	5.5565/96.39	5.5797/96.37	0.298	0.995/12	133.49
PSO	26/375.11 61/1588.47 66/245.74	5.5558/96.39	5.5558/96.39	5.86×10^{-4}	0.995/12	133.13

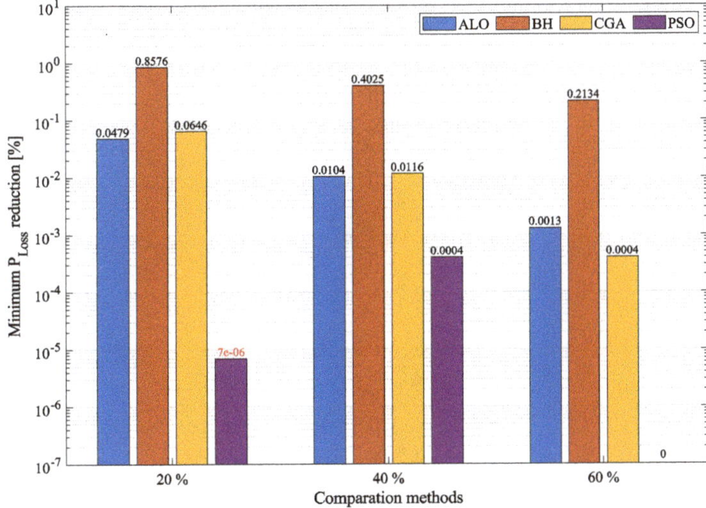

Figure 7. Percentage of reduction in minimum losses obtained by the MVO in the 69-node system compared to other methodologies.

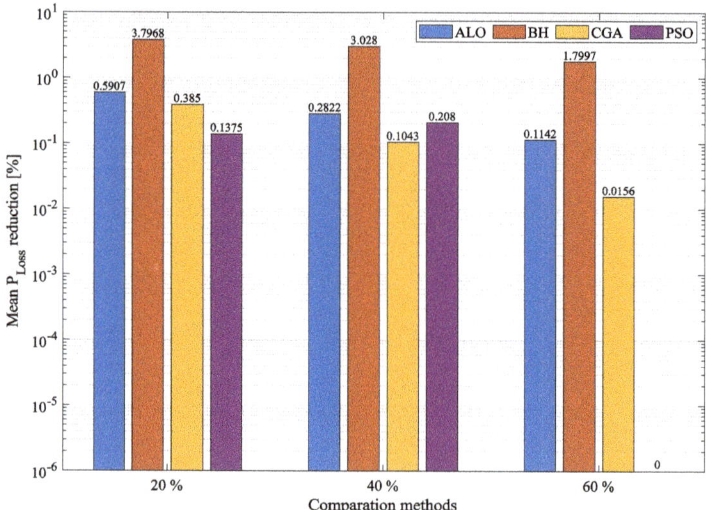

Figure 8. Percentage of reduction in average losses obtained by the MVO in the 69-node system compared to other methodologies.

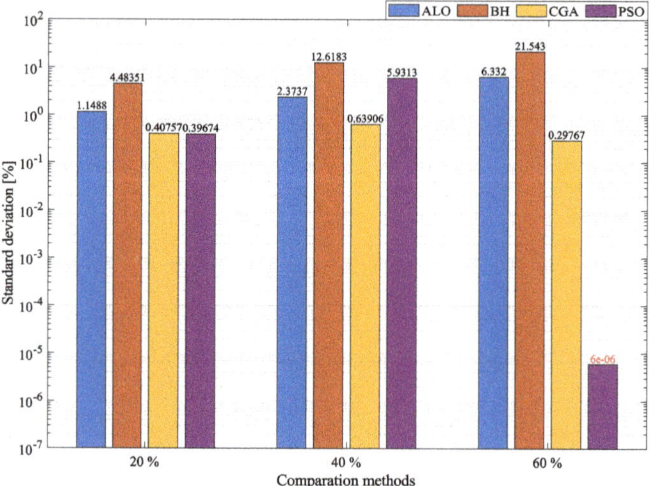

Figure 9. Percentage of standard deviation obtained by the MVO in the 69-node system compared to other methodologies.

Figures 7–9 use a logarithm with base 10 to better visualize the differences between the techniques. In Figures 7 and 9, the number in red indicates that the MVO was outperformed by PSO.

In order to clearly establish which optimization technique could provide the best solution to the OPF problem regardless of the size of the DC network, the general behavior of all the algorithms was analyzed by evaluating their results regarding minimum and average P_{loss}. Said results are reported in Figures 10–12. These figures can be used to analyze the precision of the algorithms and determine the technique that offers the best behavior in terms of the solution for each system.

Figure 10 presents the mean of the minimum P_{loss} and average P_{loss} in the 21-node system. The percentages of reduction of mean P_{loss} at 20%, 40%, and 60% penetration levels were averaged; the same process was applied to the average P_{loss}. This was the procedure followed here to obtain the results in Figure 10, which can be used to determine the algorithm that achieves the best minimum and average P_{loss} in the 21-node system. Additionally, Figure 13 presents the differences between the MVO and the other optimization algorithms in the 21-node system. In said figure, the MVO presents an adequate reduction in average minimum P_{loss}, only outperformed by PSO by 8×10^{-7}% (an almost negligible value). The MVO outperformed ALO, CGA, and BH by 0.002128%, 0.023958%, and 0.243308%, respectively. Regarding the mean values of the average P_{loss}, the MVO worked better than the other solution methodologies, outperforming PSO by 0.090245%, CGA by 0.184925%, ALO by 0.112014%, and BH by 1.881772%. This shows that the MVO is the best solution methodology for the 21-node system.

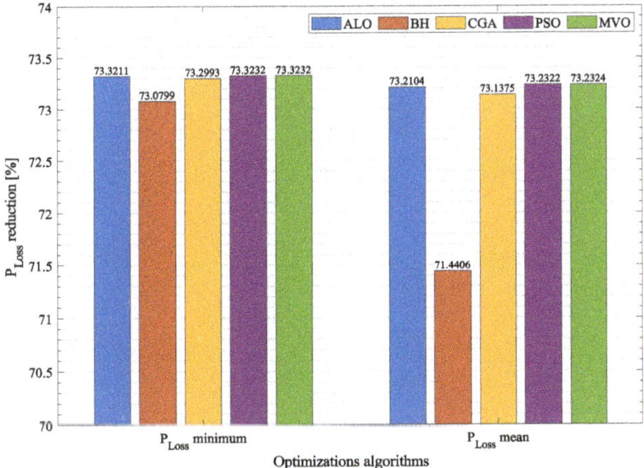

Figure 10. Average percentage of losses in the 21-node system.

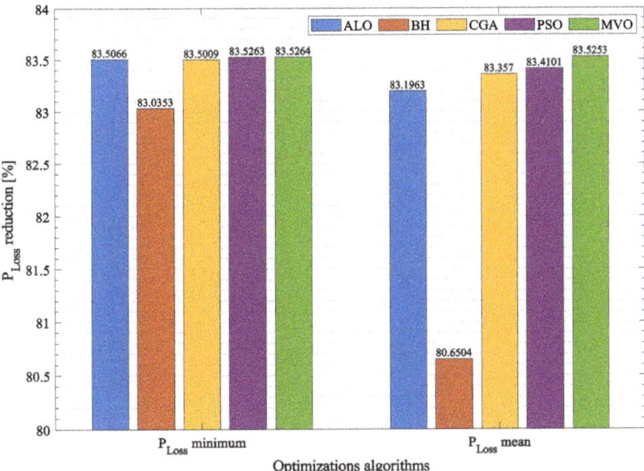

Figure 11. Average percentage of losses in the 69-node system.

Figure 11 presents the mean of the minimum P_{loss} and average P_{loss} in the 69-node system. The percentages of reduction in mean P_{loss} and average P_{loss} at 20%, 40%, and 60% injection levels obtained by each one of the optimization techniques were averaged, which produced the results reported in Figure 11. As with the 21-node system, this figure can be used to establish which algorithm can obtain the best minimum and average P_{loss}. Likewise, Figure 14 highlights the differences between the MVO and the other optimization algorithms. In the 69-node system, the MVO achieved the best percentage of reduction in average minimum P_{loss}, outperforming PSO by 0.0015%, ALO by 0.019875%, CGA by 0.025518%, and BH by 0.491145%. Regarding the mean of average P_{loss}, the MVO outperformed PSO, CGA, ALO, and BH by 0.115168%, 0.168297%, 0.329049%, and 2.874850%, respectively. This demonstrates that the MVO is the best solution methodology in the 21- and 69-node systems in terms of solution quality; furthermore, its performance improves as the network grows.

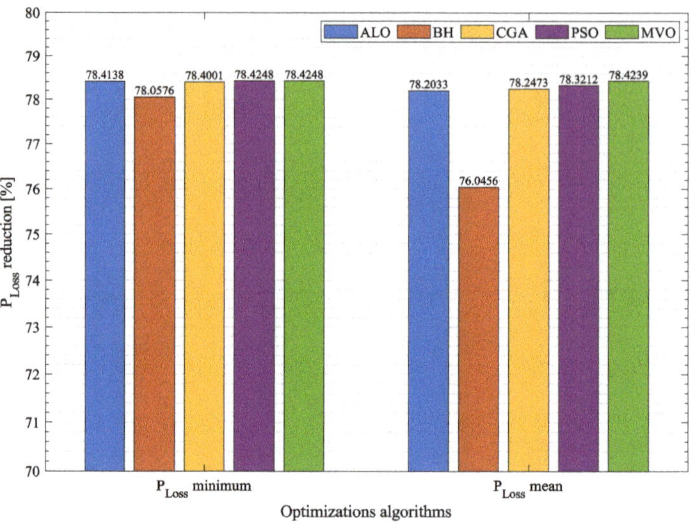

Figure 12. Robustness of the solution methodologies proposed to solve the OPF problem.

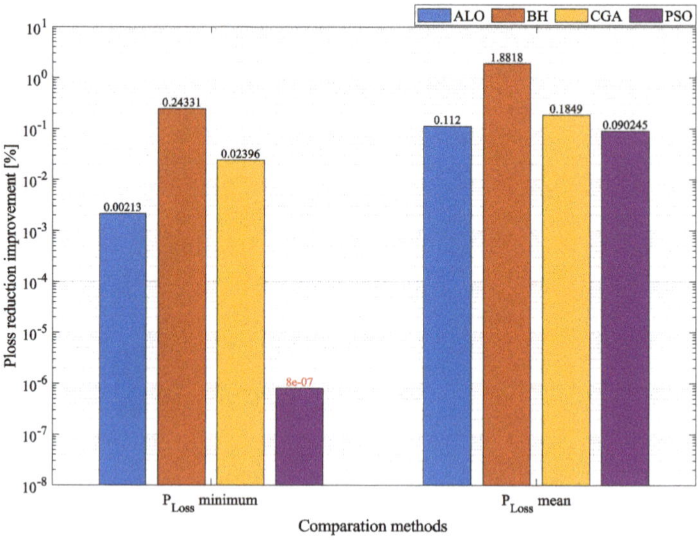

Figure 13. Percentage of improvement obtained by the MVO in the 21-node system compared to other methodologies.

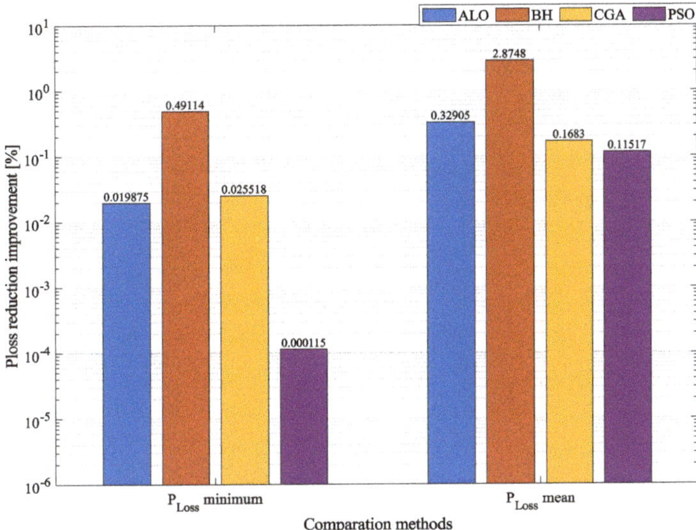

Figure 14. Percentage of improvement obtained by the MVO in the 69-node system compared to other methodologies.

Finally, to demonstrate the robustness of each one of the optimization techniques, Figure 12 reports the average minimum P_{loss} percentages and average P_{loss} percentages obtained by the solution methods in the two test systems at different DG penetration levels. These values are the result of averaging the values obtained in each scenario and test system. Therefore, the MVO presents the best minimum P_{loss} percentage, with an average reduction of 0.1008%, and the best average P_{loss} level in the two test systems, with a reduction of 0.7195% compared to the other methods tested here. Based on this analysis, it can be concluded that the MVO offers the best solution to the OPF problem in a network of any size. This demonstrates that, compared to other solution methodologies that have achieved a high performance in the specialized literature, the methodology proposed in this study can obtain the highest percentage of reduction in minimum and average P_{loss} by solving the optimal power flow problem of the DGs in the DC network.

6. Processing Time Analysis

The previous sections presented and discussed the excellent results obtained by the MVO in terms of repeatability and solution quality for solving the optimal power problem in DC grids of any size. However, these excellent results also come at a cost: the processing time they require. With respect to the 21-bus test system, the average processing time required by the solution methods were the following: MVO (11.69s), ALO (6.378s), BH (2.73s), CGA (3s), and PSO (3.78s). According to these results, the proposed method presented the longest processing time. Nevertheless, it is less than 13s, which, in real conditions of operation, is considered an excellent time to solve the optimal power flow problem because, in real life, the power flow analysis of electrical networks takes hours or, only in some cases, minutes. In the case of the 69-bus test system, all the processing times required by the solution methods increased: MVO (132s), ALO (9.67s), BH (13.91s), CGA (17.38s), and PSO (13.67s). Again, the MVO exhibited the longest processing time in the 69-bus test system, with an average value close to two minutes [33,34]. Nevertheless, said time is also adequate for solving the problem of optimal power flow in the 69-bus test system because of the same reasons explained regarding its 21-bus counterpart. These results demonstrate that, to obtain the best results in terms of repeatability and quality solution, it is necessary to sacrifice a little time in the search process conducted by the optimization algorithm. Due to this problem (i.e., the fact that the MVO takes longer to solve the optimal power flow problem), future studies should use parallel processing tools to improve the performance of the MVO in terms of processing times so that this solution method offers the best trade-off between quality solution and required processing time.

7. Conclusions

This paper proposed the implementation of a new optimization technique (MVO) to solve the OPF problem in DC networks through a master–slave methodology. In the master stage, the MVO determines the electrical power to be injected by each DG located in the DC network. In the slave stage, the numerical method known as SA establishes the impact of every solution proposed by the master stage on the objective function (reduction in power losses) and the constraints of the problem. In order to prove the efficiency and quality of the new solution methodology employed in this document, it was compared with four other optimization techniques: ALO, BH, CGA, and PSO. Each test was conducted using three levels of distributed generation penetration: 20%, 40%, and 60%. Such percentages represent the power provided by the main generator in the network (slack) in the base case (without DGs). The results and discussion here demonstrated that, when used to find a good quality solution, the MVO presents a higher percentage of reduction in average P_{loss}, as well as an excellent behavior regarding standard deviation. Thus, this technique is efficient in terms of precision and repeatability. Likewise, in several cases, the MVO obtained the best solution regarding minimum P_{loss}, only outperformed by PSO by an almost negligible difference in the 21- and 69-node systems at a 20% penetration level. Nevertheless, Figure 12 demonstrates that the MVO is the best technique to solve the OPF problem in DC networks of any size thanks to its low standard deviation and its great capacity to converge toward the best solution. The limitations of the solution methodology proposed in this paper can be overcome by (1) including an analysis of the problem of optimal location of DGs, (2) integrating and operating energy storage systems in order to improve the operation quality of the grid, (3) reducing the processing times required by the MVO and other solution methods (faster is better), and (4) using objective functions that incorporate economic and environmental indices. Hence, future studies should propose optimization strategies to solve the problem of optimal integration of DGs using the MVO-SA presented here in order to solve the sizing problem (optimal power flow problem). Moreover, further research should investigate the optimal location and operation of energy storage elements inside the DC grid, which were not included in this study. Parallel processing tools can be utilized to improve the performance of the solution methodology in terms of processing times. With respect to the indices in the objective functions, the methodology proposed here was only focused on the reduction of P_{loss}. However, said objective function can be changed to reduce the operating costs of a network or its CO_2 emissions into the environment. Finally, it should be highlighted that the MVO-SA methodology can also be used to solve the OPF problem in AC networks.

Author Contributions: Conceptualization, A.A.R.-M., L.F.G.-N., J.M., O.D.M. and A.-J.P.-M.; methodology, A.A.R.-M., L.F.G.-N., J.M., O.D.M. and A.-J.P.-M.; formal analysis, A.A.R.-M., L.F.G.-N., J.M., O.D.M. and A.-J.P.-M.; investigation, A.A.R.-M., L.F.G.-N., J.M., O.D.M. and A.-J.P.-M.; resources, A.A.R.-M., L.F.G.-N., J.M., O.D.M. and A.-J.P.-M.; writing—original draft preparation, A.A.R.-M., L.F.G.-N., J.M., O.D.M. and A.-J.P.-M. All authors have read and agreed to the published version of the manuscript.

Funding: The authors would like to thank the Instituto Tecnológico Metropolitano (ITM), two of its research groups (AEyCC and MATyER), and the University of Córdoba.

Acknowledgments: This work was supported by the Centro de Investigación y Desarrollo Científico of the Universidad Distrital Francisco José de Caldas under grant 1643-12-2020 associated with the project entitled "Desarrollo de una metodología de optimización para la gestión óptima de recursos energéticos distribuidos en redes de distribución de energía eléctrica"; the Dirección de Investigaciones of the Universidad Tecnológica de Bolívar under grant PS2020002 associated with the project entitled "Ubicación óptima de bancos de capacitores de paso fijo en redes eléctricas de distribución para reducción de costos y pérdidas de energía: Aplicación de métodos exactos y metaheurísticos"; and the Spanish Ministry of Science, Innovation and Universities under the program "Proyectos de I+D de Generacion de Conocimiento" of the National Program for Knowledege Generation and Scientific and Technological Strengthening of the R&D&i System under grant number PGC2018-098813-B-C33.

Conflicts of Interest: The authors declare no conflict of interest.

References

1. Taba, M.F.A.; Mwanza, M.; Çetin, N.S.; Ülgen, K. Assessment of the energy generation potential of photovoltaic systems in Caribbean region of Colombia. *Period. Eng. Nat. Sci.* **2017**, *5*, doi:10.21533/pen.v5i1.76 [CrossRef]
2. Gil-González, W.; Montoya, O.D.; Holguín, E.; Garces, A.; Grisales-Noreña, L.F. Economic dispatch of energy storage systems in dc microgrids employing a semidefinite programming model. *J. Energy Storage* **2019**, *21*, 1–8. [CrossRef]
3. Grisales, L.F.; Grajales, A.; Montoya, O.D.; Hincapie, R.A.; Granada, M.; Castro, C.A. Optimal location, sizing and operation of energy storage in distribution systems using multi-objective approach. *IEEE Lat. Am. Trans.* **2017**, *15*, 1084–1090. [CrossRef]
4. Montoya, O.D.; Garrido, V.M.; Gil-González, W.; Grisales-Noreña, L.F. Power flow analysis in DC grids: Two alternative numerical methods. *IEEE Trans. Circuits Syst. II Express Briefs* **2019**, *66*, 1865–1869. [CrossRef]
5. Montoya, O.D.; Grisales-Noreña, L.F.; Amin, W.T.; Rojas, L.A.; Campillo, J. Vortex Search Algorithm for Optimal Sizing of Distributed Generators in AC Distribution Networks with Radial Topology. In Proceedings of the Workshop on Engineering Applications, Santa Marta, Colombia, 16–18 October 2019; Springer: Berlin/Heidelberg, Germany, 2019; pp. 235–249.
6. Wang, W.; Barnes, M. Power flow algorithms for multi-terminal VSC-HVDC with droop control. *IEEE Trans. Power Syst.* **2014**, *29*, 1721–1730. [CrossRef]
7. Huang, G.; Ongsakul, W. Managing the bottlenecks in parallel Gauss-Seidel type algorithms for power flow analysis. *IEEE Trans. Power Syst.* **1994**, *9*, 677–684. [CrossRef]
8. Montoya, O.D.; Grisales-Noreña, L.F.; Gil-González, W. Triangular matrix formulation for power flow analysis in radial DC resistive grids with CPLs. *IEEE Trans. Circuits Syst. II Express Briefs* **2019**, *67*, 1094–1098. [CrossRef]
9. Montoya, O.D.; Grisales-Noreña, L.; González-Montoya, D.; Ramos-Paja, C.; Garces, A. Linear power flow formulation for low-voltage DC power grids. *Electr. Power Syst. Res.* **2018**, *163*, 375–381. [CrossRef]
10. Montoya, O.D. On the existence of the power flow solution in DC grids with CPLs through a graph-based method. *IEEE Trans. Circuits Syst. II Express Briefs* **2019**, *67*, 1434–1438. [CrossRef]
11. Grisales-Noreña, L.F.; Montoya, O.D.; Gil-González, W.J.; Perea-Moreno, A.J.; Perea-Moreno, M.A. A Comparative Study on Power Flow Methods for Direct-Current Networks Considering Processing Time and Numerical Convergence Errors. *Electronics* **2020**, *9*, 2062. [CrossRef]
12. Noreña, L.F.G.; Cuestas, B.J.R.; Ramirez, F.E.J. Ubicación y dimensionamiento de generación distribuida: Una revisión. *Cienc. E Ing. Neogranadina* **2017**, *27*, 157–176. [CrossRef]
13. Rendon, R.A.G.; Zuluaga, A.H.E.; Ocampo, E.M.T. *Técnicas Metaheuristicas de Optimización*; Universidad Tecnologica de Pereira: Risaralda, Colombia, 2008.
14. Garzon-Rivera, O.; Ocampo, J.; Grisales-Norena, L.; Montoya, O.; Rojas-Montano, J. Optimal Power Flow in Direct Current Networks Using the Antlion Optimizer. *Stat. Optim. Inf. Comput.* **2020**, *8*, 846–857. [CrossRef]
15. Li, J.; Liu, F.; Wang, Z.; Low, S.H.; Mei, S. Optimal power flow in stand-alone DC microgrids. *IEEE Trans. Power Syst.* **2018**, *33*, 5496–5506. [CrossRef]
16. Montoya, O.D.; Gil-González, W.; Garces, A. Sequential quadratic programming models for solving the OPF problem in DC grids. *Electr. Power Syst. Res.* **2019**, *169*, 18–23. [CrossRef]
17. Velasquez, O.S.; Montoya Giraldo, O.D.; Garrido Arevalo, V.M.; Grisales Norena, L.F. Optimal power flow in direct-current power grids via black hole optimization. *Adv. Electr. Electron. Eng.* **2019**, *17*, 24–32. [CrossRef]
18. Giraldo, J.; Montoya, O.; Grisales-Noreña, L.; Gil-González, W.; Holguín, M. Optimal power flow solution in direct current grids using Sine-Cosine algorithm. *J. Phys. Conf. Ser.* **2019**, *1403*, 012009. [CrossRef]
19. Grisales-Noreña, L.F.; Garzón Rivera, O.D.; Ocampo Toro, J.A.; Ramos-Paja, C.A.; Rodriguez Cabal, M.A. Metaheuristic Optimization Methods for Optimal Power Flow Analysis in DC Distribution Networks. 2020.
20. Moradi, M.H.; Abedini, M. A combination of genetic algorithm and particle swarm optimization for optimal DG location and sizing in distribution systems. *Int. J. Electr. Power Energy Syst.* **2012**, *34*, 66–74. [CrossRef]
21. Grisales-Noreña, L.F.; Gonzalez Montoya, D.; Ramos-Paja, C.A. Optimal sizing and location of distributed generators based on PBIL and PSO techniques. *Energies* **2018**, *11*, 1018. [CrossRef]
22. Mirjalili, S.; Mirjalili, S.M.; Hatamlou, A. Multi-verse optimizer: A nature-inspired algorithm for global optimization. *Neural Comput. Appl.* **2016**, *27*, 495–513. [CrossRef]
23. Khoury, J.; Ovrut, B.A.; Seiberg, N.; Steinhardt, P.J.; Turok, N. From big crunch to big bang. *Phys. Rev. D* **2002**, *65*, 086007. [CrossRef]
24. Tegmark, M. Parallel universes. *Sci. Am.* **2003**, *288*, 40–51. [CrossRef] [PubMed]
25. Eardley, D.M. Death of white holes in the early universe. *Phys. Rev. Lett.* **1974**, *33*, 442. [CrossRef]
26. Davies, P.C. Thermodynamics of black holes. *Rep. Prog. Phys.* **1978**, *41*, 1313. [CrossRef]

27. Morris, M.S.; Thorne, K.S. Wormholes in spacetime and their use for interstellar travel: A tool for teaching general relativity. *Am. J. Phys.* **1988**, *56*, 395–412. [CrossRef]

28. Guth, A.H. Eternal inflation and its implications. *J. Phys. A Math. Theor.* **2007**, *40*, 6811. [CrossRef]

29. Steinhardt, P.J.; Turok, N. The cyclic model simplified. *New Astron. Rev.* **2005**, *49*, 43–57. [CrossRef]

30. Lipowski, A.; Lipowska, D. Roulette-wheel selection via stochastic acceptance. *Phys. A Stat. Mech. Its Appl.* **2012**, *391*, 2193–2196. [CrossRef]

31. Garcés, A. On the convergence of Newton's method in power flow studies for DC microgrids. *IEEE Trans. Power Syst.* **2018**, *33*, 5770–5777. [CrossRef]

32. Grisales-Noreña, L.F.; Montoya, O.D.; Ramos-Paja, C.A.; Hernandez-Escobedo, Q.; Perea-Moreno, A.J. Optimal location and sizing of distributed generators in DC Networks using a hybrid method based on parallel PBIL and PSO. *Electronics* **2020**, *9*, 1808. [CrossRef]

33. Grisales-Noreña, L.F.; Montoya, O.D.; Ramos-Paja, C.A. An energy management system for optimal operation of BSS in DC distributed generation environments based on a parallel PSO algorithm. *J. Energy Storage* **2020**, *29*, 101488. [CrossRef]

34. Molina-Martin, F.; Montoya, O.D.; Grisales-Noreña, L.F.; Hernández, J.C.; Ramírez-Vanegas, C.A. Simultaneous Minimization of Energy Losses and Greenhouse Gas Emissions in AC Distribution Networks Using BESS. *Electronics* **2021**, *10*, 1002. [CrossRef]

sustainability

Article

Renewable Minigrid Electrification in Off-Grid Rural Ghana: Exploring Households Willingness to Pay

Artem Korzhenevych [1,2] and Charles Kofi Owusu [2,3,*]

[1]   Leibniz Institute of Ecological Urban and Regional Development, Weberplatz 1, 01217 Dresden, Germany; a.korzhenevych@ioer.de

[2]   Faculty of Business and Economics, Technische Universität Dresden, Helmholtzstr 10, 01062 Dresden, Germany

[3]   Kumasi Institute of Technology, Energy and Environment (KITE), Accra GA23321, Ghana

*   Correspondence: kofiowusuboateng@gmail.com

Abstract: Renewable energy minigrids hold significant prospects for Africa's energy sector and its economic development in general. The government of Ghana has established pilot renewable minigrids in five off-grid communities as a testing ground for the electrification of over 600 existing rural communities that cannot be electrified via the national grid. Although there is evidence on willingness to pay (WTP) values for renewable-generated electricity in some developing countries, little is known about households' WTP for renewable-based electricity in Ghana and, in particular, about renewable minigrids for rural electrification. This paper provides one of the first WTP estimates for renewable-based electricity for rural electrification in a developing economy context such as Ghana. Using data from a contingent valuation survey undertaken in all five pilot renewable minigrid project communities, we found that rural households are willing to pay an average of 30 GHC/month (≈5 USD/month) for high-quality renewable-powered electricity services, which is twice the amount they are currently paying based on the Uniform National Tariffs. The hypothetical bias is addressed by conducting a survey among active users of the minigrids. The starting point bias is reduced by employing random starting bids. The respondents are willing to pay between 9 and 11% of their discretionary incomes to cover the cost of accessing reliable renewable-powered electricity in the rural, off-grid communities in Ghana. The paper concludes by discussing the policy implications of these findings regarding the development of tariff regulations and business models for renewable minigrids in the rural, off-grid sector.

Keywords: willingness to pay; minigrids; rural electrification; renewable energy; Ghana

Citation: Korzhenevych, A.; Owusu, C.K. Renewable Minigrid Electrification in Off-Grid Rural Ghana: Exploring Households Willingness to Pay. *Sustainability* **2021**, *13*, 11711. https://doi.org/10.3390/su132111711

Academic Editor: Alberto-Jesus Perea-Moreno

Received: 30 September 2021
Accepted: 20 October 2021
Published: 23 October 2021

Publisher's Note: MDPI stays neutral with regard to jurisdictional claims in published maps and institutional affiliations.

1. Introduction

About one billion people in developing countries currently lack access to electricity, most of them living in sub-Saharan African and developing Asian countries [1,2]. A vast majority (87%) of these unelectrified households live in rural areas [2]. This challenge is specifically addressed by Goal 7 "Ensure access to affordable, reliable, sustainable and modern energy for all" of the Sustainable Development Goals [2]. Despite ongoing electrification projects in different jurisdictions, the current trend is likely to lead to an estimated 700 million people who will remain unelectrified in 2030, nearly all of them in sub-Saharan Africa [1].

Despite the economic feasibility of extending the electricity grid to under-served areas in some situations, minigrids may be better suited to address the low electrification rates and electrification challenges in areas with scattered households, low populations, and low demand potential [1,3,4]. A vast majority of the rural households without adequate electricity access would be better serviced with standalone systems or minigrids [5]. Alongside the existing traditional approach of electricity grid extension, off-grid renewable energy solutions, notably, solar minigrids and standalone systems, provide a modern and scalable approach to achieve universal electrification [6]. Renewable energy minigrids therefore hold enormous prospects for the African energy sector, not only by enhancing energy access, but also by enabling the increased use of low-carbon energy sources, with the benefits for sustainable rural development.

Though investment levels in the solar minigrid market remain low [6], recent years have witnessed a significant increase in interest from different stakeholders (i.e., international organizations, governments, and the private sector) in developing minigrids as cost-effective and reliable means to reach unelectrified populations [7]. Indeed, an estimated half of the investment in electrification projects in the next decades is expected to target minigrids—creating a minigrid yearly investment volume of up to USD 20 billion [7].

As donors and developing economies alone are likely unable to meet these investment levels, renewable minigrid projects must be able to attract private equity and debt financing to sustain the scale of deployment required to realize global electrification goals [6,7]. With the critical role off-grid renewable energy is expected to play in achieving universal electricity access targets [1,8], attention must be paid to how policy makers can encourage private investments into this emerging off-grid renewable sector. The respective tariffs must be able to at least generate sufficient revenues to cover operations and maintenance costs and other liabilities, generate sufficient profit, and recover minigrid investment cost to be fully commercial [9].

However, in developing economies, designing commercially viable tariffs is often not as straightforward an issue as one might expect. Electricity is generally viewed as a public good, and thus from a government perspective, equity and fairness are paramount concerns. Many African governments have established uniform national electricity tariffs in order to ensure not only fairness across customers but also affordability [7]. Often these national tariffs are set at a rate below what utilities must charge to cover their capital and operational costs.

Ghana is no exception—uniform national tariffs apply to both grid electricity and off-grid renewable projects including the five current pilot renewable minigrids developed under the World Bank-funded Ghana Energy Development and Access Project. However, these tariffs do not allow a viable business model for potential commercial investors, as the true costs are currently not passed on to the electricity consumers. Thus, there is a need to understand and model the actual households' demand for renewable-based electricity that would furnish relevant information for optimal tariff design in the rural, off-grid sector.

This study provides one of the first willingness to pay (WTP) estimates for renewable-based rural electricity provision in a developing economy context such as Ghana. Importantly, the study is new in that it was conducted among actual users of renewable minigrids, thus reducing potential bias in the WTP. Several econometric specifications making use of both dichotomous choice and open-ended survey questions are tested to increase the robustness of the results. The study is expected to inform policy makers on the amount an average rural household is willing to expend to access renewable minigrid electricity services and will consequently guide not only tariff adjustment, but also support the development of the overall business strategy for the off-grid, renewable-energy based electrification services.

To this end, this study seeks to respond to the following research questions: What is the WTP for a 24-h renewable minigrid electricity service in a rural off-grid setting in Ghana? What are the factors that influence households' WTP? What do the findings suggest regarding the choice of future business models in this sector?

The rest of the paper is organized as follows. Section 2 discusses Ghana's electricity sector in general and the off-grid electrification project development in particular. Section 3 reviews related empirical and theoretical literature. In Section 4, the methodology for the study is discussed. Analyses and discussion of study results are presented in Section 5 while Section 6 derives conclusions and policy implications.

2. Developments in Ghana's Electricity Sector and the Ghana Energy Development and Access Project

The government of Ghana launched the National Electrification Scheme (NES) in 1989, with the overall objective of providing universal access to electricity in Ghana over a 30-year horizon [10–13]. By 2009, a 65% electricity access rate was achieved from a low of 28% in 1990. A total of 4221 communities with populations of at least 500 inhabitants were initially expected to be connected by 2020, a goal accomplished to 98% [10,11].

The 2010 NES Master Plan Review showed that about 85,000 communities and 13% of the Ghanaian population remained unelectrified. The document noted that approximately 70,000 (82%) of the unserved communities have low populations, scattered settlements, and are located in rural communities far from the grid, making it prohibitively expensive to extend the national grid to serve them [14]. Moreover, some of the communities are islands and lakeside communities, and, hence, the economics and practicalities of electrifying them via the grid are unrealistic. For these unserved communities, decentralized electrification options (such as minigrids and standalone systems) have been found to be the most cost-effective way of delivering reliable energy access [1,6].

The Ghana Energy Development and Access Project (GEDAP) was launched in 2007 as part of efforts to provide the off-grid, isolated communities with alternative electrification options [12,13]. The GEDAP installs pilot photovoltaic minigrid systems (with a back-up generator) providing electricity supply to five (Pediatorkope in the Greater Accra region, Atigagome and Wayokope in the Brong-Ahafo region, Kudorkope and Aglakope in the Volta region) of these isolated rural communities on islands in the Volta Lake in Ghana. The GEDAP Project is financed with concessional funding from the World Bank, the Global Environment Fund, and the Swiss Development Agency. Ownership of the project's assets is vested in the government of Ghana. In all, a total 228 kW of photovoltaic capacity has been installed at the five minigrid sites supplying a total of 598 households. Households use this electricity typically for lighting, cell phone charging, powering their television and radio, fans, and fridges.

A dominant regulatory problem hindering the development of the minigrid market is the fact that the Uniform National Tariff policy, originally applicable to grid-connected households, has been extended to the renewable minigrids [12]. This means that consumers of electricity in the five pilot communities pay the same electricity price per kilowatt-hour (kWh) as grid-connected customers. Table 1 below shows, among others, the difference between what is deemed to be cost-reflective tariffs (only covering minigrid operational and maintenance cost) and the approved UNT in Ghana [10].

Table 1. Pricing of electricity in Ghana: Uniform National Tariff versus Cost-Reflective Tariffs.

Tariff Profile	EDA [1] (Wh/Day)	Power (kW)	Uniform National Tariff (GHC [2]/Month)	Cost-Reflective Tariff [3]—Only Operation and Maintenance (GHC/Month)
T01	275	0.5	4.20	10.00
T11	550	0.5	6.90	20.00
T21	1100	0.5	12.40	40.20
T31	1650	0.5	17.80	60.20
T42	2200	1	33.60	80.30
T53	2750	1.5	44.70	100.40

[1] Energy Daily Allowance (amount of energy per day allotted to a household per their tariff profile or category). [2] GHC = Ghana Cedi (Ghanaian currency), GHC 1 = USD 0.17 as of August 2021. [3] Given that the initial capital cost was fully funded by a grant, the cost-reflective tariff is estimated to be the minimum reference tariff that yields a positive Net Present Value (NPV > 0). The minimum reference tariff was estimated to be around 0.23 USD/kWh. This tariff will generate sufficient revenues on an annual basis to cover replacement cost of components (batteries) as well as operation and maintenance expense. Source: Data from KITE report [10].

Because of the application of the UNT, the total revenue received from minigrid customers is only sufficient to cover a fraction of the operational expenses and does not cover the investment or maintenance costs [15]. The resulting negative cash flow is a disincentive to private investments into minigrids. Thus, there is research need regarding the features of off-grid household electricity demand that would allow alternative price setting.

3. Literature Review

A number of methodological approaches exist in the valuation literature that are used to estimate people's willingness to pay (WTP) for public or non-market goods and services. The contingent valuation (CV) approach, a well-known and established valuation method [16,17] is a stated preference methodology for economic valuation characterized by the creation of hypothetical markets for non-market goods where individuals are asked how much they would be willing to pay for the good if the market really existed. This paper relies on the CV method to capture rural households' WTP for renewable-generated electricity, since it provides theoretically accurate monetary measures of utility changes as well as offers an accurate and credible estimate of the respondent's full non-market value of a good [16,17].

Specifically for Ghana, there are a handful of empirical studies on WTP for electricity. Twerefou [18] used a CV method to assess WTP for improved electricity supply in Ghana from a survey of 1000 households. The survey captured the northern, coastal, and middle zones of the country. The study used a combination of dichotomous choice and open-ended question elicitation methods, and from the author's ordered probit estimations, the results showed that households in Ghana are willing to pay an average of GHC 2.7 for a kilowatt-hour of electricity supply, about one and a half times more than what they were actually paying. However, respondents in the study were asked to state their WTP estimates based on an amount (in kWh) of electricity consumed. Analysis based on kWh is a bit technical (and might not elicit the right value placed on electricity consumption by households) as compared to what most users are accustomed to: the average amount they pay in a month for power consumed within that month. As noted in previous studies [19–22], energy is abstract, invisible, and measured in kWh, a unit hard to deal with for most consumers. The study also did not capture any heterogeneity, such as the rural–urban distinction in the WTP figures or the north–south divide within Ghana in terms of income profiles and living standards.

Using a tobit regression technique, Taale and Kyeremeh [23] showed that urban households in Ghana are willing to pay 44% (GHC 6.8) more, compared to their current average monthly electricity bill, in order to access improved electricity services. The study showed that prior notice of power outages, monthly income, education level, and household size are among the factors that significantly affect households' willingness to pay for reliable electricity in Ghana. The authors, in their econometric model, however, did not account for differences in geographical location and in economic circumstances among communities, which could affect the household's WTP for electricity.

A number of other electricity-related WTP studies have been conducted in other developing countries. There also have been studies on willingness to pay for renewable-generated electricity in both developing and developed countries. A summary of the relevant literature is provided in Table 2 below. To the best of the authors' knowledge, the literature so far does not include a WTP study on renewable-generated electricity in Ghana and, in particular, on renewable minigrids for rural electrification.

Table 2. Summary of relevant literature.

	Author (s)	Country	Good/Service Valued	Study Method	Econometric Estimation Method
1	Twerefou [18]	Ghana	Improved electricity	CV:WTP	ordered probit
2	Taale and Kyeremeh [23]	Ghana	Reliable electricity	CV:WTP	tobit
3	Abdullah and Jeanty [24]	Kenya	Renewable energy for rural electrification	CV:WTP	parametric/non-parametric models
4	Abdullah and Mariel [25]	Kenya	Electricity services	Choice modeling	mixed logit
5	Alam and Bhattacharyya [26]	Bangladesh	Renewable minigrid electricity	CV: WTP	logit, OLS
6	Ayodele et al. [27]	Nigeria	Renewable energy minigrid/Renewable electricity	CV:WTP	ANOVA test
7	Deutschmann et al. [28]	Senegal	Reliable electricity	CV:WTP	probit, OLS
8	Dogan and Muhammad [29]	Turkey	Renewable electricity	CV:WTP	tobit/probit/logit
9	du Preez et al. [30]	South Africa	Wind farm	CV:WTA	logit
10	Entele [31]	Ethiopia	Solar PV vs. Grid electricity	CV:WTP	probit
11	Graber et al. [32]	India	Solar microgrids	Choice modeling	mixed logit
12	Gunatilake et al. [33]	India	24 h electricity supply	CV:WTP	probit, OLS
13	Harajli and Chalak [34]	Lebanon	Energy efficient appliances	CV:WTP	multivariate tobit
14	Kim et al. [35]	South Korea	Renewable electricity	CV:WTP	spike model
15	Kim et al. [36]	South Korea	Reliable electricity	CV:WTP	spike model
16	Oseni [37]	Nigeria	Reliable electricity	CV:WTP	double-bounded (interval) model
17	Scarpa and Willis [38]	United Kingdom	Renewable electricity	Choice modeling	multinomial logit
18	Zhang and Wu [39]	China	Green electricity	CV:WTP	multinomial logit

Note: CV—contingent valuation, WTP—willingness to pay, WTA—willingness to accept, OLS—ordinary least squares. Source: Authors.

4. Study Methodology

4.1. Study Area and Selection of Survey Households

In order to estimate the willingness to pay for renewable-powered electricity service in rural Ghana, a contingent valuation survey was undertaken in all five renewable minigrid project communities in Ghana, located in 3 (Greater Accra, Volta, and Brong-Ahafo) of the 16 regions of Ghana. All of them are located on islands in the Volta River.

The communities are mainly accessible by water and are predominantly rural, with mud houses and thatched roofs. Fishing and farming are the predominant occupations and the source of income for most households. Fish trading, clothing making, hairdressing, livestock breeding, and small retail stores also provide income for households. Only a few households are employed in petty trade and public service (e.g., district assembly employees and teachers).

Prior to the minigrid electrification project, there was no electricity in the communities. All traditional sources of energy and the respective equipment such as storm lamps, kerosene, dry batteries, diesel generators, etc., were purchased at very high prices from surrounding towns, increasing energy costs and overall household expenditures. In the absence of electricity to run cold storage equipment, households were forced to sell their fish harvest in the market at cheap prices. Processing of agricultural products was also problematic owing to the high cost of diesel to operate the existing mills. The minigrid electrification project can therefore be considered a very important infrastructure that will help meet the social, health, and economic needs of the communities.

The conceptual framework for WTP analysis and contingent valuation is consistent with consumer demand theory and captures both use and non-use values of a commodity.

The contingent valuation method is deeply rooted in microeconomic welfare theory, where households or individuals minimize their expenditure under utility constraints or maximize their utility subject to income or budget constraints [40,41].

Households in the project communities (refer to Table 3) were selected for interview using a combination of a cluster sampling approach and simple random sampling. Cluster sampling was applied because of the scattered nature of the settlements in the project communities. The number of households picked from each cluster was set in proportion to the cluster population. Inside any cluster, the households interviewed were selected randomly. The number of households selected per community for the face-to-face interviews was in proportion to the total number of households in the community. The survey took place between 28 October 2020 and 14 November 2020, and a total of 200 households (respondents) were interviewed (see Table 3). Four field researchers participated in the main survey after being trained with a pilot survey.

Table 3. List of communities and number of households interviewed per community.

Study Community	Region	Number of Clusters in the Community	Number of Households Interviewed
Pediatorkope	Greater Accra	10	49
Atigagome	Brong-Ahafo	7	25
Aglakope	Volta	5	46
Wayokope	Brong-Ahafo	3	17
Kudorkope	Volta	4	63
Total		**29**	**200**

Source: Authors.

4.2. Questionnaire Design and WTP Elicitation Process

The survey was structured in three main sections. These included (i) respondent's socioeconomic characteristics, (ii) utility-related information, and (iii) questions on WTP for renewable minigrid electricity. Renewable minigrid electricity access and bills paid were captured under a second, utility-related information, section.

Contingent valuation questions were asked in the third part of the questionnaire using the double-bounded dichotomous choice (DBDC) and open-ended techniques. The DBDC method implies that two different monetary payments are subsequently suggested to survey respondents. The second amount proposed to respondents is contingent on their response to the first proposed monetary payment [42–44]. The DBDC technique was adopted for its several advantages. This elicitation method is robust to poorly-designed bids [45,46] and is incentive-compatible [47]. It is also efficient [45] and robust to strategic and cognitive biases [48]. The open-ended question asked directly for the maximum WTP after the first two questions of yes/no type. Both elicitation approaches were used to increase construct validity of the WTP estimates [43].

In each of the minigrid project communities, there are currently the same 6 electricity tariff levels (with corresponding monthly payments) based on the Uniform National Tariffs, as reported in Table 1. The majority (82%) of the households in the communities have signed onto the T11, T21 and T31 tariff bands and are paying GHC 7, GHC 12.4 and GHC 17.8, respectively, as average monthly electricity bills. For the dichotomous choice WTP questions, this study randomized the starting bid values in order to control for the anchoring effect or the starting point bias. For each household, the starting bid (b^0) was randomly picked from a set of six possible monthly tariff tiers ranging from GHC 15 to 40 (this corresponds to the middle level of the cost-reflective tariff calculations for the most popular current tariff categories, see Table 1).

The initial question was formulated as "Assume your household is provided with a 24-h, reliable renewable minigrid electricity supply, which is able to power all your electrical equipment. Are you willing to pay amount x per month to cover the cost of power production?" It was followed by another similar dichotomous choice question, where the starting bid was adjusted upwards or downwards, depending on the first answer (see Figure 1).

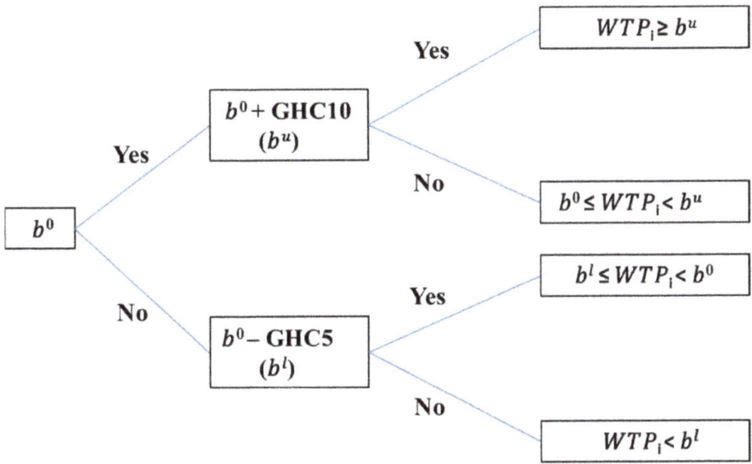

Figure 1. Elicitation of willingness to pay in the survey (b^0 = starting bid, b^l = lower bid, b^u = upper bid). Source: Authors.

The final question in the survey was open-ended: "State the maximum amount you are willing to pay, to cover cost of power production, assuming your household is provided with a 24-h, reliable renewable minigrid electricity supply, which is able to power all your electrical equipment." The DBDC results and the maximum WTP were used separately in the econometric analyses to check the validity of the results.

Despite the fact that contingent valuation methodology has been widely applied in research, some researchers have raised concerns over its validity. According to some authors [49–52] as reported in Hanemann [53], results from a contingent valuation study may be inconsistent with economic theory. According to Hausman [54], the approach is plagued by three main issues, namely: willingness to pay—willingness to accept dichotomy, hypothetical response bias, and scope effect (which renders it an ineffective tool in terms of policy formulation). Hausman [54] averred that people do not do what they say; a 'yes' response to a hypothetical question, as happens in contingent valuation studies, does not signify economic power neither can it be suggested to mean that survey respondents would do exactly in reality. However, other authors [55–57] have adduced counterarguments to the views of the critics.

Evidence can be used to justify application of the contingent valuation approach in this study along the lines of the criticism. First, according to Rowe et al. [58] (p. 6), hypothetical bias is "the potential error induced by not confronting any individual with the real situation." The renewable minigrid electricity, which is being valued, is not new to the survey households; the commodity is not a hypothetically described market good and hence cannot be so predisposed to the hypothetical bias. In terms of scope sensitivity, Morey et al. [59] affirmed that, "economic theory suggests that in general WTP will depend on income, justifying the inclusion of income in the utility difference model." This study estimated the WTP of households subject to their income, which is consistent with consumer demand theory.

4.3. Econometric Estimation

4.3.1. Dichotomous Choice Models Estimation

The answers of respondents to the discrete choice questions of the survey were employed to construct two models estimated using maximum likelihood techniques. First, a simple *probit* model was estimated using only the answers to the first dichotomous choice question. Second, the answers to both dichotomous choice questions were used together in the estimation of a *double-bounded (interval) model* using the doubleb STATA command developed by Lopez-Feldman [60]. The interval model used the combinations of the two answers (Yes–Yes, Yes–No, No–Yes, No–No) to a certain extent to limit the individual WTP values to the bands within the known bids (Figure 1).

Selected households' socioeconomic characteristics included in the regressions were informed by previous studies discussed in Section 3. They included the initial bid, the current electricity bill, household income, marital status, gender and educational level of the respondent, household size and a dummy taking a value of 0 if the household was using electricity for all activities/energy services and 1 otherwise. All monetary variables were converted into logs. Community-specific dummy variables were added to all models.

As a measure of household income, the household's discretionary income was used, which was the remaining portion of the household's income after committed expenditures on clothing, housing, food, transportation, and other market and non-market goods were taken into account. According to Laitila [61], this is a relevant measure of household income, which is supported by economic theory, because the household's maximum WTP for any good should be restricted by their ability to pay.

4.3.2. OLS Estimation of the Maximum WTP

In the definition of elicitation methods, when open-ended questions are posed and a continuous bid variable is obtained, ordinary least squares (OLS) can be an appropriate estimation method [44]. The OLS model uses the stated maximum WTP values as the dependent variable. The explanatory variables employed were the same as in the dichotomous choice models estimation, including community-specific dummy variables. All regressions were run in STATA.

5. Results and Discussion

5.1. Descriptive Analysis

From the summary statistics shown in Table 4, the mean household size was 6.6, which was higher than the national average of 5.5 for rural dwellers [62]. Out of the 200 respondents interviewed, approximately 70% were males, which is characteristic of male dominance in Ghanaian households, affirmed by the national average of 72% in the rural areas [62]. On the average, respondents reported a monthly discretionary income of GHC 323 per household, which was below the national estimate of GHC 422 for rural dwellers [62]. Furthermore, 63% of the respondents were married as compared to an estimated 48% for rural inhabitants in Ghana [62]. From the survey, an average of 59% of households were not able to meet all their energy service needs as they wished, compared with 41% who did not have those capacity constraints. With respect to education, a majority (60%) of the respondents had attained a basic education while 13% had not acquired any form of education. A fifth (20%) of the sample had a secondary education, while 7% were schooled up to the tertiary level. For comparison, according to national statistics, about a fifth (20%) of all rural dwellers have never been to school while approximately 47% have some basic education. Additionally, about 15% of the rural population have acquired a secondary or higher level of education [62].

As for the WTP, 29% of the respondents were willing to pay at most GHC 25 per month, the majority (60%) were willing to pay between GHC 25 and GHC 35 while the remaining 11% were willing to pay above GHC 35. However, their current average monthly electricity bill equaled approximately GH 15. A recorded average WTP of GHC 29 was almost twice the respondents' current average electricity expenditure.

Table 4. Descriptive statistics of variables used in the WTP model.

Variable	Classification	Expected Sign	Obs.	Mean	Std. Dev.	Min	Max
Maximum WTP (GHC)	Continuous		200	29.79	6.284	15	50
First bid response (Yes = 1; No = 0)	Dummy		200	0.65	0.478	0	1
Second bid response (Yes = 1; No = 0)	Dummy		200	0.50	0.501	0	1
Electricity bill (GHC)	Continuous	+	200	14.91	7.355	7	45
Starting bid (GHC)	Discrete	+	200	26.80	8.237	15	40
Monthly discretionary income (GHC)	Continuous	+	200	322.65	105.747	95	705
Marital status (Married = 1; Otherwise = 0)	Dummy	+	200	0.63	0.484	0	1
Gender (Male = 1, Female = 0)	Dummy	+	200	0.70	0.462	0	1
Use of electricity for all activities (No = 1; Yes = 0)	Dummy	+	200	0.59	0.493	0	1
Household size	Continuous	+	200	6.61	3.346	1	18
No education	Dummy	−	200	0.125	0.331	0	1
Basic education	Dummy	−	200	0.61	0.490	0	1
Secondary	Dummy	+	200	0.20	0.401	0	1
Tertiary	Dummy	+	200	0.07	0.255	0	1

Source: Authors.

5.2. Factors Influencing the Willingness to Pay

The WTP was estimated using three models: a probit model using the first round of dichotomous choice questions, a double-bounded (interval) model using two rounds of dichotomous choice questions, and an OLS model using the stated maximum WTP. Figure 2 compares the estimated WTP from three methods at three points of the distributions: 25th, 50th, and 75th percentiles. The OLS model using the maximum WTP produced the most reliable estimates, and the results of the interval model were quite similar. This is why later in this section the WTP determinants are discussed based on these two models. Other estimation results are presented in the Appendix A, Table A1.

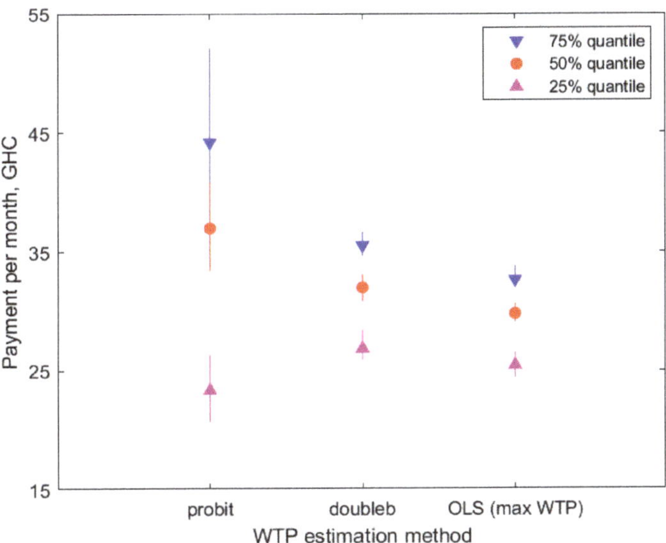

Figure 2. Comparison of WTP estimates from the three methods (25th, 50th, and 75th percentile WTP estimates with respective 95% confidence intervals). Source: Authors.

The econometric estimation results are shown in Table 5. Community dummies were used to control for community-specific effects. The Variance Inflation Factor (VIF) was applied to test for the presence of multicollinearity. The results from the test showed that

VIFs for all the regressors were less than 10, which deemed multicollinearity unproblematic. Furthermore, the models were estimated with robust standard errors owing to the problem of heteroskedasticity associated with cross-sectional data. The F-tests (OLS) undertaken to evaluate the validity and significance of the model parameters showed the estimated models were highly significant at the 1% significance level.

Table 5. Estimation results for the maximum willingness to pay.

Explanatory Variables	OLS Model	Double-Bounded (Interval) Model
Starting bid (Log)	−0.038	−0.170 **
	−0.027	−0.067
Household monthly income (Log)	0.463 ***	0.572 ***
	−0.045	−0.055
Electricity bill (Log)	−0.010	−0.009
	−0.021	−0.03
Respondent marital status	0.063 ***	0.037
(Married = 1)	−0.021	−0.028
Respondent gender (Male = 1)	−0.009	0.003
	−0.018	−0.028
Use of electricity for all activities	0.060 ***	0.071 **
(No = 1)	−0.021	−0.028
Household size	0.007 ***	0.007 *
	−0.003	−0.004
Respondent education level,	0.036	−0.002
basic = 1	−0.024	−0.043
Respondent education level,	0.047 *	0.001
secondary = 1	−0.027	−0.048
Respondent education level,	−0.006	−0.012
tertiary = 1	−0.04	−0.059
Constant	0.752 ***	0.663 **
	−0.249	−0.274
Community dummies	*Yes*	*Yes*
Observations	200	200
Expected Mean	29.14	31.21
R-squared	0.713	
F-test	26.13 ***	
Wald chi2(14)		195.01 ***

Note: Robust standard errors in parentheses. *** $p < 0.01$, ** $p < 0.05$, * $p < 0.1$. Source: Authors.

The starting bid variable was insignificant in the OLS estimation, which was important and demonstrated the absence of a starting point bias in this estimation. In the interval model, the starting bid was significant and negative, which is an often-observed effect in dichotomous choice models, where a high initial bid is more likely to be rejected.

Household income was highly significant and carried the expected positive sign. Furthermore, the coefficient decreased when the sample was reduced toward low-income households (see Appendix). This was consistent with economic theory and showed that an increase in household income will lead to an increased WTP for renewable-powered electricity in rural areas. The coefficient was in the order of 0.5, further implying that renewable-powered electricity was regarded as a normal good or even a necessity. This result confirmed the findings of other studies in developing countries [18,23–26,31,33,34,63], which found that income is an important variable in determining the amount households are willing to pay for electricity.

Marital status was significant in the OLS regression with a positive impact on households' WTP for clean electricity services. This could be due to the fact that married couples were more likely to have their own children and corresponding increased energy needs. Previous studies [19,23,26] found similar results, which showed that married couples are more likely to pay for electricity connection services relative to the unmarried.

The size of a household was also important in determining the WTP for renewable electricity. The coefficient of household size was positive. This was also in accordance with previous findings [19,25,33,36]. The positive sign could be attributed to the fact that people attached status to large household sizes and the show of economic strength, especially in the rural areas. Hence, with an increasing household size, families may be forced to live in line with this status symbol. However, other studies [18,64] found a negative relationship between the size of a household and willingness to pay for electricity.

Households that wanted the opportunity to use the electricity for all the activities and energy services they require were generally inclined toward paying more for electricity services. This study revealed similar trends. Households that did not have enough electrical capacity to meet all their current and potential energy needs were willing to pay more for renewable-powered electricity services. This result confirmed the findings in other studies [23,48,65], which found a positive association between duration of power outage and willingness to pay for improved electricity service. This was summed up in Otegbulu [65] and Oseni [37], who found that a majority of Nigerian households, irrespective of their socioeconomic status, valued reliable electricity supply and were thus willing to pay more to access it.

Secondary education was the only education dummy, which was positive and significant in the OLS regression. No significant negative effects of higher education were observed. It suggested the importance that household heads with some level of schooling attached to electricity in general. The assumption was that households with some level of education understood the benefits of electricity access, including convenience, income generation opportunities, and quality of life in general relative to the uneducated. Findings from several studies in developing countries [39,66] affirmed that education is a key socioeconomic variable that positively impacts the adoption and WTP for renewable technology electrification. Zarnikau [67] also found that the education levels of households positively impact their WTP for electricity efficiency investments. Other studies [18,23,34,35] found similar results indicating higher WTP with the levels of education of a household head.

5.3. Mean WTP Levels

This section discusses the mean maximum WTP values as obtained from the sample prior to estimating the regression and the predicted mean WTP estimates using the OLS estimation. The sample means represent the observed WTP for the renewable-powered electricity services while the predicted estimates reflect the impact of the socioeconomic variables. Table 6 captures both the observed and empirical mean estimates for the full and respective sub-samples.

Table 6. Mean WTP from the OLS estimation.

Sample	Observations	Mean Stated WTP	Mean Income	Estimated Mean WTP
Full sample	200	29.79 (6.35) [28.90–30.68]	322.65 (105.75)	29.14 (1.24) [28.97–29.31]
Sub-sample 1 (Tariff level T11 = GHC 7)	37	29.81 (5.89) [27.84–31.77]	328.40 (111.98)	28.25 (1.21) [27.50–29.02]
Sub-sample 2 (Tariff level T21 = GHC 12.4)	90	29.25 (6.5) [27.88–30.61]	311.78 (111.38)	28.55 (1.24) [28.29–28.81]
Sub-sample 3 (Tariff level T31 = GHC 17.8)	55	30.76 (5.67) [29.23–32.29]	335.65 (89.30)	29.96 (1.21) [26.63–30.28]

Note: Standard deviation in parentheses, 95% confidence intervals in square brackets. Source: Authors.

In Table 6, the mean WTP in the full sample is GHC 29.79. This constituted an estimated 9% of the rural households' discretionary income. This proportion of income to electricity expenditure was consistent with previous contingent valuation results [23,68].

The average WTP were further evaluated based on sub-samples belonging to current tariff categories (of the Uniform National Tariff) in the minigrid project communities. As explained earlier (see Table 1), there were six main tariff categories. The respective numbers of households surveyed were as follows: 37 (18.5%) households in T11, 90 (45%) households in T21, 55 (22.5%) households in T31, 16 (8%) households in T42, and 2 (1%) households subscribed to the T53 tariff. None of the households surveyed subscribed to the T01 tariff category. Results of the regressions based on the three large sub-samples are included in the Appendix B as Table A2.

Table 6 shows the observed and predicted average WTP of households within these tariff bands. The predicted average WTP for the T11 tariff band sub-sample was GHC 28.25. This estimated amount corresponded to about 9% of those households' monthly income and was more than three times their current electricity expenditure (GHC 7). In the same vein, the predicted average WTP for the T21 tariff band sub-sample was GHC 28.55 as compared to the current amount of GHC 12.4 per month. This estimated WTP amount (GHC 28.55) was more than twice their current expenditure on electricity and corresponded to about 9% of the households' monthly income. A similar WTP result was obtained for the T31 tariff band sub-sample. The expected mean WTP for this group of households was GHC 29.96, compared with their current monthly electricity expenditure of GHC 17.8, representing about a 68% increment over their current electricity expenditure and corresponding to 9% of the average monthly income from this subsample.

Thus, households were willing to pay substantially more than currently if a stable renewable energy supply was guaranteed. Oseni [37] similarly showed that Nigerian households are similarly willing to pay up to 86% above their current electricity tariffs for an enhanced power supply. According to ESMAP [69], rural households in Ghana want electricity more than they want low electricity tariffs, because in the absence of electricity access, households resort to paying higher proportions of their income on inferior energy forms. This also suggests that any form of business model to be considered and adopted by the government in the long term must seriously consider households' willingness to enjoy more and better available electricity and the fact that rural households are ready to discharge appropriate financial commitment to support sustainability of minigrids. However, poorer households currently signed on to low tariffs would be confronted with the largest increase in electricity expenditure, if they were charged according to their stated WTP. This calls for caution in the implementation of business models that might replace the uniform tariffs.

6. Conclusions and Policy Implications

Universal access to reliable and sustainable energy services requires expanding access to electricity, a key precondition for achieving the Sustainable Development Goals. Many governments in Africa set the agenda of meeting the universal electrification goals by electrifying remote and off-grid communities in rural areas with renewable minigrids. These minigrid technologies require huge capital outlays and therefore would need the backing of government, private sector, and households living in isolated, rural communities to achieve electrification goals and more so to ensure minigrid systems scalability and sustainability. This study relied on the contingent valuation method to estimate households' willingness to pay for renewable-generated electricity in the rural, off-grid communities in Ghana.

The results from the study indicated that rural households are willing to pay an average of about GHC 30 (USD 5) per month for renewable-powered electricity services, which is on average twice the amount they are currently paying, based on the Uniform National Tariffs. The surveyed households are thus willing to pay around 9% of their discretionary incomes for renewable-powered electricity. The results also showed that the elasticity of willingness to pay with regards to household income is 0.46. Given the economic growth rate of Ghana, at approximately 1.7% in 2020 (a decline from 6.5% in 2019 due to the COVID-19 shock), households' willingness to pay for electricity is expected to significantly increase in the future.

Household income, household size, basic education level of household head and marital status of respondents were noted to be significant factors that impact households' willingness to pay for renewable minigrid services. Another important finding was that households that do not have enough electrical capacity to meet all their current and potential energy needs are willing to pay more for renewable-powered electricity services.

The benefits that come with electricity access are evident to all the minigrid-connected communities, as even households without adequate electrical capacity indicated their readiness and willingness to pay a "premium" price for the minigrid electricity services. This should serve as a major signal to the policy makers: first, of the households' readiness to embrace new forms of alternative energy sources, and second, of the need to fast-track access provision for the energy have-nots and under-served areas, bringing into sharp focus the importance of the minigrid business model.

The results suggested that a private sector model could be considered and adopted for minigrid electrification in the future, as off-grid rural households' financial circumstances can support the sustainability of this business model. A hybrid minigrid business model (Public–Private Partnership) could also be explored. To this end, the government must develop the relevant regulatory and policy frameworks that support sustainable tariff approaches and minigrid business models, in order to de-risk investments and attract private developers into the off-grid renewable sector.

Another policy implication of the study is for the government and district/municipal authorities to support the minigrid communities with the development of productive uses of clean energy. Such initiatives have the potential of generating income for households from light industrial and agro-processing activities and thus enhancing the wealth of families in the rural areas. Poor households will then be better positioned to withstand shocks that may come with the abolishment of the Uniform National Tariff policy.

The study exclusively surveyed the five minigrid communities (which were the only renewable minigrid-electrified communities at the time of the study), and the sample was thus representative of the population covered by rural minigrids. Although the surveyed communities generally shared similar socioeconomic characteristics with the rural poor in Ghana (and hence results are generalizable), these minigrid communities have had the benefit of already enjoying renewable electricity access relative to the other rural population with little or no electricity access. Thus, perceptions and attitudes about alternative energy sources and the level of willingness to pay for these energy sources may differ from the general population. It will be a task for future research to evaluate whether there are significant differences in WTP values for renewable minigrid electricity services across already electrified communities and unelectrified rural locations.

Author Contributions: C.K.O.: Conceptualization, methodology, formal analysis, writing—original draft, funding acquisition. A.K.: Methodology, writing—review and editing, supervision, formal analysis, funding acquisition. All authors have read and agreed to the published version of the manuscript.

Funding: The Alexander von Humboldt Foundation provided support for conducting the research and preparing the article (Ref 3.5-1208742-GHA-IKS). They did not have any role in the study design, data collection, analysis and report writing, and in the decision to submit the article for publication.

Institutional Review Board Statement: Not applicable.

Informed Consent Statement: Not applicable.

Data Availability Statement: The data presented in this study may be made available from the authors upon request.

Acknowledgments: The authors wish to thank Selorm Kpoh (and the entire KITE field team) for the technical assistance provided in the survey design and instrumentation and for facilitating the field survey. We express our appreciation to the participants of the RSAI 2021 Conference, the GfR Summer Conference 2021, and the Seminar Series at the Faculty of Business and Economics, TU Dresden, for their insightful comments. The usual disclaimer applies.

Conflicts of Interest: The authors declare no conflict of interest. The funders had no role in the design of the study; in the collection, analyses, or interpretation of data; in the writing of the manuscript, or in the decision to publish the results.

Appendix A. Discrete Choice Model Results (Double-Bounded and Probit Models)

Table A1. Maximum WTP regression results.

Explanatory Variable	Double-Bounded Model	Probit Model
Starting bid (log)	−0.170 **	−11.970 ***
	(0.067)	(−2.321)
Electricity bill (log)	−0.009	0.179
	(0.030)	(−0.479)
Monthly discretionary income (log)	0.572 ***	6.507 ***
	(0.055)	(−1.099)
Marital status (Dummy)	0.037	0.327
	(0.028)	(−0.406)
Gender (Male)	0.003	0.065
	(0.028)	(−0.356)
Use of electricity for all activities (No = 1)	0.071 **	0.876 **
	(0.028)	(−0.404)
Household size	0.007	0.117 *
	(0.004)	(−0.065)
Educational level, basic	−0.002	−0.264
	(0.043)	(−0.554)
Educational level, secondary	0.001	−0.292
	(0.048)	(−0.622)
Educational level, tertiary	−0.012	−1.842 *
	(0.059)	(−0.947)
Constant	0.663 **	3.725
	(0.274)	(−3.963)
Community dummies	*Yes*	*Yes*
Regression's estimated standard error	0.114 ***	
	(0.012)	
Observations	200	200
Hosmer and Lemeshow goodness-of-fit (prob > chi2)		0.75

Note: Standard errors in parentheses; *** $p < 0.01$, ** $p < 0.05$, * $p < 0.1$. Source: Authors.

Appendix B. Further OLS Regression Results

Table A2. OLS regression results—full vs. sub-samples (tariff categories).

Variable	(1) Full Sample	(2) Sub-Sample 1 (Tariff Level = GHC 12.4)	(3) Sub-Sample 2 (Tariff Level = GHC 17.8)	(4) Sub-Sample 3 (Tariff Level = GHC 7)
Starting bid (Log)	−0.038	−0.017	−0.011	−0.161 *
	(0.027)	(0.038)	(0.039)	(0.086)
Monthly income (Log)	0.463 ***	0.450 ***	0.534 ***	0.397 ***
	(0.045)	(0.067)	(0.059)	(0.097)
Marital status (Dummy)	0.063 ***	0.090 ***	0.012	0.057
	(0.021)	(0.028)	(0.032)	(0.077)

Table A2. *Cont.*

Variable	(1) Full Sample	(2) Sub-Sample 1 (Tariff Level = GHC 12.4)	(3) Sub-Sample 2 (Tariff Level = GHC 17.8)	(4) Sub-Sample 3 (Tariff Level = GHC 7)
Gender (Male)	−0.009	0.012	−0.017	−0.037
	(0.018)	(0.029)	(0.027)	(0.053)
Use of electricity for all	0.060 ***	0.060 *	0.001	0.097 *
activities (No)	(0.021)	(0.031)	(0.038)	(0.051)
Household size	0.007 ***	0.009 *	0.003	0.017 **
	(0.003)	(0.005)	(0.006)	(0.008)
Education level, basic	0.036	0.065 *	−0.054	0.052
	(0.024)	(0.038)	(0.046)	(0.049)
Education level,	0.047 *	0.050	0.024	−0.016
secondary	(0.027)	(0.044)	(0.038)	(0.067)
Education level,	−0.006	0.020	0.032	
tertiary	(0.040)	(0.057)	(0.046)	
Constant	0.752 ***	0.640 *	0.382	1.411 **
	(0.249)	(0.324)	(0.360)	(0.629)
Community dummies	*Yes*	*Yes*	*Yes*	*Yes*
Observations	200	90	55	37
Expected Mean	29.14	28.55	29.96	
R-squared	0.713	0.746	0.801	0.735
F-test	26.13 ***	14.17 ***	24.42 ***	11.52 ***

Dependent Variable for WTP: WTP Amount (Final); Robust standard errors in parentheses. *** $p < 0.01$, ** $p < 0.05$, * $p < 0.1$. Source: Authors.

References

1. International Energy Agency. *WEO-2017 Special Report: Energy Access Outlook*; IEA: Paris, France, 2017. Available online: https://www.iea.org/reports/energy-access-outlook-2017 (accessed on 17 August 2021).
2. International Energy Agency; International Renewable Energy Agency; United Nations; World Bank Group; World Health Organization. *Tracking SDG7: The Energy Progress Report 2018*; World Bank: Washington, DC, USA, 2018.
3. International Energy Agency. *Energy for All*; IEA: Paris, France, 2011. Available online: https://www.iea.org/reports/energy-for-all (accessed on 17 August 2021).
4. Peters, J.; Sievert, M.; Toman, M.A. Rural electrification through mini-grids: Challenges ahead. *Energy Policy* **2019**, *132*, 27–31. [CrossRef]
5. Antonanzas-Torres, F.; Antonanzas, J.; Blanco-Fernandez, J. State-of-the-Art of Mini Grids for Rural Electrification in West Africa. *Energies* **2021**, *14*, 990. [CrossRef]
6. UNDP; ETH Zurich. *Derisking Renewable Energy Investment: Off-Grid Electrification*; United Nations Development Program: New York, NY, USA; ETH Zurich, Energy Politics Group: Zurich, Switzerland, 2018.
7. Reber, T.J.; Booth, S.S.; Cutler, D.S.; Li, X.; Salasovich, J.A. *Tariff Considerations for Micro-Grids in Sub-Saharan Africa*; National Renewable Energy Lab: Golden, CO, USA, 2018.
8. World Bank; Foster, V.; Azuela, G.; Bazilian, M.; Sinton, J.; Banergee, S.; De Wit, J.; Ahmed, A.; Portale, E.; Angelou, N.; et al. *Sustainable Energy for All 2015: Progress toward Sustainable Energy*; The World Bank: Washington, DC, USA, 2015.
9. Deshmukh, R.; Carvallo, J.P.; Gambhir, A. *Sustainable Development of Renewable Energy Mini-Grids for Energy Access: A Framework for Policy Design*; Lawrence Berkeley National Laboratory: Berkeley, CA, USA, 2013.
10. KITE (Kumasi Institute of Technology, Energy and Environment). *An Assessment of Renewable Energy Technologies as Off-Grid Power Solution in Ghana, Final Report Submitted to the African Development Bank*; KITE: Accra, Ghana, 2018.
11. Kemausuor, F.; Ackom, E. Toward universal electrification in Ghana. *Wiley Interdiscip. Rev. Energy Environ.* **2016**, *6*, e225. [CrossRef]
12. Bukari, D.; Kemausuor, F.; Quansah, D.A.; Adaramola, M.S. Towards accelerating the deployment of decentralised renewable energy mini-grids in Ghana: Review and analysis of barriers. *Renew. Sustain. Energy Rev.* **2021**, *135*, 110408. [CrossRef]
13. Kumi, E.N. *The Electricity Situation in Ghana: Challenges and Opportunities*; Center for Global Development: Washington, DC, USA, 2017.
14. Ghana Ministry of Energy. *National Electrification Scheme Master Plan Review (2011–2020)*; MoE: Accra, Ghana, 2010.
15. Trama TecnoAmbiental. *Final O&M Technical and Financial Report Submitted to the GEDAP Secretariat*; Trama TecnoAmbiental: Barcelona, Spain, 2018.
16. Arrow, K.; Solow, R.; Portney, P.R.; Leamer, E.E.; Radner, R.; Schuman, H. Report of the NOAA Panel on Contingent Valuation. *Fed. Regist.* **1993**, *58*, 4601–4614.

17. Mitchell, R.C.; Carson, R.T. *Evaluating the Validity of Contingent Valuation Studies*; Venture Publishing: State College, PA, USA, 1988.
18. Twerefou, D.K. Willingness to Pay for Improved Electricity Supply in Ghana. *Mod. Econ.* **2014**, *5*, 489–498. [CrossRef]
19. Amoah, A. *Estimating Demand for Utilities in Ghana: An Empirical Analysis*; University of East Anglia: Norwich, UK, 2016.
20. Buchanan, K.; Russo, R.; Anderson, B. The question of energy reduction: The problem(s) with feedback. *Energy Policy* **2015**, *77*, 89–96. [CrossRef]
21. Burgess, J.; Nye, M. Re-materialising energy use through transparent monitoring systems. *Energy Policy* **2008**, *36*, 4454–4459. [CrossRef]
22. Hargreaves, T.; Nye, M.; Burgess, J. Making energy visible: A qualitative field study of how householders interact with feedback from smart energy monitors. *Energy Policy* **2010**, *38*, 6111–6119. [CrossRef]
23. Taale, F.; Kyeremeh, C. Households' willingness to pay for reliable electricity services in Ghana. *Renew. Sustain. Energy Rev.* **2016**, *62*, 280–288. [CrossRef]
24. Abdullah, S.; Jeanty, P.W. Willingness to pay for renewable energy: Evidence from a contingent valuation survey in Kenya. *Renew. Sustain. Energy Rev.* **2011**, *15*, 2974–2983. [CrossRef]
25. Abdullah, S.; Mariel, P. Choice experiment study on the willingness to pay to improve electricity services. *Energy Policy* **2010**, *38*, 4570–4581. [CrossRef]
26. Alam, M.; Bhattacharyya, S. Are the off-grid customers ready to pay for electricity from the decentralized renewable hybrid mini-grids? A study of willingness to pay in rural Bangladesh. *Energy* **2017**, *139*, 433–446. [CrossRef]
27. Ayodele, T.; Ogunjuyigbe, A.; Ajayi, O.; Yusuff, A.; Mosetlhe, T. Willingness to pay for green electricity derived from renewable energy sources in Nigeria. *Renew. Sustain. Energy Rev.* **2021**, *148*, 111279. [CrossRef]
28. Deutschmann, J.W.; Postepska, A.; Sarr, L. Measuring willingness to pay for reliable electricity: Evidence from Senegal. *World Dev.* **2021**, *138*, 105209. [CrossRef]
29. Dogan, E.; Muhammad, I. Willingness to pay for renewable electricity: A contingent valuation study in Turkey. *Electr. J.* **2019**, *32*, 106677. [CrossRef]
30. Du Preez, M.; Menzies, G.; Sale, M.; Hosking, S. Measuring the indirect costs associated with the establishment of a wind farm: An application of the contingent valuation method. *J. Energy S. Afr.* **2012**, *23*, 2–7. [CrossRef]
31. Entele, B.R. Analysis of households' willingness to pay for a renewable source of electricity service connection: Evidence from a double-bounded dichotomous choice survey in rural Ethiopia. *Heliyon* **2020**, *6*, e03332. [CrossRef]
32. Graber, S.; Narayanan, T.; Alfaro, J.; Palit, D. Solar microgrids in rural India: Consumers' willingness to pay for attributes of electricity. *Energy Sustain. Dev.* **2018**, *42*, 32–43. [CrossRef]
33. Gunatilake, H.; Maddipati, N.; Patil, S. Willingness to Pay for Electricity Supply Improvements in Rural India. *J. Resour. Energy Dev.* **2013**, *10*, 55–78. [CrossRef]
34. Harajli, H.; Chalak, A. Willingness to Pay for Energy Efficient Appliances: The Case of Lebanese Consumers. *Sustainability* **2019**, *11*, 5572. [CrossRef]
35. Kim, J.-H.; Kim, S.-Y.; Yoo, S.-H. Public Acceptance of the "Renewable Energy 3020 Plan": Evidence from a Contingent Valuation Study in South Korea. *Sustainability* **2020**, *12*, 3151. [CrossRef]
36. Kim, J.-H.; Lim, K.-K.; Yoo, S.-H. Evaluating Residential Consumers' Willingness to Pay to Avoid Power Outages in South Korea. *Sustainability* **2019**, *11*, 1258. [CrossRef]
37. Oseni, M.O. Self-Generation and Households' Willingness to Pay for Reliable Electricity Service in Nigeria. *Energy J.* **2017**, *38*, 165–194. [CrossRef]
38. Scarpa, R.; Willis, K. Willingness-to-pay for renewable energy: Primary and discretionary choice of British households' for micro-generation technologies. *Energy Econ.* **2010**, *32*, 129–136. [CrossRef]
39. Zhang, L.; Wu, Y. Market segmentation and willingness to pay for green electricity among urban residents in China: The case of Jiangsu Province. *Energy Policy* **2012**, *51*, 514–523. [CrossRef]
40. Cullen, R.; Hanley, N.; Spash, C.L. *Cost-Benefit Analysis and the Environment*; Edward Elgar Publishing Limited: Hants, UK, 1993.
41. Spash, C.L. *The Contingent Valuation Method: Retrospect and Prospect*; CSIRO Sustainable Ecosystems: Canberra, Australia, 2008.
42. Boyle, K.J.; Bishop, R.C. Welfare Measurements Using Contingent Valuation: A Comparison of Techniques. *Am. J. Agric. Econ.* **1988**, *70*, 20–28. [CrossRef]
43. Whitehead, J.C.; Huang, J.-C.; Blomquist, G.C.; Ready, R.C. Construct Validity of Dichotomous and Polychotomous Choice Contingent Valuation Questions. *Environ. Resour. Econ.* **1998**, *11*, 107–116. [CrossRef]
44. Rietbergen-McCracken, J.; Abaza, H. *Environmental Valuation: A Worldwide Compendium of Case Studies*; Earthscan: London, UK, 2000.
45. Hanemann, M.; Loomis, J.; Kanninen, B. Statistical Efficiency of Double-Bounded Dichotomous Choice Contingent Valuation. *Am. J. Agric. Econ.* **1991**, *73*, 1255–1263. [CrossRef]
46. Scarpa, R.; Bateman, I. Efficiency Gains Afforded by Improved Bid Design versus Follow-up Valuation Questions in Discrete-Choice CV Studies. *Land Econ.* **2000**, *76*, 299. [CrossRef]
47. Carson, R.T.; Groves, T. Incentive and informational properties of preference questions. *Environ. Resour. Econ.* **2007**, *37*, 181–210. [CrossRef]
48. Carlsson, F.; Martinsson, P. Does it matter when a power outage occurs?—A choice experiment study on the willingness to pay to avoid power outages. *Energy Econ.* **2008**, *30*, 1232–1245. [CrossRef]

49. Diamond, P.A.; Hausman, J.A. On contingent valuation measurement of nonuse values. In *Health Econometrics*; Emerald Publishing Limited: Bentley, UK, 1993; Volume 220, pp. 3–38.
50. Diamond, P.A.; Hausman, J.A. Contingent Valuation: Is Some Number Better than No Number? *J. Econ. Perspect.* **1994**, *8*, 45–64. [CrossRef]
51. Milgrom, P. Is sympathy an economic value? Philosophy, economics, and the contingent valuation method. In *Contributions to Economic Analysis*; Elsevier BV: Amsterdam, The Netherlands, 1993; pp. 417–441.
52. Mcfadden, D.; Leonard, G.K. Issues in the contingent valuation of environmental goods: Methodologies for data collection and analysis. In *Contributions to Economic Analysis*; Elsevier BV: Amsterdam, The Netherlands, 1993; pp. 165–215.
53. Hanemann, W.M. Willingness To Pay and Willingness To Accept: How Much Can They Differ? Reply. *Am. Econ. Rev.* **2003**, *93*, 464. [CrossRef]
54. Hausman, J. Contingent Valuation: From Dubious to Hopeless. *J. Econ. Perspect.* **2012**, *26*, 43–56. [CrossRef]
55. Carson, R. Contingent Valuation: A Practical Alternative when Prices Aren't Available. *J. Econ. Perspect.* **2012**, *26*, 27–42. [CrossRef]
56. Haab, T.C.; Interis, M.G.; Petrolia, D.R.; Whitehead, J.C. From Hopeless to Curious? Thoughts on Hausman's "Dubious to Hopeless" Critique of Contingent Valuation. *Appl. Econ. Perspect. Policy* **2013**, *35*, 593–612. [CrossRef]
57. Kling, C.L.; List, J.A.; Zhao, J. A dynamic explanation of the willingness to pay and willingness to accept disparity. *Econ. Inq.* **2011**, *51*, 909–921. [CrossRef]
58. Rowe, R.D.; D'Arge, R.C.; Brookshire, D.S. An experiment on the economic value of visibility. *J. Environ. Econ. Manag.* **1980**, *7*, 1–19. [CrossRef]
59. Morey, E.R.; Shaw, W.; Rowe, R.D. A discrete-choice model of recreational participation, site choice, and activity valuation when complete trip data are not available. *J. Environ. Econ. Manag.* **1991**, *20*, 181–201. [CrossRef]
60. Lopez-Feldman, A. Introduction to Contingent Valuation Using Stata. 2012. Available online: https://mpra.ub.uni-muenchen.de/41018/2/intro_CV.pdf (accessed on 29 August 2021).
61. Bateman, I.J.; Carson, R.T.; Day, B.; Hanemann, M.; Hanley, N.; Hett, T.; Jones-Lee, M.; Loomes, G.; Mourato, S.; Özdemiroglu, E.; et al. Economic Valuation with Stated Preference Techniques: A Manual. *Ecol. Econ.* **2004**, *50*, 155–156. [CrossRef]
62. Ghana Statistical Service. Ghana Living Standards Survey Report of the 6th Round. 2014. Available online: https://www2.statsghana.gov.gh/publications.html (accessed on 18 August 2021).
63. Kateregga, E. The Welfare Costs of Electricity Outages: A Contingent Valuation Analysis of Households in the Suburbs of Kampala, Jinja and Entebbe. *J. Dev. Agric. Econ.* **2009**, *1*, 1–11.
64. Ito, N.; Takeuchi, K.; Tsuge, T.; Kishimoto, A. Applying threshold models to donations to a green electricity fund. *Energy Policy* **2010**, *38*, 1819–1825. [CrossRef]
65. Otegbulu, A.C. A contingent valuation model for assessing electricity demand. *J. Financ. Manag. Prop. Constr.* **2011**, *16*, 126–146. [CrossRef]
66. Lay, J.; Ondraczek, J.; Stoever, J. Renewables in the energy transition: Evidence on solar home systems and lighting fuel choice in Kenya. *Energy Econ.* **2013**, *40*, 350–359. [CrossRef]
67. Zarnikau, J. Consumer demand for 'green power' and energy efficiency. *Energy Policy* **2003**, *31*, 1661–1672. [CrossRef]
68. Whittington, D.; Lauria, D.T.; Mu, X. A study of water vending and willingness to pay for water in Onitsha, Nigeria. *World Dev.* **1991**, *19*, 179–198. [CrossRef]
69. Energy Sector Management Assistance Program (ESMAP). *Mini Grids for Timely and Low-Cost Electrification in Ghana: Exploring Regulatory and Business Models for Electrifying the Lake Volta Region*; World Bank: Washington, DC, USA, 2017.

sustainability

Article

Importance and Performance of Value-Based Maintenance Practices in Hospital Buildings

Wai Fang Wong [1,*], AbdulLateef Olanrewaju [2] and Poh Im Lim [3]

1   Faculty of Built Environment, Tunku Abdul Rahman University College, Setapak, Kuala Lumpur 53300, Malaysia
2   Department of Construction Management, Universiti Tunku Abdul Rahman, Jalan Universiti, Bandar Barat, Kampar 31900, Malaysia; olanrewaju@utar.edu.my
3   Department of Architecture & Sustainable Design, Lee Kong Chian Faculty of Engineering & Science, Sungai Long Campus, Universiti Tunku Abdul Rahman, Kajang 43000, Malaysia; limpi@utar.edu.my
*   Correspondence: wongwf@tarc.edu.my

Abstract: After two decades of privatization of building maintenance service in government hospitals in Malaysia, evidence of under-maintained hospital buildings suggests a need to raise the level of hospital maintenance service delivery. This study identified the critical success factors to enhance the value outcomes of hospital maintenance service. A total of 66 questionnaire survey responses from maintenance personnel in public hospitals were analyzed using the Importance-Performance Matrix Analysis (IPMA) in the SmartPLS3.0 software. The Importance versus the Performance of value-based practices was mapped to identify the critical areas that require greater considerations to improve maintenance service delivery. The findings revealed four critical success factors: Responsive to Needs, Integrated Service Solutions, Innovative Improved Practices, and Value for Money. These practices were found to be the impetus that can bring significant enhancement to hospital building maintenance service delivery. Although the findings are based on data derived from public hospitals in Malaysia, the outcomes are applicable to private hospitals both in and outside of Malaysia.

Keywords: hospital building maintenance; critical success factor; value-based practices; importance-performance matrix analysis

Citation: Wong, W.F.; Olanrewaju, A.; Lim, P.I. Importance and Performance of Value-Based Maintenance Practices in Hospital Buildings. *Sustainability* **2021**, *13*, 11908. https://doi.org/10.3390/su132111908

Academic Editor: Alberto-Jesus Perea-Moreno

Received: 10 September 2021
Accepted: 25 October 2021
Published: 28 October 2021

Publisher's Note: MDPI stays neutral with regard to jurisdictional claims in published maps and institutional affiliations.

1. Introduction

Hospitals perform essential functions concerning human lives, health, and well-being. Maintenance of hospital buildings is challenging due to their engineering plants and systems [1,2]. The healthcare industry is subjected to high standards and regulation compliance and has a high operational risk of failure [3]. Besides safety, comfort, and security considerations [4], hospitals are also expected to attain energy efficiency requirements [5]. Failure in healthcare facilities can cause catastrophic impacts [3], and poor service quality affects patient satisfaction [6,7]. It is crucial for hospital buildings to always perform at an optimal state so that the delivery of medical functions is not compromised. Thus, maintenance works are necessary to support the functionality and continuity of hospitals' healthcare service [8].

There are 144 public hospitals in Malaysia with 42,424 beds and 210 private hospitals with 15,957 beds [9]. The government privatized the maintenance service in 1997 under Facilities Management (FM) to standardize and improve service delivery nationwide [10]. Concession agreements were signed with concession companies to provide various supports including the maintenance of M&E engineering, civil engineering, and architecture works. Despite the privatization of maintenance service, there is still room for improvement. Problems such as fungal attack, building defects, lift breakdown, and moisture problems were frequently reported. Poor maintenance of fire safety was pointed out as a frequent incident [11]. In 2016, a fatal fire incident in a public hospital killed six patients and resulted in the evacuation of 487 patients and staff [12]. The same hospital had seven fire incidents before that fatal incident [13]. Other issues include a woman giving birth in a lift [14], a pre-mature baby trapped in a lift [15], and a collapsed ceiling [16]. Empirical evidence on building condition assessment found that facilities for persons with disabilities in public hospitals were critical [17]. These issues of poor performance in hospitals could be prevented if maintenance service were delivered effectively.

Extant research revealed weaknesses in maintenance management [10,18,19] due to multifaceted causes such as low service level by contractors [19], lack of understanding of user's needs [20,21], and less focus on collaborative working. The maintenance process involves constant interactions of various systems and different parties, i.e., the maintenance contractors, building users, and maintenance personnel and hospital management. Hence, issues of maintenance should be investigated from the practices of the parties involved and the effect of collaboration among them. Value concepts emphasizing user involvement [22,23], value-adding practices of contractors [24–26], and value co-creation through collaborative work [25,27–31] can potentially mitigate the problems in maintenance management.

However, there is a lack of theoretical and empirical justification for implementing a value-based approach in healthcare building maintenance [20,21]. Past research in hospital building maintenance was fragmented and diversified in maintenance efficiency [32], benchmarking [33], cost [34,35], audit assessment [19], defects [36], strategy [37], and effectiveness [2]. Recent research trends focus on energy-saving [38], green hospitals [39], service quality in hospital FM [6,40], lean six sigma [41], performance [42], and fire safety [12]. Research in value-based maintenance is limited to the case study by Okoroh et al. [24] and recent studies by Olanrewaju et al. [21] and Wong et al. [43].

In our previous work on the value-based building maintenance model for hospitals [43], the causal relationships between value factors and value outcomes of building maintenance were established, where value-adding and value co-creation were found to have a positive influence on achieving value outcomes. The previous study provided information about the main factors and their respective sets of practices in general. However, it has limitations where the specific sub-factors or indicators that are critical were not known. Information about the main areas that directly or indirectly impact the performance of maintenance service delivery are crucial for systemic decision-making. Hence, this study continues our previous work to address its limitation [43] by mapping the Importance versus the Performance of value-based indicators of the model with the aim of establishing the critical success factors (CSFs) of value-based building maintenance. CSFs are the set of criteria that facilitate the achievement of the objective of the project or services.

2. Literature Review

Value is referred to as "value-in-use" in the service-dominant (S-D) logic model developed by Vargo and Lusch [28] and Vargo et al. [29]. Under the S-D logic model, value is determined in the usage, rather than "value-in-exchange" in the traditional exchange of goods and money from goods-dominant (G-D) logic [29]. The "value-in-use" is defined as "a customer's outcome, purpose or objective that is achieved through service" [31]. The term "service" is defined as the application of competences by one party for the benefit of another [29]. In this study, the outsourced maintenance is a form of service, where maintenance contractors offer their competencies in the form of technology, knowledge, resources, and innovation to benefit the hospitals. In such an arrangement, value is determined in use; hence, the S-D logic model is relevant in this study.

Value is perceived and determined by the customer, instead of embedded in goods and determined by the producer [28]. Similarly, Gummerus [44] suggested that the beneficiary determines value. Hence, in this study, value outcomes of hospital maintenance are dictated by the hospitals and their users.

2.1. Value Outcomes

Based on the value-in-use concept, value outcomes for public hospital maintenance can be two-fold, with an emphasis on daily operational outcomes and strategic ones. The operational value outcomes are the daily work processes [30] such as shorter response time by the contractor [27], reduced risk and quality of output [27], and basic maintenance requirements such as health and safety [26]. In the longer term, strategic value outcomes result from parties' collaborative effort in skill, knowledge, and technology advancement and transfer [27]. Through synergy, contractors are treated as partners and therefore entrusted with more diverse roles [30]. In the long run, corporate image [24] and user satisfaction [24,27,45,46] can be achieved.

2.2. Value-Based Factors

2.2.1. User Involvement

User involvement explores the demand side of the maintenance arrangement in terms of user expectations, user involvement, and user satisfaction. Users are classified as hospital staff who perform day-to-day operations in the hospital to deliver healthcare services. They are the medical or clinical staff, and administrative and supporting staff. A recent survey conducted on doctors and nurses in public hospitals in Malaysia revealed their dissatisfaction over unresolved complaints on maintenance issues [47]. The findings indicated a low attainment of users' needs even though their roles have shifted from passive recipients to knowledgeable and active participants.

Gathering and knowing the expectations of main stakeholders will facilitate better-informed decision-making and evaluation, as suggested by Jensen and Maslesa [22] in value-based building renovation. By understanding users' expectations, the gap between users and maintenance providers can be minimized or closed [48]. Organizations should co-opt customer competency to increase their competitiveness [49]. Hence, this study postulated the need for inclusion of users from three aspects. Firstly, organizations must understand user expectations [22,23,50,51]. Secondly, it is beneficial to involve users in the maintenance process [22,23,50]. Thirdly, user satisfaction needs to be measured to gauge the achievement of their expectations [23,40,52].

2.2.2. Value-Adding Practices

The concept of added value explores the supply side of the maintenance arrangement. Maintenance service providers are expected to deliver more than essential transactional maintenance functions. It was propounded that public sector real estate has moved beyond satisfying customers based on the traditional time–cost–quality trilogy [53]. There is an urgency to provide a relationship or partnership with their customers in the value-adding service offering [25,26,54]. On top of that, FM research conducted in Nordic countries [46,50,55] postulated that the focus of value has shifted from a customer focus towards a value-adding notion.

The contracted service providers of hospital support are bound by conditions stipulated in the concession agreement. Contractors may appoint their sub-contractors to carry out these tasks. However, there are areas of concern in terms of the contractor's efficiency in the outsourced FM [19], heavy reliance on the sub-contractors [10], and a lack of top management support for the contractor [10]. Issues of rising operation costs [34] also indicate the lack of value for money. To mitigate problems faced by maintenance contractors, the concept of added value was explored. Innovative practices, value for money, and cost reduction/cost savings from contractors [24] can potentially add value to the service provider's provision. Ali-Marttila et al. [26] also identified the service partner's ability to solve problems and provide a service solution as essential factors in value creation. Besides, responsiveness to needs is also a necessary value that customers look for in their partners [23,25].

2.2.3. Value Co-Creation

The S-D model emphasized the importance of value co-creation, where service systems such as people, information, and technology engage with other service systems for adaptability and survival [29]. Value is co-created jointly or reciprocally among the provider/producer and the consumer or beneficiaries, through physical and tangible resources [29]. Value co-creation is also defined as "the joint, collaborative, concurrent, peer-like process of producing new value, both materially and symbolically" [56]. The business focus should be shifted to co-create unique value with customers rather than the traditional company-centric value creation [57]. Grönroos and Voima [58] suggested a "joint sphere" where value is co-created when service providers and customers interact jointly, directly or indirectly. Malaysian public hospitals currently rely on a fee deduction system to ensure conformance and performance [10], which drains the manpower in contractor supervision and monitoring [10] when hospitals are already reported to be under-staff [59]. Improvement in the collaborative working environment rather than a punitive system could potentially help to enhance service delivery. Previous studies shown that collaborative working improved the construction supply chain [60] and infrastructure asset maintenance [61]. Past literature acknowledged that value co-creation could be achieved through intensive cooperation [51], sharing of information [29], knowledge transfer [29,30] and effective communication [51]. Additionally, openness and honesty; mutual trust and confidence [25] and relationship synergies [26], strategic alignment [30], strategic integration [27], and strong governance [25] were also explored as value co-creation factors. Hence, it is postulated that value co-creation practices can potentially improve the delivery of maintenance service to hospitals, therefore enhancing the value outcomes.

In this study, three causal relationships between the value-based factors and value outcomes are hypothesized as follows:

Hypothesis 1 (H1). *User involvement positively influences value outcomes.*

Hypothesis 2 (H2). *Value-adding practices positively influence value outcomes.*

Hypothesis 3 (H3). *Value co-creation positively influences value outcomes.*

Value-based factors and their indicators are presented in Table 1.

Table 1. Value-based factors and indicators.

	Value-Based Factors	Indicators
1	User Involvement	User Expectation
		User Involvement
		User Satisfaction
2	Value Added	Integrated Service Solutions
		Innovative Improved Practices
		Value for Money
		Cost Reduction/Saving
		Responsive to Needs
3	Value Co-Creation	Sharing of Information
		Operational Integration
		Intensive Cooperation
		Knowledge Transfer
		Effective Communication
		Transparency of Internal Information
		Openness and Honesty
		Shared Risks
		Mutual Trust and Confidence
		Relationship Synergies
		Strategic Integration
		Strategic Alignment
		Strong Governance
		Sharing of Information
		Operational Integration

3. Methods

A questionnaire survey was administered from January to July 2019 using the online platform. The target respondents were the engineers of public hospitals in Malaysia who are appointed by the ministry to monitor, supervise, and inspect the privatized support service [18]. They were the most suitable respondents since they represent the maintenance department in tasks involving hospital users and maintenance contractors, and they report to the hospitals' top management. Only five concession companies were contracted to provide maintenance service to public hospitals in the entire country [9]. Hence, this study attempted to collect data from all 139 public hospitals (excluding medical institutions/centers) listed in the Ministry of Health Malaysia website. The census method was selected due to the well-defined, accessible, and small population [62].

The first section of the questionnaire gathers the background information of respondents and the hospitals, whereas the second section measures the value-based factors (User Involvement, Value-Adding Practices, Value Co-Creation) and value outcomes. The constructs' measurement items were developed from the literature review based on the synthesis and integration of past studies of value concepts. The constructs and number of items are shown in Table 2. This study employed an even-numbered 6-point Likert scale to measure respondents' experience of practices and outcomes. The 6-point Likert items were chosen to give a greater discriminant and reliability value [63].

Table 2. Measurement items.

Constructs	Items	Scale
Value Outcomes	11 (reflective)	6-points Likert scale: 1 = Strongly Disagree
User Involvement	3 (reflective)	2 = Disagree
Value Add	5 (reflective)	3 = Slightly Disagree 4 = Slightly Agree
Value Co-Creation	15 (reflective)	5 = Agree 6 = Strongly Agree

Two parts of the analyses were performed in this study. In Part 1, PLS-SEM Smart-PLS 3.0 software [64] was used to perform the structural equation modeling (SEM) to test the three proposed hypotheses; the detailed analysis and outcomes were presented in Wong et al. [43]. PLS-SEM was chosen due to the nature of the study being more exploratory than confirmatory [65,66] and it is recommended for non-parametric data [65].

This paper focuses on Part 2 of the analysis, which applied the same software to perform the Importance-Performance Matrix Analysis (IPMA) at the construct level and indicator level. The IPMA function was used to analyze the importance versus performance of constructs and indicators resulting from Part 1 of the SEM analysis. The IPMA analysis was conducted based on steps outlined by Ringle and Sarstedt [67] and Ramayah et al. [65]. The first step involved a requirement check where latent variable scores were re-scaled from 0 to 100, while all indicator codes must be in the same scale direction. The second step was to compute the performance values, where the average value of the latent variable score represented the average Performance. Next, the Importance value was computed. This represented the total effect of the relationship between two constructs. Lastly, the Importance–Performance map was created. IPMA was conducted at the construct and indicator levels to identify specific areas of improvement required.

To interpret the Importance–Performance map, the revised IPA grid by Abalo et al. (2006), as cited in Abalo et al. [68], is referred to. Any point on the map with an importance rating above its corresponding Performance rating was identified as an area that requires improvement effort ("Concentrate Here" category). Subsequently, the points that were high in both the Importance and Performance ratings were the areas to be maintained ("Keep Up the Good Work"). Using the same principle, points that were low in both the Importance and Performance ratings were considered "Low Priority". In contrast, those with low Importance but high in the Performance rating were categorized as "Possible Overkill". In this study, factors that fell in the "Concentrate Here" and "Keep Up the Good Work" categories were established as the CSFs. The IPMA technique has been adopted in numerous studies in various disciplines to obtain detailed insights into factors investigated (for example, Ong and Bahar [69], Su and Cheng [70], Ting et al. [71], Valaei et al. [72], and Tailab [73]).

4. Results

A total of 66 usable responses were collected out of 139 public hospitals. The distribution of respondents comprised 41% engineers and 59% assistant engineers; the average number of years of experience was five years. The profile of the respondents is shown in Table 3.

Table 3. Respondent profiles.

Description	Frequency (66 Samples)	%
Years of Experience		
Mean	5.19	
Standard deviation	2.593	
Range	1–14	
Position		
Engineer	27	40.9
Assistant engineer	39	59.1
Total	66	100
Education (Level)		
Diploma	32	48.5
Bachelor's degree	31	47.0
Master's degree	3	4.5
Others	0	0.0
Total	66	100

The sample size was sufficient for PLS-SEM analysis, where only a minimum of 30 cases of observations was required [65]. Based on the rule of thumb of 10 times of structural paths directed to a construct [74], the minimum sample size required was 30. Besides, it also fulfilled the requirement of the minimum R-squared method, whereby for a model with maximum three arrows pointing to a construct and an R^2 of 0.423, with a 5% probability of error, the minimum sample size required should be within the range of 16 to 37 [66].

4.1. Part 1: Structural Equation Modeling (SEM)

This section reports the essence of the SEM analysis as previously presented in Wong et al. [43]. The reflective measurement model was assessed in terms of internal consistency reliability, convergent validity, and discriminant validity. Composite reliability (CR) for all constructs in the range of 0.756 to 0.927 met the criteria of above 0.7 [66] and below the maximum CR value of 0.95 to avoid indicator redundancy [75]. Overall, six indicators were deleted due to low factor loading and within the caveat of the 20% limit of overall number indicators [76]. The average variance extracted (AVE) of all constructs was above the minimum acceptable level of 0.5. Both the Fornell–Larcker criterion and the cross-loading pattern confirmed sufficient discriminant value validity [65].

The structural model validation was conducted based on the five essential steps to assess lateral collinearity, the path coefficient, the coefficient of determination, the effect size to R^2, and the Stone–Geisser Q^2 predictive relevance [65]. All three exogenous latent variables had a variance inflator factor (VIF) value of below 3.3 [77], cited in Ramayah et al. [65], which indicates no collinearity problem. The three hypotheses H1 (User $\rightarrow$ Value Outcomes), H2 (Val Add $\rightarrow$ Value Outcomes), and H3 (Co-Creation $\rightarrow$ Value Outcomes) were tested. The results show that Value Added had a significant influence on the value outcomes ($p = 0.000$, $t = 3.476$) with a medium effect size, whereas Value Co-Creation significantly influenced the value outcomes ($p = 0.014$, $t = 2.214$) with small effect size. However, Hypothesis 1 is not supported, as the User Involvement construct was not found to have a significant effect on the value outcomes. The result of the SEM is presented in Figure 1.

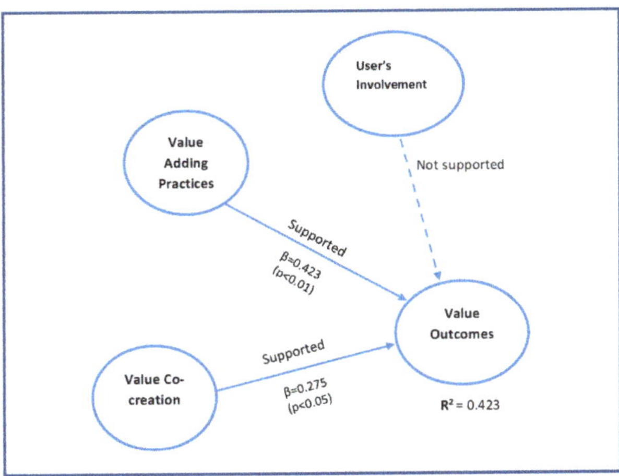

Figure 1. Result of structural equation modeling (SEM).

4.2. Part 2: Importance-Performance Matrix Analysis (IPMA)

Based on the outcome of the SEM, further analysis using IPMA function in SmartPLS3.0 was carried out. Table 4 shows that Value Added had the highest Importance but lowest Performance rating. This indicates that even though Value Added is the most important construct, it was given the least attention to achieve the desired performance. Subsequently, the Value Co-Creation construct ranked second in both Importance and Performance ratings, whereas the User Involvement construct was lowest in Importance, but highest in Performance, which means it was the least critical area.

Table 4. Importance and performance (constructs).

Construct	Importance	Performance
Co-Creation	0.318	78.696
User Involvement	0.071	82.309
Value Added	0.417 (highest)	74.413 (lowest)

Further analysis was extended on the total 18 indicators. From Table 5, the indicator VAL5 (Responsive to Needs) had the highest Importance rating (0.129) but was relatively lower in terms of its Performance (76.061), which ranked 14th out of 18. Figure 2 depicts VAL5 falling within the "Concentrate Here" category. Subsequently, VAL1 (Integrated Service Solutions), VAL2 (Innovative Improved Practices), and VAL3 (Value for Money), which ranked second, third, and fourth in terms of Importance, all fell in the "Keep Up the Good Work" category. Other indicators fell below the 50% continuum of the Importance axis but above the 50% continuum of the Performance axis, indicating their Performance was higher than their relative Importance, or in the category "Possible Overkill". Hence, these indicators are not the main focus for hospital maintenance improvement compared to other areas. The four indicators that fell in the "Concentrate Here" and "Keep up the Good Work" categories were established as the CSFs of the value-based building maintenance in this study (see Table 6).

Table 5. Importance and Performance of indicators.

Code	Indicators	Indicator Importance	Ranking of Importance	Indicator Performance	Ranking of Performance
VAL5	Responsive to needs	0.129	1	76.061	14
VAL1	Integrated service solutions	0.081	2	80.303	6
VAL2	Innovative improved practices	0.078	3	73.636	16
VAL3	Value for money	0.069	4	67.879	18
VAL4	Cost reduction/saving	0.061	5	71.515	17
STR1	Strategic integration	0.043	6	76.364	13
WWW3	Relationship synergies	0.040	7	81.212	5
STR3	Strong governance	0.035	8	78.182	9
USE3	Measure user satisfaction	0.034	9	84.848	1
COM3	Openness and honesty	0.033	10	77.273	12
STR2	Strategic alignment	0.032	11	77.879	10
WWW2	Mutual trust and confidence	0.031	12	74.848	15
OPE2	Intensive cooperation	0.030	13	80.303	6
JOR3	Sharing of information	0.028	14	78.788	8
COM1	Effective communication	0.027	15	82.424	3
USE2	User involvement	0.023	16	77.879	10
OPE3	Knowledge transfer	0.021	17	81.515	4
USE1	User expectation	0.014	18	83.333	2

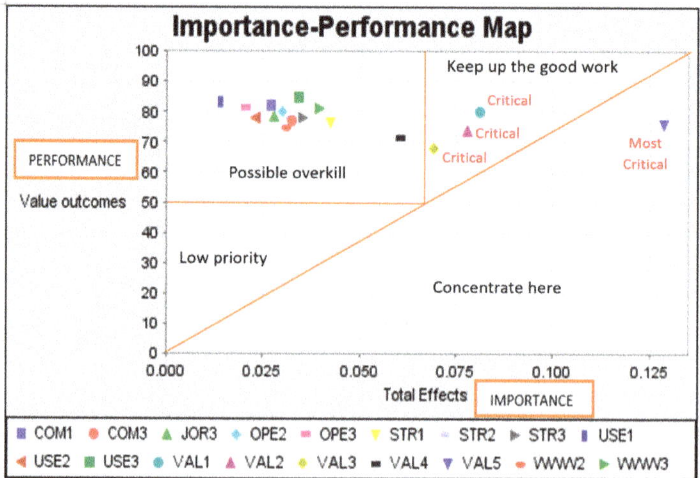

Figure 2. Importance versus Performance of value-based maintenance practices.

Table 6. Critical success factors.

Category	Indicators	Decision
Concentrate Here	VAL5	Critical success factor
Keep Up the Good Work	VAL1, VAL2, VAL3	Critical success factor
Possible Overkill	COM1, COM3, JOR3, OPE2, OPE3, STR1, STR2, STR3, WWW2, WWW3, VAL4, USE1, USE2, USE3	-
Low Priority	-	-

5. Discussions

From the SEM outcomes, Value-Adding Practices and Value Co-Creation were found to positively influence the value outcomes in hospital maintenance. User Involvement was not supported to have influence on value outcomes, which merits further investigation. Further analysis on 18 indicators using IPMA found Responsive to Needs, Integrated Service Solutions, Innovative Improved Practices, and Value for Money were critical, and hence were established as the CSFs for value-based hospital maintenance. Even though there are no direct comparable CSFs on value-based maintenance in past research, comparison with the closest work by Ab Ghani [78] on CSFs for FM in the Malaysian healthcare sector found two related CSFs, which are "value for money" and "integrated process". However, their research focused on generic FM in the healthcare sector, and not specific to a value-based approach. In comparison, Amaratunga et al.'s [79] case study on CSFs for FM in NHS facilities in the UK found three related CSFs, which are "Timeliness", "Service Delivery Innovation", and "Value for Money". From their study, corresponding measures for Timeliness are patient environment assessment, Service Delivery Innovation measured by the effectiveness of service planning, and Value for Money' measured using estate returns measures, budget variance, absenteeism, and benchmarking tools. Their research is not value-based; however, the examples of measurement tools can be adapted to develop KPI for CSFs in this study.

In a review on CSFs for healthcare FM by Ahmad Pakrudin et al. [3], top management commitment and support was found to be the top-ranked cited factor. In contrast, our results show that the indicator "Strong Governance" falls under the "Possible Overkill" category, which indicates the Performance rating exceeded the Importance rating, hence it is not as critical in the Malaysian public hospital context.

The identified critical success factors of the value-based building maintenance model are presented in Figure 3.

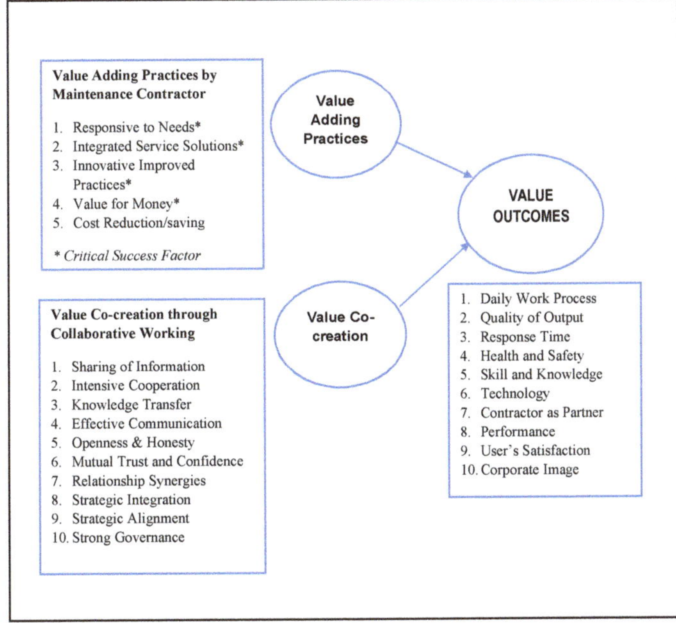

Figure 3. Critical success factors of the value-based building maintenance model.

6. Conclusions

Our previously established value-based building maintenance model [43] identified value-adding practices and value co-creation as main factors having positive impacts on value outcomes in an integrated model. The previous study recommended for maintenance contractors to provide value-adding practices in their delivery and justified the needs for collaborative working in hospital maintenance arrangements. However, the general sets of practices are not sufficient to guide practitioners and policymakers to concentrate explicitly on the key aspects that could give the most impact to the desired outcomes. This study extended analysis on all the indicators from the previous model [43] to evaluate their level of importance and their respective performance to obtain more valuable and specific information. From the evaluation, the importance and performance were weighed so that the important areas that were lacking in terms of performance could be identified. The outcome of this study complemented the previous model with four identified critical success factors. The new findings add value to the model by providing deeper insights for practitioners to concentrate on the vital areas, i.e., Responsive to Needs, Integrated Service Solutions, Innovative Improved Practices, and Value for Money, to heighten the value of maintenance service.

Notably, all four critical success factors identified from this study attributed to value-adding practices, which were derived from the supply side of the maintenance arrangement. The essential finding highlights the crucial role of the maintenance contractors. Hospital management and policymakers could prioritize contractors' selection based on their track records in the aforesaid critical areas. These criteria can also be extended to develop key performance indicators (KPIs) to monitor their performance for contract renewals.

Specifically, contractors could keep up their good work by providing integrated service solutions such as providing loans of equipment during breakdowns and providing technical advice. Innovative improved practices such as predictive maintenance, technology, and innovation are critical attributes expected from contractors. Value for money through initiatives such as energy-saving can add value to hospitals in terms of sustainability. As more responsibilities are placed on concession companies by the government in achieving sustainability goals [5], it becomes essential for contractors to pro-actively add value to their services in this respect.

Lastly, this study identified Responsive to Needs as the most critical success factor. Further action and research can focus on contractors' responsiveness to close the gap of Importance and Performance, such as on-site productivity measurement to eliminate non-value-added tasks or further supply chain analysis extended to the subcontractor and suppliers' level.

Author Contributions: Conceptualization, W.F.W., A.O. and P.I.L.; methodology, W.F.W., A.O. and P.I.L.; software, W.F.W.; validation, W.F.W.; formal analysis, W.F.W.; investigation, W.F.W., A.O. and P.I.L.; resources, W.F.W., A.O. and P.I.L.; data curation, W.F.W.; writing—original draft preparation, W.F.W.; writing—review and editing, W.F.W., A.O. and P.I.L.; visualization, W.F.W., A.O. and P.I.L.; supervision, A.O. and P.I.L.; project administration, W.F.W. All authors have read and agreed to the published version of the manuscript.

Funding: This research received no external funding.

Informed Consent Statement: Informed consent was obtained from all subjects involved in the study.

Data Availability Statement: The data presented in this study are available on request from the corresponding author. The data are not publicly available due to privacy reasons.

Acknowledgments: This research was supported by the Centre for Construction Research, Faculty of Built Environment, Tunku Abdul Rahman University College, Malaysia.

Conflicts of Interest: The authors declare no conflict of interest.

References

1. Enshassi, A.A.; El Shorafa, F. Key performance indicators for the maintenance of public hospitals buildings in the Gaza Strip. *Facilities* **2015**, *33*, 206–228. [CrossRef]
2. Omar, M.F.; Ibrahim, F.A.; Wan Omar, W.M.S. Key performance indicators for maintenance management effectiveness of public hospital building. In *MATEC Web of Conferences*; EDP Sciences: Les Ulis, France, 2017; Volume 97, pp. 1–6. [CrossRef]
3. Ahmad Pakrudin, N.A.; Abdullah Mohd Asmoni, M.N.; Mei, J.L.Y.; Jaafar, M.N.; Mohammed, A.H. Critical success factors for facilities management implementation in the healthcare industry. *Int. J. Real Estate Stud.* **2017**, *11*, 69–83.

4. Lavy, S.; Shohet, I.M. Integrated maintenance management of hospital buildings: A case study. *Constr. Manag. Econ.* **2004**, *22*, 25–34. [CrossRef]

5. Abdullah, M.S.I.; Abd Rahman, N.M.; Ahmad Zaidi, T.Z.; Kamaluddin, K.A. Latest development on sustainability programme initiatives in Malaysian healthcare facility management. In Proceedings of the 37th Conference of the ASEAN Federation of Engineering Organisations, Jakarta International Expo, Jakarta, Indonesia, 11–15 September 2019; pp. 1–6.

6. Amankwah, O.; Choong, W.W.; Mohammed, A.H. Modelling the influence of healthcare facilities management service quality on patients satisfaction. *J. Facil. Manag.* **2019**, *17*, 267–283. [CrossRef]

7. Boadi, E.B.; Wang, W.; Bentum-micah, G.; Asare, I.K.J.; Bosompem, L.S. Impact of service quality on customer satisfaction in Ghana hospitals: A PLS- SEM approach. *Can. J. Appl. Sci. Technol.* **2019**, *7*, 503–511.

8. Yousefli, Z.; Nasiri, F.; Moselhi, O. Healthcare facilities maintenance management: A literature review. *J. Facil. Manag.* **2017**, *15*, 352–375. [CrossRef]

9. Ministry of Health Malaysia. *Annual Report Ministry of Health Malaysia 2017*; Ministry of Health Malaysia: Putrajaya, Malaysia, 2018.

10. Fan, H.P. *Privatization of Facility Management in Public Hospitals: A Malaysian Perspective*; Patridge Publishing: Singapore, 2016.

11. Ab Ghani, M.Z.; Aripin, S. Comparative review of design requirements for natural smoke ventilation in hospital buildings. *J. Malays. Inst. Plan.* **2018**, *16*, 334–344. [CrossRef]

12. Muhamad Salleh, N.; Agus Salim, N.A.; Jaafar, M.; Sulieman, M.Z.; Ebekozien, A. Fire safety management of public buildings: A systematic review of hospital buildings in Asia. *Prop. Manag.* **2020**, *38*, 497–511. [CrossRef]

13. Anon. Seven Fires at HSA in past Four Years. *Star Online*. 2016. Available online: https://www.thestar.com.my/news/nation/2016/10/27/seven-fires-at-hsa-in-past-four-years (accessed on 3 December 2017).

14. Anon. Woman Gives Birth while Trapped in Sarawak General Hospital's Lift. *Malay. Mail Online*. 2016. Available online: https://www.malaymail.com/news/malaysia/2016/10/07/woman-gives-birth-while-trapped-in-sarawak-general-hospitals-lift/1222701 (accessed on 3 December 2017).

15. Lai, C. Hospital elevators need better maintenance. *Star Online*. 2012. Available online: https://www.thestar.com.my/news/community/2012/06/07/hospital-elevators-need-better-maintenance (accessed on 3 December 2017).

16. Anon. Authorities Aware of Problems at Ampang Hospital. *Star Online*. 2007. Available online: https://www.thestar.com.my/news/nation/2007/03/14/authorities-aware-of-problems-at-ampang-hospital (accessed on 12 March 2017).

17. Awang, N.A.; Chua, S.J.L.; Ali, A.S. Building condition assessment focusing on persons with disabilities' facilities at hospital buildings. *J. Des. Built Environ.* **2017**, *17*, 73–84. [CrossRef]

18. National Audit Department Malaysia. *Auditor General's Report 2015: Activities of the Federal Ministries Departments and Management of the Government Companies—Series 2*; National Audit Department Malaysia: Putrajaya, Malaysia, 2016.

19. Ali, M.; Wan Mohamad, W.M.N.W. Audit assessment of the facilities maintenance management in a public hospital in Malaysia. *J. Facil. Manag.* **2009**, *7*, 142–158. [CrossRef]

20. Olanrewaju, A.L.A.; Wong, W.F.; Seong, Y.T. Hospital building maintenance management model. *Int. J. Eng. Technol.* **2018**, *7*, 747–753. [CrossRef]

21. Olanrewaju, A.L.A.; Wong, W.F.; Nik Yahya, N.N.H.; Lim, P.I. Proposed research methodology for establishing the critical success factors for maintenance management of hospital buildings. *AIP Conf. Proc.* **2019**, *2157*, 020036. [CrossRef]

22. Jensen, P.A.; Maslesa, E. Value based building renovation—A tool for decision-making and evaluation. *Build. Environ.* **2015**, *92*, 1–9. [CrossRef]

23. Olanrewaju, A.L.A.; Abdul Aziz, A.R. *Building Maintenance Processes and Practices: The Case of a Fast Developing Country*; Springer: Singapore, 2015. [CrossRef]

24. Okoroh, M.I.; Gombera, P.P.; John, E.; Wagstaff, M. Adding value to the healthcare sector—A facilities management partnering arrangement case study. *Facilities* **2001**, *19*, 157–164. [CrossRef]

25. Dibley, A.; Clark, M. *How to Implement Best Practice in Strategic Partnerships: An Outsource Supplier and Client Perspective*; The Henley Centre for Customer Management: London, UK, 2011.

26. Ali-Marttila, M.; Marttonen-Arola, S.; Kärri, T.; Pekkarinen, O.; Saunila, M. Understand what your maintenance service partners value. *J. Qual. Maint. Eng.* **2017**, *23*, 144–164. [CrossRef]

27. Bititci, U.S.; Martinez, V.; Albores, P.; Parung, J. Creating and managing value in collaborative networks. *Int. J. Phys. Distrib. Logist. Manag.* **2004**, *34*, 251–268. [CrossRef]

28. Vargo, S.L.; Lusch, R.F. Evolving to a new dominant logic for marketing. *J. Mark.* **2004**, *68*, 1–17. [CrossRef]

29. Vargo, S.L.; Maglio, P.P.; Akaka, M.A. On value and value co-creation: A service systems and service logic perspective. *Eur. Manag. J.* **2008**, *26*, 145–152. [CrossRef]

30. Joshi, K.P.; Chebbiyyam, M. Determining value co-creation opportunity in B2B services. In Proceedings of the 2011 Annual SRII Global Conference (SRII 2011), San Jose, CA, USA, 29 March–2 April 2011; pp. 674–684. [CrossRef]

31. Macdonald, E.K.; Wilson, H.; Martinez, V.; Toossi, A. Assessing value-in-use: A conceptual framework and exploratory study. *Ind. Mark. Manag.* **2011**, *40*, 671–682. [CrossRef]

32. Lavy, S.; Shohet, I.M. On the effect of service life conditions on the maintenance costs of healthcare facilities. *Constr. Manag. Econ.* **2007**, *25*, 1087–1098. [CrossRef]

33. Li, Y.; Cao, L.; Han, Y.; Wei, J. Development of a conceptual benchmarking framework for healthcare facilities management: Case study of Shanghai municipal hospitals. *J. Constr. Eng. Manag.* **2020**, *146*, 05019016. [CrossRef]

34. Mustapa, F.D.; Mustapa, M.; Ismail, F.; Ali, K.N. Outsourcing in Malaysian healthcare support services: A study on the causes of increased operational costs. In Proceedings of the International Conference in Construction Industry, Universitas Bung Hatta, Padang, Indonesia, 21–24 June 2006; pp. 1–10.

35. Vanzanella, C.; Fico, G.; Arredondo, M.T.; Delfino, R.; Viggiani, V.; Triassi, M.; Pecchia, L. Interactive management control via analytic hierarchy process: An empirical study in a public university hospital. *J. Int. Bus. Entrep. Dev.* **2015**, *8*, 144–159. [CrossRef]

36. Othman, N.L.; Jaafar, M.; Wan Harun, W.M.; Ibrahim, F. A case study on moisture problems and building defects. *Procedia Soc. Behav. Sci.* **2015**, *170*, 27–36. [CrossRef]

37. Abd Rani, N.A.; Baharum, M.R.; Nizam Akbar, A.R.; Nawawi, A.H. Perception of maintenance management strategy on healthcare facilities. *Procedia Soc. Behav. Sci.* **2015**, *170*, 272–281. [CrossRef]

38. Kamaluddin, K.A.; Abdullah, M.S.I.; Yang, S.S. Development of energy benchmarking of Malaysian government hospitals and analysis of energy savings opportunities. *J. Build. Perform.* **2016**, *7*, 72–87.

39. Sahamir, S.R.; Zakaria, R. Green assessment criteria for public hospital building development in Malaysia. *Procedia Environ. Sci.* **2014**, *20*, 106–115. [CrossRef]

40. Pheng, L.S.; Rui, Z. *Service Quality for Facilities Management in Hospitals*; Springer: Singapore, 2016. [CrossRef]

41. Bawab, F.; Baxter, L. The relationships between lean six sigma strategic, operational and tactical factors and organizational performance in hospitals: A proposed model. In Proceedings of the Seventh International Conference on Lean Six Sigma, Dusit Thani Hotel, Dubai, United Arab Emirates, 7–8 May 2018; pp. 38–47.

42. Jandali, D.; Sweis, R. Factors affecting maintenance management in hospital buildings: Perceptions from the public and private sector. *Int. J. Build. Pathol. Adapt.* **2019**, *37*, 6–21. [CrossRef]

43. Wong, W.F.; Olanrewaju, A.L.A.; Lim, P.I. Value-based building maintenance practices for public hospitals in Malaysia. *Sustainanility* **2021**, *13*, 6200. [CrossRef]

44. Gummerus, J. Value creation processes and value outcomes in marketing theory: Strangers or siblings? *Mark. Theory* **2013**, *13*, 19–46. [CrossRef]

45. Olanrewaju, A.L.A. Quantitative analysis of criteria in university building maintenance in Malaysia. *Australas. J. Constr. Econ. Build.* **2010**, *10*, 51–61. [CrossRef]

46. Jensen, P.A.; van der Voordt, T.J.M.; Coenen, C.; Sarasoja, A.L. Reflecting on future research concerning the added value of FM. *Facilities* **2014**, *32*, 856–870. [CrossRef]

47. CodeBlue. Water, Power Cuts, Collapsing Ceilings: Survey Bemoans State of Malaysia's Public Hospitals and Clinics. Available online: https://codeblue.galencentre.org/2019/12/20/water-power-cuts-collapsing-ceilings-survey-bemoans-state-of-malaysias-public-hospitals-and-clinics/ (accessed on 3 December 2017).

48. Olanrewaju, A.L.A.; Khamidi, M.F.; Idrus, A. Validation of building maintenance performance model for Malaysian universities. *Int. J. Educ. Pedagog. Sci.* **2011**, *5*, 1031–1035. [CrossRef]

49. Prahalad, C.K.; Ramaswamy, V. Co-opting customer competence. *Harv. Bus. Rev.* **2000**, *78*, 79–90.

50. Jensen, P.A. The facilities management value map: A conceptual framework. *Facilities* **2010**, *28*, 175–188. [CrossRef]

51. Coenen, C.; Alexander, K.; Kok, H. Facility management value dimensions from a demand perspective. *J. Facil. Manag.* **2013**, *11*, 339–353. [CrossRef]

52. Zulkarnain, S.H.; Ahmad Zawawi, E.M.; Rahman, M.Y.A.; Mustafa, N.K.F. A review of critical success factor in building maintenance management practice for university sector. *Int. J. Civ. Environ. Struct. Constr. Archit. Eng.* **2011**, *55*, 215–219. [CrossRef]

53. Wilson, C.; Leckman, J.; Cappucino, K.; Pullen, W.; Wilson, C.; Leckman, J.; Cappucino, K.; Pullen, W. Towards customer delight: Added value in public sector corporate real estate. *J. Corp. Real Estate* **2014**, *3*, 215–222. [CrossRef]

54. Toossi, A.; Lockett, H.L.; Raja, J.Z.; Martinez, V. Assessing the value dimensions of outsourced maintenance services. *J. Qual. Maint. Eng.* **2013**, *19*, 348–363. [CrossRef]

55. Jensen, P.A.; van der Voordt, T.J.M.; Coenen, C. (Eds.) *The Added Value of Facilities Management: Concepts, Findings and Perspectives*; Polyteknisk Forlag: Lyngby, Denmark, 2012.

56. Galvagno, M.; Dalli, D. Theory of value co-creation: A systematic literature review. *Manag. Serv. Qual.* **2014**, *24*, 643–683. [CrossRef]

57. Prahalad, C.K.; Ramaswamy, V. Co-creating unique value with customers. *Strateg. Leadersh.* **2004**, *32*, 4–9. [CrossRef]

58. Grönroos, C.; Voima, P. Critical service logic: Making sense of value creation and co-creation. *J. Acad. Mark. Sci.* **2013**, *41*, 133–150. [CrossRef]

59. Aliman, K.H. Audit Finds Malaysian Hospitals Understaffed, Underfunded and Overcrowded. *Edge Mark.* 2019. Available online: https://www.theedgemarkets.com/article/audit-finds-malaysian-hospitals-understaffed-underfunded-and-overcrowded (accessed on 11 October 2019).

60. Kwofie, T.E.; Aigbavboa, C.O.; Matsane, Z.S.S. Key drivers of effective collaborative working in construction supply chain in South Africa. *Int. J. Constr. Supply Chain Manag.* **2019**, *9*, 81–93. [CrossRef]

61. Munro, T.; Childerhouse, P. Construction supply chain integration: Understanding its applicability in infrastructure asset maintenance and renewal programmes. *Int. J. Constr. Supply Chain Manag.* **2018**, *8*, 1–18. [CrossRef]

62. Mooi, E.; Sarstedt, M. *Concise Guide to Market Research: The Process, Data and Methods Using IBM SPSS Statistics*; Springer: Berlin/Heidelberg, Germany, 2011. [CrossRef]
63. Chomeya, R. Quality of psychology test between Likert Scale 5 and 6 points. *J. Soc. Sci.* **2010**, *6*, 399–403.
64. Ringle, C.M.; Wende, S.; Becker, J.M. *SmartPLS 3*; SmartPLS GmbH: Bönningstedt, Germany, 2015.
65. Ramayah, T.; Cheah, J.; Chuah, F.; Ting, H.; Memon, M.A. *Partial Least Squares Structural Equation Modeling (PLS-SEM) Using SmartPLS 3.0*, 2nd ed.; Pearson Malaysia Sdn. Bhd.: Kuala Lumpur, Malaysia, 2018.
66. Hair, J.F.; Hult, G.T.M.; Ringle, C.M.; Sarstedt, M. *A Primer on Partial Least Squares*, 2nd ed.; SAGE Publications, Inc.: Los Angeles, CA, USA, 2017.
67. Ringle, C.M.; Sarstedt, M. Gain more insight from your PLS-SEM results: The importance-performance map analysis. *Ind. Manag. Data Syst.* **2016**, *116*, 1865–1886. [CrossRef]
68. Abalo, J.; Varela, J.; Manzano, V. Importance values for importance-performance analysis: A formula for spreading out values derived from preference rankings. *J. Bus. Res.* **2007**, *60*, 115–121. [CrossRef]
69. Ong, C.H.; Bahar, T. Factors influencing project management effectiveness in the Malaysian local councils. *Int. J. Manag. Proj. Bus.* **2019**, *12*, 1146–1164. [CrossRef]
70. Su, C.H.; Cheng, T.W. A sustainability innovation experiential learning model for virtual reality chemistry laboratory: An empirical study with PLS-SEM and IPMA. *Sustainability* **2019**, *11*, 1027. [CrossRef]
71. Ting, S.H.; Yahya, S.; Tan, C.L. Importance-performance matrix analysis of the researcher's competence in the formation of university-industry collaboration using Smart PLS. *Public Organ. Rev.* **2020**, *20*, 249–275. [CrossRef]
72. Valaei, N.; Nikhashemi, S.R.; Javan, N. Organizational factors and process capabilities in a KM strategy: Toward a unified theory. *J. Manag. Dev.* **2017**, *36*, 560–580. [CrossRef]
73. Tailab, M.M.K. Using importance-performance matrix analysis to evaluate the financial performance of American banks during the financial crisis. *Sage Open* **2020**, *10*, 1–17. [CrossRef]
74. Barclay, D.W.; Higgins, C.A.; Tompson, R. The partial least squares approach to causal modeling: Personal computer adoption and use as illustration. *Technol. Stud.* **1995**, *2*, 285–309.
75. Hair, J.F.; Risher, J.J.; Sarstedt, M.; Ringle, C.M. When to use and how to report the results of PLS-SEM. *Eur. Bus. Rev.* **2019**, *31*, 2–24. [CrossRef]
76. Hair, J.F.; Black, W.C.; Babin, B.J.; Anderson, R.E.; Tatham, R. *Multivariate Data Analysis*, 7th ed.; Pearson Prentice Hall: Hoboken, NJ, USA, 2010.
77. Diamantopoulos, A.; Siguaw, J.A. Formative versus reflective indicators in organizational measure development: A comparison and empirical illustration. *Br. J. Manag.* **2006**, *17*, 263–282. [CrossRef]
78. Ab Ghani, M.Z.; Abd, Z.; Ibrahim, I.; Musa, Z. Defining the critical success factor in FM Malaysian healthcare sector. In Proceedings of the 3rd International Building Control Conference, Hotel Royale Chulan, Kuala Lumpur, Malaysia, 21 November 2013; pp. 1–10.
79. Amaratunga, D.; Haigh, R.; Sarshar, M.; Baldry, D. Application of the balanced scorecard concept to develop a conceptual framework to measure facilities management performance within NHS facilities. *Int. J. Health Care Qual. Assur.* **2002**, *15*, 141–151. [CrossRef]

sustainability

MDPI

Article

Energy Evolution: Forecasting the Development of Non-Conventional Renewable Energy Sources and Their Impact on the Conventional Electricity System

Vadim A. Golubev [1], Viktoria A. Verbnikova [2], Ilia A. Lopyrev [3], Daria D. Voznesenskaya [3], Rashid N. Alimov [2], Olga V. Novikova [4] and Evgenii A. Konnikov [3,*]

[1] Laboratory for Digital Modeling of Underground Oil and Gas Reservoirs and Well-Test-Analysis, Peter the Great St. Petersburg Polytechnic University, 195251 Saint Petersburg, Russia; doveva@mail.ru
[2] Graduate School of High Voltage Energy, Peter the Great St. Petersburg Polytechnic University, 195251 Saint Petersburg, Russia; verbvika@mail.ru (V.A.V.); alimov.rn@edu.spbstu.ru (R.N.A.)
[3] Graduate School of Industrial Economics, Peter the Great St. Petersburg Polytechnic University, 195251 Saint Petersburg, Russia; ilyalo1808@mail.ru (I.A.L.); vdaria0722@gmail.com (D.D.V.)
[4] Graduate School of Nuclear and Thermal Power Engineering, Peter the Great St. Petersburg Polytechnic University, 195251 Saint Petersburg, Russia; novikova-olga1970@yandex.ru
* Correspondence: konnikov.evgeniy@gmail.com; Tel.: +7-961-808-4582

Citation: Golubev, V.A.; Verbnikova, V.A.; Lopyrev, I.A.; Voznesenskaya, D.D.; Alimov, R.N.; Novikova, O.V.; Konnikov, E.A. Energy Evolution: Forecasting the Development of Non-Conventional Renewable Energy Sources and Their Impact on the Conventional Electricity System. *Sustainability* **2021**, *13*, 12919. https://doi.org/10.3390/su132212919

Academic Editor: Alberto-Jesus Perea-Moreno

Received: 1 September 2021
Accepted: 17 November 2021
Published: 22 November 2021

Publisher's Note: MDPI stays neutral with regard to jurisdictional claims in published maps and institutional affiliations.

Abstract: The development of the world's electric power systems goes back over a century. During this period, the overwhelming majority of states have formed stable, typically centralized systems for generation, transmission, and distribution of electrical energy. At the same time, technologies, primarily for energy generation, are steadily developing, which leads to the emergence of potentially effective technological solutions based on fundamentally new energy sources. The most rapidly expanding group at the moment are renewable energy sources (RES). This fact is due to the significant coverage of the potential environmental and economic benefits of using technologies based on RES in the information environment. At the same time, the process of transformation of traditional electric power systems, by integrating generation technologies based on the use of renewable energy sources, is extremely resource-intensive, and also potentially reducing the level of sustainability and efficiency of the entire system functioning as a whole. This thesis is primarily true for exclusively centralized power systems. The purpose of this study is to create a forecasting model for the development of non-conventional renewable energy sources (NCRES) for short, medium, and long term, which makes it possible to form an action plan to ensure a reliable and uninterrupted supplying of consumers, taking into account the existing electric power system. The developed model made it possible to identify the most promising directions of NCRES from the integration point of view, and for them the quantification and clustering of the information environment was carried out, which made it possible to identify key trends and the specifics of the development of technological solutions for these directions of renewable energy sources. The developed tool and systemic conclusions formulated on the basis of its application make it possible to develop mathematically sound solutions in the direction of managing the development of traditional electric power systems based on the integration of NCRES.

Keywords: renewable energy sources; non-conventional renewable energy sources; RES; NCRES; electric power system; information environment

1. Introduction

The world is currently on the verge of a fourth energy transition to the widespread use of renewable energy sources and the displacement of fossil fuels. The rate of these changes, the speed of transition, and the impact on the already established electricity system are associated with high uncertainty [1]. Any global changes in technologies require risk assessment within the framework of existing systems, since the sustainability and continuity of energy supply determine not only energy security, but also the stability of economic development of individual regions and the country as a whole. Modern electric power systems in many countries have been formed for more than a century and their key elements are the fuel supply system, electric power generation, and the transmission and distribution systems of electrical energy. Modern trends encourage decentralization, on the one hand, stimulating the development of small generation facilities, including renewable energy sources, and, on the other hand, require taking into account the specifications of each technology being introduced [1,2]. The presence of significant negative consequences from underestimating the impact of an increase in the share of renewable energy sources in the balance of the power system's capacity is emphasized by the facts of massive disruption of power supply that consumers experience in different countries.

Non-conventional renewable energy sources (NCRES) include wind power; solar energy; small-scale hydropower; wave energy and energy of ebb and flow; geothermal energy; and bioenergy. Consider the dynamics of the development of renewable energy sources in different countries and in the world as a whole. This increment affects traditional energy systems which can cause significant issues, such as blackouts, due to instability of output for renewables. Such issues could influence the expansion of NCRES in some developing countries. Therefore, a model is required to predict the capacity of energy sources and predict possible problems within energy systems. Based on data provided by BP Statistical Review of World Energy [1] and the MGBM model described in the article, which is supposed to be the most accurate model [2], we made a trend model of the renewable's development (Figure 1). As is shown on a graph, there is a significant difference in the development of renewable energy sources in different regions of the world; however, there is a trend of increasing shares of renewables. On the downside, the model has a significant issue. It is a trend model which is based on statistical data, which leads to high errors during forecast periods in which rare events take place. For example, in 2014 there was a significant spike in renewable development, which can be explained by scientific exploration. Analyzing the 2014 spike [3] led us to conclusion that there were technology developments which led to a breakthrough and numerous projects in the renewables sphere. A trend model could not forecast these kinds of events, so our research is based on an originally developed mathematical model which includes developing projects and innovation trends within articles, so it makes prediction more valuable and accurate. Moreover, our model can be used for long-term forecasting because it is possible to evaluate future projects and some scientific trends, which makes our prediction more precise in long-term predictions, in contrast to traditional trend models which are influenced by cumulative error.

The purpose of this study is a predictive analysis of the prospects for the development of renewable energy sources, taking into account the existing energy system in the short, medium, and long term. In order to achieve this goal, it is necessary to accomplish the following tasks:

1. Investigate the specifics of the development of NCRES in relation to a universal object;
2. Propose a mathematical model that can be used to assess the development of NCRES and their mutual influence on the existing energy system;
3. Justify the features of the application of the forecasting model for the short, medium, and long term;
4. Investigate the degree of reliability of the application of the proposed model for certain technologies of NCRES;
5. Determine the most promising NCRES in the context of the identified predictive dynamics;
6. Identify technological trends typical for the most promising NCRES.

Non-conventional renewable energy sources world development forecast is shown in Figure 1.

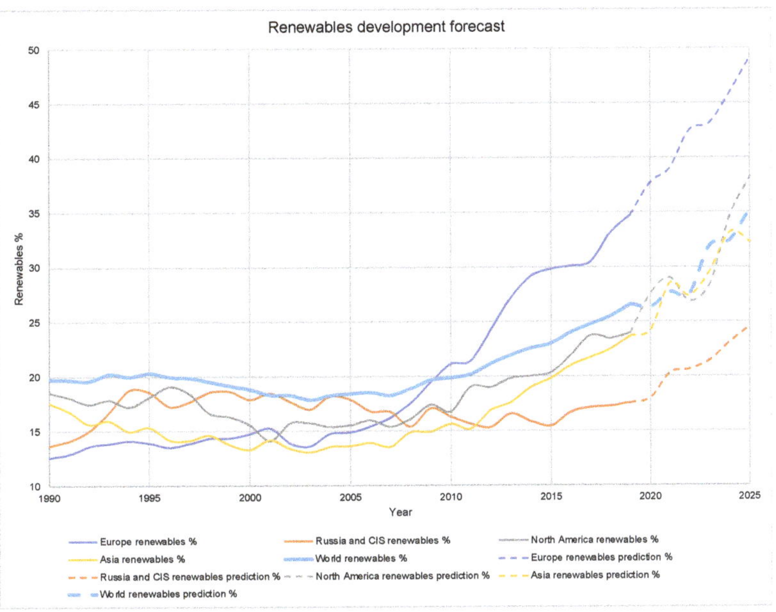

Figure 1. Non-conventional renewable energy sources world development forecast [1–3].

The established purpose of this study is not unique to the current scientific environment. This statement is due to the fact that the traditional energy system has been undergoing a rapid transition in the past 10 years with two most notable features, one of which is the high penetration of renewable energy generators using discontinuous renewable sources such as wind and solar. This transition is also causing a high degree of penetration of power electronic devices in generation, such as wind turbine converters and solar inverters, transmission, such as flexible AC or DC transmission system converters, and distribution/utilization systems, such as electric vehicles and microgrids. The development of modern power systems with multidirectional high penetration, i.e., with high penetration of renewable energy sources and power electronic devices, significantly affects the dynamics of the power systems and causes new sustainability problems [4].

At the same time, the importance of conventional power plants is increasing in parallel with the development of generation based on the penetration of renewable energy sources, its primary cause being the unsustainability of the generation process [5]. At the same time, a 10% increase in power generation using solar panels and wind could reduce the annual CO_2 emissions of the average thermal plant by about 4% [6]. Thus, researchers distinguish two key vectors of the impact of NCRES on the traditional power system: reduction of CO_2 emissions and increasing the level of unsustainability.

It should be noted that the electrical grid infrastructure is undergoing unprecedented transformations, mainly due to external factors: policies by government and regulatory authorities aimed at combating climate change; increasing opportunities and requirements of consumers; the development of distributed energy and the growth of electric vehicles quantity; the digitalization of grids with increasing integration of information and operational technologies; increased risks of cyberattacks; and energy market reforms, paving the way for completely new forms of competition. Attention should be paid to the fact that the impact of NCRES development is indirect. Transmission and distribution participants tend to interact more with each other in order to cope with the increased decentralized stochastic generation and changes in operating rules and procedures aimed at improving load regulation. These processes become the consequences of increasing the share of NCRES in traditional power systems [7].

In the confines of combating climate change, international experts generally consider only SPP (Solar Power Plant), WPP (Wind Power Plant) and some other renewable sources as an alternative to traditional sources [8]. At the same time, the prospects of using the traditional power system with reduced environmental impact and without the direct participation of NCRES should be noted. Technologies based on hydrogen and conversion of electricity into gas may become worldwide leaders [9,10]. Despite the prevailing trend toward electrification of energy sources and end-use, a balanced and reliable energy system is likely to require simple ways to transmit and store gas—possibly decarbonized gas. Consequently, energy-gas technology has enormous potential to provide synergistic coupling of sectors, which essentially means transmitting electricity for end-use in non-electric form. Gas infrastructure powered by greener gas can help ensure security of supply by bridging the mismatch between levels of peak power generation (most of the time by irregular renewable sources such as wind and solar) and demand [11].

Research of storage systems to enable the operation of NCRES in accordance with the mode of consumption is undoubtedly relevant [12,13]. Breakthroughs in electrochemical energy storage technologies—such as lithium or sodium ion batteries and supercapacitors—were used to create small-sized mobile electronic devices, medium-sized vehicles, portable and stationary devices, as well as for energy storage in large electric grids, paving the way to a new market with unlimited potential [14–16].

Energy recycling also has significant potential. Municipal solid waste-global production of 3.6 million tons per day gives an energy potential of 178 GW; hazardous waste—1.2 million tons per day, 43 GW; bio-waste—14 million tons per day, 685 GW; car tires—28,000 tons per day, 1.4 GW. The combined total is 907.4 GW, which compares to the entire U.S. installed capacity of 1100 GW [17]. As a different example, plasma recycling of municipal solid waste can provide about 5% of U.S. electricity needs. The most promising for the nearest future seems to be a complex approach which consists of waste management system development for each region [18,19].

NCRES has a variety of directions for development. In addition to the already common SPP technologies, chemical energy can be produced as a byproduct of utilizing sunlight [20]. Solar energy is a key element for many different ways to produce chemical fuel: production of biofuel from biomass; production of hydrogen in the process of microalgae; and production of biofuel by photocatalysis, performed by artificial devices [21–23]. Each of these methods has its own advantages and disadvantages. Successful development of these methods and overcoming the existing drawbacks requires further research in each of these sectors. Currently, the most popular solar fuel is biofuel derived from plant biomass. At the same time, molecular hydrogen is considered to be the fuel of the future, since it is a carbon-free chemical compound enriched with energy [24,25].

Thus, we can conclude that the prospects for the development of NCRES are extremely diverse, which is also multiplied by their level of impact on the traditional energy system. Parallel developing technological solutions, which cannot yet have a significant impact, but unambiguously contribute to the harmonization of the development of traditional energy together with renewable energy, are noted. However, the presented studies practically do not consider the problems of integrating generation systems based on renewable energy sources into existing traditional energy systems. Possible problems are extremely diverse and ignoring them can lead to both significant financial losses and the loss of energy security of entire regions. One of the most significant ways is to determine the most promising from the point of view of complex development of NCRES. It is necessary to effectively balance traditional and renewable sources. Only in such a case will the electric power industry develop towards minimal environmental risk and energy security [26]. Thus, the main gap in the current level of knowledge is the extreme inconsistency of research in the field of integration of modern energy generation technologies based on renewable energy sources into existing energy systems. The existing energy systems have been formed for decades and their current state and structure is determined both by the needs of energy consumers and by technological, natural, and infrastructural constraints of the environment.

Eliminating this gap requires systematization of the current trends in the development of technologies based on renewable energy sources and the formation the forecasts of development of this industry. This will make a significant contribution to the sustainability management of energy systems. This study raises the question of assessing the prospects of development and areas of impact of NCRES on the functioning of traditional power system, which can be largely investigated using statistical information processing. The scientific significance of this study and its difference from all those presented lies in a systematic approach to the analysis of the development directions of NCRES in the context of their integration with existing traditional energy systems. A systematic analysis of this issue is based on the analysis of objective statistical information and the results of quantification of the information environment, and does not use expert assessments, which increases the objectivity of the conclusions.

Thus, the main contribution of this research can be described by the following points:

1. Formation of a universal forecasting model for the development of basic NCRES technologies.
2. Identification of the most dynamically and steadily developing NCRES technologies.
3. Description of the specifics of the development of identified emerging NCRES technologies in the context of their integration into existing energy systems and increasing their sustainability.

This article, presenting the results of the current research, is divided into 5 main sections. The first section provides an overview of the current scientific basis and articulates a key gap in existing research. In the second section, the research methodology is presented in detail and the key methods for constructing a forecasting model for the development of the main NCRES technologies and an information environment analysis model describing the state of research of identified technologies are formed. The results section presents a forecasting model for the development of main NCRES technologies and identifies the most dynamically and steadily evolving NCRES technologies. The discussion section explores the information environment describing the development of the main NCRES technologies in the context of their integration into existing energy systems and identifies the key vectors for the development of renewable energy in general, taking into account the current scientific basis. Finally, the results are aggregated, and a brief summary of the study is presented.

2. Materials and Methods

The above-mentioned goal of the study determines the need for a consistent structural description and quantification of the development process of renewable energy in conjunction with traditional power systems. At the same time, the analysis of theoretical and methodological frameworks has established that the directions of NCRES development are extremely differentiated, both from a technological and economic point of view. Thus, the initial phase in this case should be the quantification and hierarchical classification of key renewable energy technologies, considering the dynamics of perspective development. However, in addition to the global technological development of NCRES, it is necessary to consider the importance of natural, territorial, and social factors that invariably affect the prospects for the use of the described technologies within a particular state or region. Consequently, the results of the initial stage should be specified taking into account these limitations. For the purpose of this study, the Russian Federation is chosen as a subject of analysis. There are several key reasons for this:

1. The Russian Federation is extremely vast from a territorial point of view, which determines a significant differentiation of natural and climatic conditions of NCRES-based projects development.
2. The Russian Federation has significant reserves of traditional energy resources, which, in turn, determined the formation of a developed traditional power system, that is totally centralized and sensitive to technological transformations.

3. The Russian Federation has a significant scientific and economic potential for the development of non-conventional renewable energy sources, which is being confirmed by the projects implemented in this area.

Thus, the primary stage of this study is aimed at identifying the most perspective NCRES influencing traditional energy system. The effect will be more significant in developing energy systems, which start energy transition to renewables, so it is possible to use the Russian energy system as an example to prove the model. However, the identified areas of development are extremely multidimensional; thus, they need to be clarified. For these purposes, it is proposed to conduct thematic clustering of the informational scientific environment focusing on the identified promising spheres, and to identify key areas of research in these segments. System analysis of links between the identified key areas of research will allow us to highlight the specifics of development of non-conventional renewable energy sources in the context of traditional power systems.

As part of the research, a mathematical model was developed, which was used to evaluate the development of renewable energy sources and their mutual influence on the existing power system of the Russian Federation. The analysis began with a review of historical data on the dynamics of capacity of non-conventional renewable energy sources in the Russian power system (Figure 2) [27].

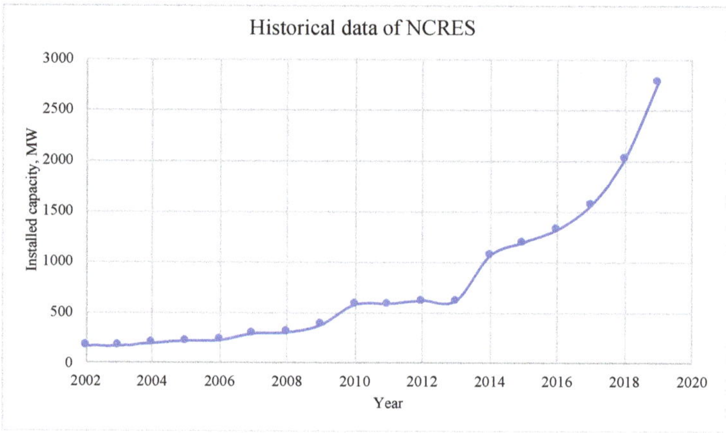

Figure 2. Historical data of NCRES development in Russia.

The graph has a characteristic non-linearity, which does not allow using simple polynomial time trends to assess the development of NCRES. Usage of such trend models can lead to a poor result shown in Table 1. Because of that we suggest developing a more advanced model which will be described below.

Table 1. Time-based trends in the development of renewable energy sources.

	R-Squared	MSE
Linear trend	0.56263715	308.4257
Quadratic trend	0.82746759	200.0688
Cubic trend	0.89081251	164.7449

None of the results obtained can be called satisfactory, because in the case of the cubic trend the number of variables becomes too large compared to the sample size, which makes this analysis irrelevant. On the other hand, the temporal influence on the development of the industry cannot be neglected.

In the analysis of technological development, inventions in non-conventional renewable energy technologies were analyzed, after which a correlation was found that all peaks in renewable energy growth are preceded by major discoveries within NCRES industries, as well as the completion of major projects. For example, in 2013, Sharp made a breakthrough in solar photovoltaic cells, increasing their efficiency to 44% [28]. In 2014, a project to build a biopower plant was implemented. In 2017, a large plant for the production of innovative solar panels "HEVEL" was built.

As for the current state of NCRES industries in Russia, solar energy is 0.72% of all energy capacity. Wind energy accounts for around 0.56% of the energy system. Other types of NCRES industries amount to about 0.3%. In its current state the influence is rather low, but will increase in the future development.

Due to the fact that non-conventional renewable energy is a knowledge-intensive industry, it is proposed to address the relationship between the creation of renewable energy facilities and the development of technologies of NCRES industries. However, in the development of technology there are a large number of innovations that do not receive any follow-up. In this regard, it is proposed to introduce a link between the development of technology and projects that have actually been implemented. A similar idea of linking the growth of emerging industries was expressed in the Thomas L. Heath series of books "The Thirteen Books of the Elements" [29].

Based on these judgments, it is suggested that the correlation between articles written on technology in any of the NCRES industries and actual projects using the technology described in the articles or using the result of the article in an actual project has to be evaluated. In addition, it is proposed to assess the general trend of projects within the selected NCRES industry through the value of the projects being developed. The last criterion will be the probability of discovery. This criterium is based on the correlation between the technologies of real projects and those described in highly significant articles, which can be considered a kind of fundamental work for the industry. Model scheme and results within every model step is presented in Figure 3.

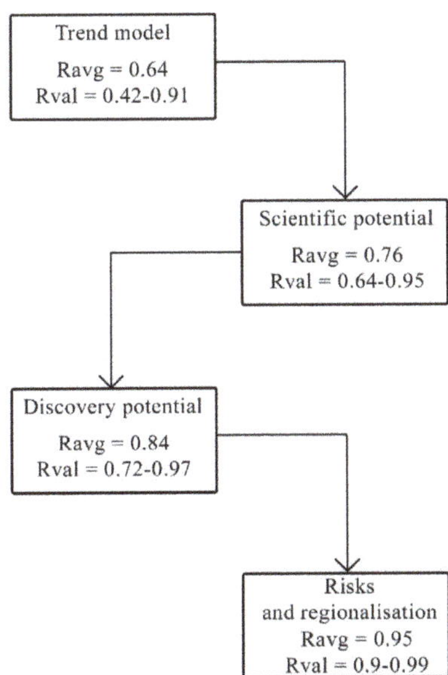

Figure 3. Scheme of mathematic model for prediction of NRCES development.

Ravg—Average R-squared value across all significant NCRES industries.

Rval—Is a range between lowest and highest R-squared value across all significant NCRES industries.

Based on the identified dependencies, three percentage values were calculated:

Estimated growth rate from developments-assuming an annual increase in the share in the balance of power of the industry, due to the projects implemented in it.

Scientific potential, showing the percentage of scientific activity that has real application within the industry.

Discovery potential, a percentage value reflecting the percentage of works of high importance with technologies that are implemented in projects of different companies.

As a final step, the model involves risks and regionalization which are supposed to make our prediction more precise.

For this purpose, algorithms were created in the Python 3 programming language for the article aggregators E-Library and Science Direct, as well as the procurement aggregators. With the help of the developed programs, articles and projects were uploaded after being sorted by keywords. Next, the technologies in the articles were analyzed and their correlation with the projects was confirmed. The Table 2 shows the results of the correlation analysis for 2020.

The result of the application of the described methodology is a model that allows us to identify the most promising dynamically developing types of NCRES. However, for the purposes of a predictive analysis of the prospects of renewable energy sources, considering the existing energy system in the short-term, it is necessary to describe the specifics of the most promising dynamically developing types of NCRES. To achieve this goal, it is necessary to analyze the information environment of each of the selected NCRES types. The automated algorithm for analyzing the information environment is shown in Figure 4.

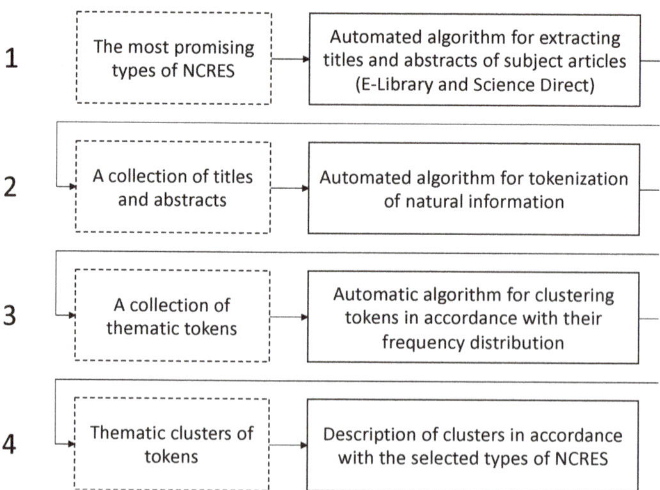

Figure 4. Information environment analysis algorithm.

The details of this methodology are presented in the study [30]. Based on the results of the application of this algorithm, a set of conclusions is formed regarding the specifics of the dynamically developing types of NCRES, which will make it possible to detail the specifics of the evolution of the energy sector. The complete algorithm of this study is shown in Figure 5.

Table 2. Calculated parameters for the model for 2020.

Type of NCRES	Keywords	Company Purchases	Number of Articles	Estimated Growth Rate from Development	Scientific Potential	Discovery Potential	Highly Significant Points	Share in NCRES
WPP	WPP (ВЭУ-Ветроэнергоустановка) *	1	1086	7.17%	0.38%	1.73%	16	5.76%
	WES (ВЭС-ветроэлектростанция) *	34	666				5	
SPP	Solar panel	22	2103	0.998%	5.89%	8.01%	5	45.27%
	Solar module	4	1934				17	
	Solar battery	79	5488				20	
	Charge controller	2	382				0	
	Inverter	132	16996				55	
	Batteries	Not relevant					-	
SHPP (Small Hydropower Plant)	Hydroelectric turbine	9	1978	21.29%	0.64%	0.91%	6	5.22%
	Hydroelectric generator	2	755				3	
	Small hydropower	22	203				2	
GEOEP (Geopower Energy Plant)	Turbounit	32	1206	7.52%	0.26%	0.66%	8	4.46%
Tidal power plant	Tidal	0	2045	0.00%	0.45%	3.55%	43	0.06%
BIOEP (Biofuel Power Plant)	Bioreactor	5	2891	0.04%	0.63%	2.39%	29	30.91%
Waste	Waste incineration	210	535	9.80%	0.12%	0.08%	1	8.23%
Wave power plant	Wave stations	0	438	0.00%	0.10%	0.25%	3	0.00%
Landfill gas disposal	PSU—Pressure Steam Unit	5	27	11.98%	0.06%	0.08%	0	0.08%
	Landfill gas	2	232				1	

* Research was made within Russian Science Citation Index, so keywords were used in Russian.

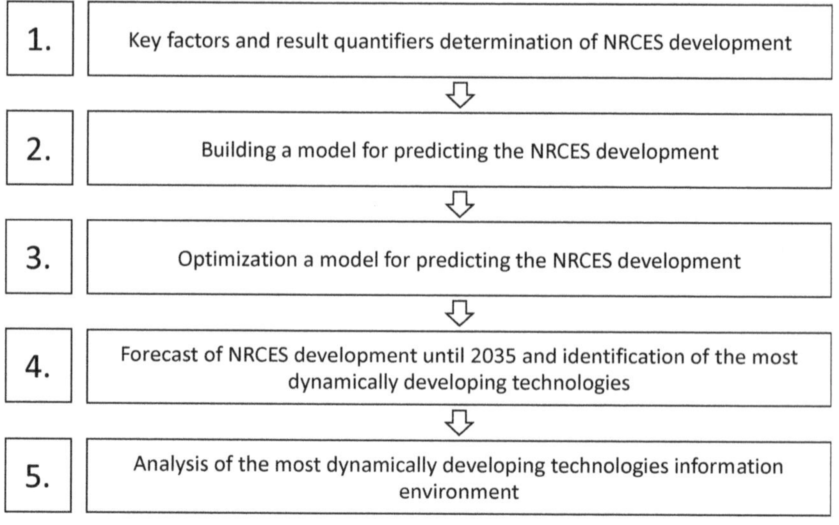

Figure 5. Research algorithm.

3. Results

Within the framework of this study, it was revealed that the main areas of development of non-conventional renewable energy sources are SPP, WPP, and BIOEP (Biofuel Power Plant). At the same time there is a significant increase in the number of SPP-based projects under development, which is again due to the recent development of technology in this area. Tidal power plants are promising in terms of development, but at the moment there are no implemented and functioning projects.

After the research, the linear trend model was extended by percentage coefficients. The final formula is as follows:

$$PO_i = PO_{i-1} \times \left(1 + \frac{K_r^{i-1}}{1.5}\right) + \left(1 + \frac{K_r^{i-1} + K_s^i}{100\%}\right) \times T_{tr} + P(V_{ex}) \times \left(1 + \frac{\sum_{n=0}^{i} K_s^n}{100\%}\right) \times T_{tr} \tag{1}$$

where

1. PO_i—Industry volume in period i, expressed in installed capacity, in MW.
2. T_{tr}—Trend annual growth coefficient of the industries calculated on the basis of historical data on the development of the industry.
3. K_r^{i-1}—The estimated growth rate from developments in the previous period.
4. K_s^i—Current scientific potential.
5. V_{ex}—Discovery potential.
6. $P(V_{ex})$—Is a function returning 1 with probability V_{ex} and returning 0 with probability $(1 - V_{ex})$.

Part of the calculations was the approximation that all projects in the procurements are implemented within 1.5 years of acceptance; this approximation was validated by the Pearson's Chi-squared criterion and the hypothesis that projects are implemented in 1.5 years with a 95% probability was accepted.

For any predictive model it is the predictive ability that is significant, which can be evaluated by calculating statistical parameters. To validate the model, we took data from 2002 up to 2020 with a lag of 2 years. The lag is caused by the fact that in order to use the linear regression algorithm a certain minimum sample is required, otherwise it will be impossible to construct a linear trend. In this regard, we took the data on the branches of NCRES and calculated the model. This dataset was taken as a validation to ensure that model is suitable for predictions within different states in energy evolution. In its current state, renewables in Russia account for a small share of such energy sources in energy system, but its influence increases as the time goes by. Next, statistical indicators such as coefficient of determination, standard deviation, and coefficient of variation were estimated. The results are presented in the table below for some typical industries, in which the dependence is very different from the linear one. These industries had the highest deviations from the time trend, which prevented it from being used in the calculations (Table 3).

Table 3. Statistical parameters of the developed model.

Parameter	WPP	SPP	BIOEP
R squared	99.79%	99.72%	99.94%
MSE	2.22	16.47	5.18
Coefficient of variation	4.16%	5.67%	1.84%

As we can see, the coefficient of determination indicates that the model accurately describes the real dependence. The low values of the mean square error and the low coefficient of variation also indicate the high accuracy of the model. For the remaining industries, the coefficient of determination ranges from 98 to 99%, and the coefficients of variation do not exceed 7%, which indicates the versatility of this model for the non-conventional renewable energy industries. A visual comparison of the power NCRES graphs is presented below (Figures 6–8).

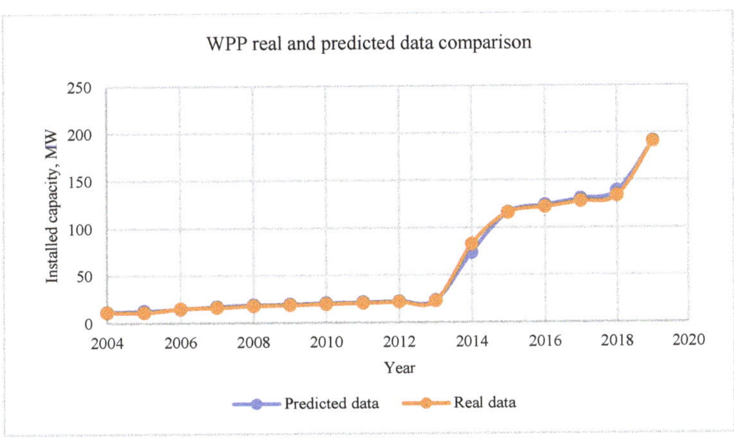

Figure 6. Comparative analysis of wind energy industry.

As we can see, the model has high accuracy, which makes it possible to make predictions on its basis and create certain patterns. It can be noted that for solar power plants prediction is in significance interval and it still can be used for a prediction modeling and forecasting, but has lowest accuracy across prediction of NRCES that is caused by significant non-linearity in data. This is due to the fact that a linear trend deriving from previous periods is used for each calculation. In turn, the linear trend becomes quite stable with a large sample, then the effect of outliers is not so great, in this case, the sample consists of five values, which are not enough for the trend to stabilize.

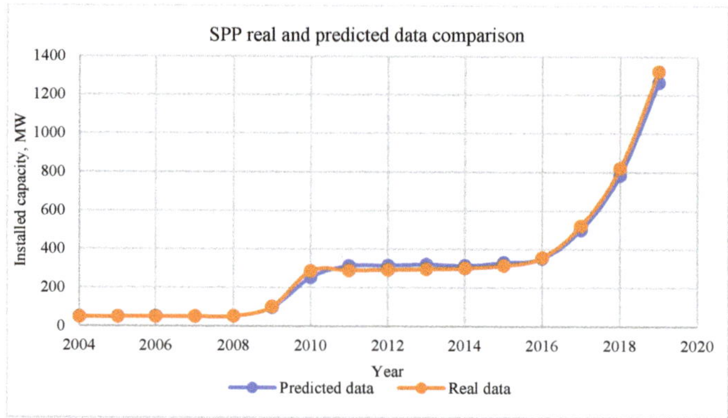

Figure 7. Comparative analysis of solar energy industry.

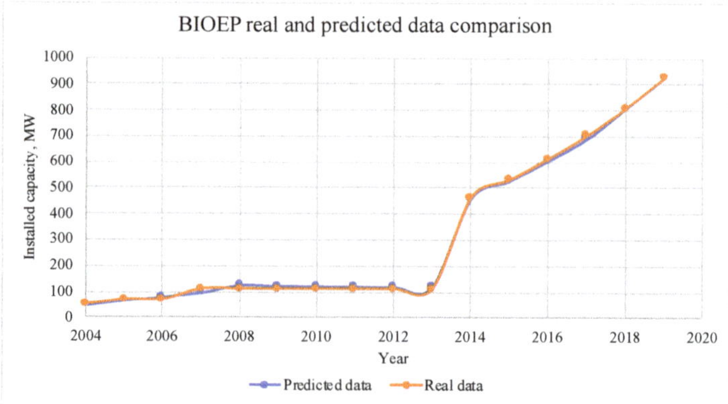

Figure 8. Comparative analysis of biological energy industry.

The presented model does not extensively highlight the specific sectors of renewable energy sources prevailing in each region of Russian Federation, focusing more on the rate of development of these technologies on the whole. This need derives from the inexpediency of placing certain types of renewable energy sources in certain regions due to their physical, economic, or other features. For example, all types of hydroelectric power plants need to be located in regions with an abundance of water resources due to the technical features of electricity generation. In this regard, regions with large volumes of water resources are more likely to build hydroelectric power plants. In order to take into account regional features in the model, the regions were ranked according to the development potential of the non-conventional renewable energy sectors [31,32]. For forecasting purposes, it is necessary to estimate the parameters of the model for future periods:

1. Scientific potential;
2. Potential of discovery;
3. Estimated growth rate of the industry from project development.

As part of data analysis on publications, a characteristic linearity was revealed in the percentage of publications whose technologies were applied in projects. On the basis of this, a linear trend was built and the scientific potential in the future period was calculated using the trend. The potential for discovery was estimated using the average percentage of high value articles that correlated with actual completed projects. From the point of view of this indicator, it does not change over time, remaining rather low in any sector of non-conventional renewable energy sources.

The growth rate of the industry from projects in development is proposed to be estimated using the average value of the growth of industries over the recent years, since the influence of this indicator has a high impact in the short term. In the medium and long term, the key factor will be the scientific potential and the capacity for discovery, which allows us to use a rough estimate of this indicator due to the specificities of the model. Afterwards, based on the above conditions established for the model, the forecast was divided by time intervals into short-term (up to 5 years), medium-term (up to 10 years), and long-term (up to 15 years) forecasts. Due to the structure of the model, a certain segment of the formula presented above will affect the formation of each forecast. The first one is responsible for the short-term perspective, which does not take into account the potential of scientific discoveries, assuming that the development of the industry in this case is brought only by projects launched into implementation, since any innovative pilot project requires investment and is most often implemented in a much longer time frame. In the medium-term, for which the second item is responsible, scientific potential appears, showing that some designs that are in the pilot status can become full-fledged projects in the end and significantly affect the non-conventional renewable energy sector. In the long term, some discoveries are possible in each of the non-conventional renewable energy sectors, which was taken into account in the form of a probabilistic function of the opening potential. Within the framework of the baseline scenario for the development of non-conventional renewable energy sources, the probability had the pattern of a Gaussian distribution, and the number of cycles according to the Monte Carlo method was 1000 iterations. Figure 9 illustrates the rates of NCRES development over the years.

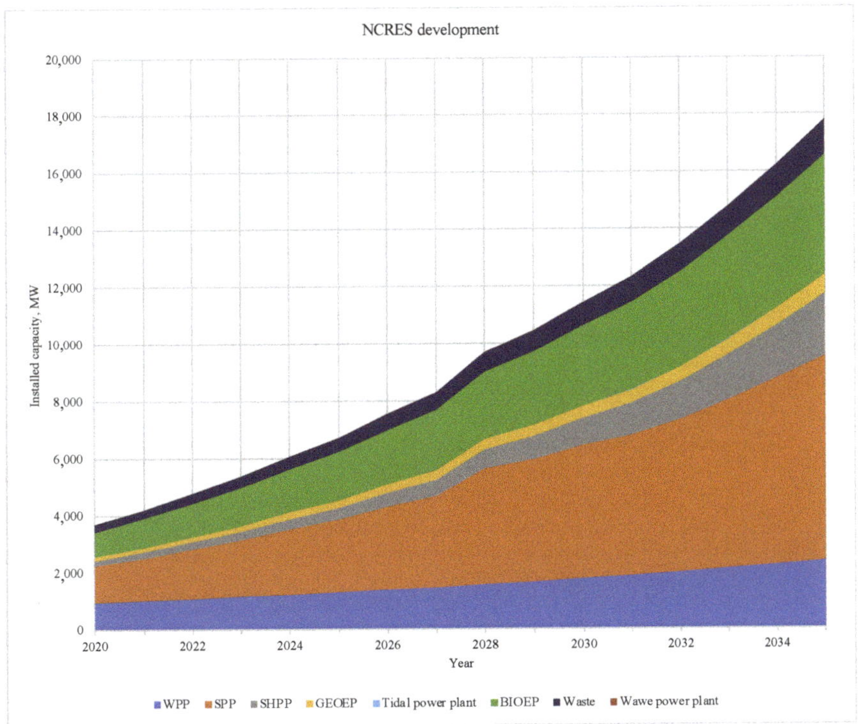

Figure 9. Forecast of RES development until 2035.

This graph reveals that some industries are developing more monotonously than others; for example, the SPP industry has a strong dependence on the emergence of new technologies. In our country, the industry has received extensive development, including scientific, which allows us to conclude the possibility of new discoveries in this sector of non-conventional renewable energy sources. Moreover, for all industries mentioned above as a whole, exponential development is typical, which may be a consequence of a favorable climate for the development of these areas in the long term, but this forecast is not completely accurate, since at the moment the risks of drastic structural change are not taken into account, which can seriously affect the further development of the industry. For forecasting purposes, it is not enough to rely solely on the current parameters of the system, since unforeseen events may occur in the future that may affect the development of a particular sector of non-conventional renewable energy sources. In this regard, it is necessary to introduce an additional risk correction parameter, estimated by PESTLE analysis using a fuzzy logic tool, which makes it possible to obtain a quantitative risk value, which will be taken into account in the presented model by shifting the distribution in the generator relative to normal, which seriously affects the change in the results after calculation by the Monte Carlo method. It should be noted that this study does not take into account the influence of solar cycles and global warming when predicting the development of solar and wind energy generation technologies.

Thus, it was revealed that the solar and wind energy industries are promising, as they have high potential for development at the moment. Within the SPP industry, a large development potential was identified, which allows us to expect high growth rates in the next 7–10 years. The industries of wave and tidal power plants will not become widespread due to the lack of large-scale projects being implemented, as well as new technologies that could help increase their efficiency. The growth rate of other industries (small hydroelectric power plants, geoPPs, bioelectric power plants, and the production of electricity from waste) will remain at an average level, due to their low level of development and the specificity of use of similar non-conventional renewable energy technologies. However, it is worth noting the systemic importance of BioEP, as a fundamental development vector from a technological point of view. The resource specificity of these technologies makes them universal, which largely determines their prospects. Therefore, it is necessary to study the information scientific environment of the solar, wind, and bioenergy industries.

4. Discussion

The conducted research allows us to assert that the dynamics of the development of existing NCRES in many respects confirms the conclusions of previous studies, and the primary driver of development is indeed environmental issues [6]. The conclusions of the study [8] were fully confirmed—the key technologies at the moment are solar and wind power plants. This fact indicates the extremely high importance of the climate change factor in the mediation of the development of renewable energy sources. Moreover, it is these technological areas that are currently being actively integrated into existing energy systems. At the same time, the dynamics of the development of technologies based on the use of hydrogen [9,10] or gas [11] is much less intense. The reason for this may be the logistic specifics of using these technologies, as well as the risk specifics. At the same time, only the analysis of the information environment will make it possible to establish the significance of the development of digital technologies within the framework of the development of NCRES [7].

In order to study the areas and topics related to non-conventional renewable energy sources, the most promising in the context of the study, it was decided to analyze the articles and research papers published over the past 3 years. Sciencedirect.com, an aggregator of scientific works, was chosen as the fundamental source of information. This site has several significant advantages over its analogue: only works from reputable journals and periodicals are published on the resource; it presents the results of research from around the world; and it is possible to fine-tune the scope of the search, which was used to simplify the process of collecting and processing information. It is also necessary to note that articles aggregated within the framework of this source require mandatory detailed reviewing, which in turn does not exclude generalized and low-quality materials from the analytical array. It is also necessary to take into account the openness of these materials. This model was created to predict renewable capacity in an energy system, using a system theory to describe such a complex concept. Calculation of exact values for Russia's united energy system requires some simplifications in the data that are estimated. Citation rate was used to calculate science potential, because it represents the paper's popularity and relevance among other scientists. To account for the connection between articles and real-life projects, authors used correlation between technology in articles and real projects of companies that includes this technology. Citation rate was also used to calculate probability of discovery. However, in this case only the most valuable articles were considered (which have a citation rate of more than 50; value was calculated according to articles that were most influential in renewables in the past). To confirm the connection between new discoveries and their practical state, correlation between developed technologies and articles was used. This data were collected across the whole dataset and the model proved to be accurate according to series of statistical tests that were mentioned in 2nd block.

Since the key task of this stage of the research is the cluster analysis of tokens obtained during the processing of scientific papers on several main topics related to non-conventional renewable energy sources, a set of keywords should be chosen for each topic assessed. Search topics are selected according to their industry relevance. Consequently, tokens on the topics of solar, wind, and bioenergy will be researched. The fundamental difference between aggregators of scientific articles in comparison with search engines familiar to a modern person is that, due to the variety of articles, one list of keywords can get results from completely different areas of scientific research, which will be clearly demonstrated below. Accordingly, when choosing a set of keywords to form a sample of scientific articles, it requires accuracy in terminology and compliance with a strict limit on the number of selected words. The situation is also complicated by two additional factors: first, differences in the terminology of the Russian and English languages in the energy industry significantly limit the choice that allows you to maintain the accuracy of the query; secondly, when using automated methods of data processing, it is rather difficult to weed out scientific papers that do not correspond to the meaning, if their keywords match the request. As a result, the following keywords were chosen for data collection: for SPP—"solar power", for WPP—"wind power", for BioEP—"biogas energy".

As for the methods of processing the obtained data, for this work we chose an automated analysis of information using the Python 3 programming language, since, thanks to the almost unlimited possibilities of creating functions and programs, Python 3 allows you to create the most convenient analytic tools. This language was chosen because of its simplicity, the ability to run and test the program's performance in the process of writing code, as well as wide support from the community and the abundance of additional libraries that are perfect for analyzing both quantitative and qualitative information.

The program code has been divided into three parts for ease of editing and executing. The first part performs an automated entry to the site, selection of keywords, and loading the titles of all found articles into a separate *.xlsx file. For these purposes, in addition to the basic Python functions, the selenium, BeautifulSoup and time libraries were used. The selenium library, originally created to test the operation of browsers, today is often used for scraping—the process of automated collection and processing of data from the Internet. BeautifulSoup allows you to turn a page on the Internet into a set of text for the subsequent selection of the desired data by "tags"—elements of the code of a web page written in the html language. Time library—for setting delay timers for some program actions.

The next step was the transformation of the obtained data into a set of words for analyzing the frequency of mentioning various terms. For this, the following libraries were imported: nltk with different extensions—for tokenization (dividing text into words), lemmatization (highlighting the original form of a word to simplify aggregation and analysis), and removing words without independent meaning (prepositions, conjunctions, articles)); the re library for extracting the title and individual words from the corresponding parts of the web page code; and pandas—for the formation of datasets and their subsequent study. Since language analysis requires fine tuning for automated functioning, several variables were introduced containing a list of designations for the key parts of speech and the choice of a list of stop words (words that have no independent meaning).

The last part of the code is the calculation of the frequency distribution of each token. For this, the collections library was used, which allows one, in particular, to count the total number of mentions of each word in the titles of articles. Thus, the final distribution file was generated.

The final stage of data processing before analysis was sorting the received tokens by frequency and excluding insignificant parts of the sample. For research purposes, those tokens are considered insignificant if the frequency of their mention is less than 13 times. Additionally, the tokens with the highest frequency of mention, as well as the keywords used for the search, are removed from the selection, since they have no meaning. Thus, for each area of data collection, a list of words was formed, with a volume of about 100 units, each of which was divided into clusters, according to the meaning of each of the tokens.

Based on the analysis results, the following conclusions can be drawn:

1. The most popular research topic related to all analyzed non-conventional renewable energy technologies is electricity generation and its technologies. For SPP, the corresponding tokens occupy about 6% of the sample (127 references), for WPP—about 2% of the sample, BioEP—4%. However, each of the industries has its own characteristics: articles on solar technologies place the greatest emphasis on photocatalytic decomposition as the main technology and works on wind energy place emphasis on a comparative analysis of existing technologies of wind power plants and solar power plants. In terms of bioenergy research, it mentions different ways to obtain fuels from biomaterials, accounting for 3.6% of the sample. Additionally, in the works analyzed, great emphasis is placed on the disclosure of different types of fuel resources or resources suitable for the mass production of biofuels. So, 6.7% (241 references) of the sample are tokens associated with this aspect.

2. Hydrogen energy is an integral part of almost all discussions related to the topic of non-conventional renewable energy sources. This fact is reflected in the research papers studied: each of the three samples contains references to hydrogen energy in varying degrees—from 1 to 1.5% of the sample. However, articles on bioenergy also contain descriptions of several technologies for hydrogen production: steam conversion, reforming, and gasification. Thus, it can be concluded that the discussion of bioenergy in the world scientific community is closely related to the production of hydrogen.

3. One of the important points in research on the topic of non-conventional renewable energy sources is also the ecological component of the relevant technologies. So, articles about SPP contain references to ecological tokens in the amount of 2% of the sample, WPP—3.2%, BioEP—an outstanding 5.4%, which is not surprising, given the lower environmental friendliness of bioenergy from the public point of view.

4. Another common theme for the three industries is accumulation systems, which occupy 2–3% of the sample for WPP, SPP, and BioEP. Based on this fact, it can be concluded that scientific community is highly interested in the issues of leveling and stabilizing the generation of non-conventional renewable energy sources using various storage devices, such as pumped storage power plants, battery complexes, and other technologies and resources, such as, for example, hydrogen, which was mentioned above.

5. It is also worth noting that scientific works on wind energy stand out from the rest because of the abundant interest in the issues of forecasting production. In particular, it is proposed to do this using neural networks, the mentions of which occupy about 1.5% of the sample. This trend was noticed only in this industry, which suggests the development of research directions, the results of which allow expanding the list of economically efficient wind generation regions in the future.

Thus, according to the results of this study, it can be concluded that the development process of the selected technological areas is extremely differentiated, which indicates fundamentally different technological development problems. The current gap in scientific knowledge is associated not with the lack of a unified position regarding the prospects for the development of renewable energy sources, but with the technological uniqueness of each of the selected areas. The technologies of solar and wind power plants are the most researched, which led to their dynamic development. At the moment, it is the technologies of solar and wind power plants that are developing in the context of practical use and, as a result, integration into existing energy systems.

5. Conclusions

Based on the conducted study using the tools of regression analysis, as well as quantification and clustering of the information environment, the features of the development of technological trends characteristic of the development of NCRES were identified. The main technologies that can influence the traditional electric power system are WPP, SPP, and bioenergy. The proposed mathematical model, with the help of which it is possible to assess the development of NCRES and its mutual influence on the existing energy system, considers the results of assessing the development of scientific potential and the potential for discovery. Since all the identified factors differ in time in terms of their impact, it is proposed to take into account the specificities of applying the forecasting model for the short, medium, and long term, namely projects at the stage of implementation, scientific potential, and the potential of discoveries. A high degree of application reliability of the proposed model for certain technologies of NCRES was demonstrated.

Based on the results of the study, models have been created for predicting the development of NCRES for the short, medium, and long term, which makes it possible to form an action plan to ensure a reliable and uninterrupted supply of consumers, taking into account the existing electric power system. This goal has not previously been considered separately using data on scientific potential and potential for discovery. The authors consider it important to note the versatility of the approach to forecasting the development of NCRES in a specific power system with the simultaneous planning of measures to ensure energy security. The obtained results contribute to the elimination of the current knowledge gap associated with non-systematic research in the field of integration of modern energy generation technologies based on renewable energy sources into existing energy systems and form the basis for further research in this area, potentially allowing us to significantly increase the sustainability of energy systems. As key limitations, it is worth noting that this study does not take into account the influence of solar cycles and global warming when predicting the development of solar and wind energy generation technologies, and also that the results of the analysis of the information environment can be interpreted in different ways. However, despite the existing limitations, the results obtained are of a multilevel theoretical and applied nature. The results can be used to develop plans for the development of the electric power system of the regions and the country as a whole.

Author Contributions: E.A.K. and V.A.G. designed the model and the computational framework and analyzed the data. V.A.G. and I.A.L. carried out the implementation. E.A.K. performed the calculations. V.A.V. and D.D.V. wrote the manuscript with input from all authors. O.V.N. and R.N.A. conceived the study and were in charge of overall direction and planning. All authors have read and agreed to the published version of the manuscript.

Funding: The research is partially funded by the Ministry of Science and Higher Education of the Russian Federation under the strategic academic leadership program 'Priority 2030' (Agreement 075-15-2021-1333 dated 30 September 2021).

Acknowledgments: The research was supported by the Peter the Great St. Petersburg Polytechnic University.

Conflicts of Interest: The authors declare no conflict of interest.

References

1. BP Statistical Review of World Energy. 2020. Available online: http://www.bp.com/statisticalreview (accessed on 15 June 2021).
2. Tsai, S.B.; Xue, Y.; Zhang, J.; Chen, Q.; Liu, Y.; Zhou, J.; Dong, W. Models for forecasting growth trends in renewable energy. *Renew. Sustain. Energy Rev.* **2017**, *77*, 1169–1178. [CrossRef]
3. 2014 Renewable Energy Data Book (Book), U.S. Department of Energy (DOE), Energy Efficiency & Renewable Energy URL. Available online: https://www.nrel.gov/docs/fy16osti/64720.pdf (accessed on 22 June 2021).
4. Shair, J.; Li, H.; Hu, J.; Xie, X. Power system stability issues, classifications and research prospects in the context of high-penetration of renewables and power electronics. *Renew. Sustain. Energy Rev.* **2021**, *145*, 111111. [CrossRef]
5. Singh, A.; Eser, P.; Chokani, N.; Abhari, R. High Resolution Modeling of the Impacts of Exogenous Factors on Power Systems—Case Study of Germany. *Energies* **2015**, *8*, 14168–14181. [CrossRef]
6. Graf, C.; Marcantonini, C. Renewable energy intermittency and its impact on thermal generation. *Energy Econ.* **2017**, *66*, 421–430. [CrossRef]
7. ELicwPML. *Nist Framework and Roadmap for Smart Grid Interoperability Standards, Release 2.0*; Office of the National Coordinator for Smart Grid Interoperability and I. T. Laboratory. Available online: https://nvlpubs.nist.gov/nistpubs/SpecialPublications/NIST.SP.1108r2.pdf (accessed on 3 May 2021).
8. The Future of Hydrogen–Analysis IEA. Available online: https://www.iea.org/reports/the-future-of-hydrogen (accessed on 3 May 2020).
9. Marchese, M.; Giglio, E.; Santarelli, M.; Lanzini, A. Energy performance of Power-to-Liquid applications integrating biogas upgrading, reverse water gas shift, solid oxide electrolysis and Fischer-Tropsch technologies. *Energy Convers. Manag. X* **2020**, *6*, 100041. [CrossRef]
10. Daggash, H.A.; Patzschke, C.F.; Heuberger, C.F.; Zhu, L.; Hellgardt, K.; Fennell, P.S.; Bhave, A.N.; Bardow, A.; Mac Dowell, N. Closing the carbon cycle to maximise climate change mitigation: Power-to-methanolvs.power-to-direct air capture. *Sustain. Energy Fuels* **2018**, *2*, 1153–1169. [CrossRef]
11. Fasihi, M.; Efimova, O.; Breyer, C. Techno-economic assessment of CO_2 direct air capture plants. *J. Clean. Prod.* **2019**, *224*, 957–980. [CrossRef]
12. Yang, D.; Liu, C.; Rui, X.; Yan, Q. Embracing High Performance Potassium-Ion Batteries with Phosphorus-Based Electrodes: A Review. *Nanoscale* **2019**, *11*, 15402–15417. [CrossRef] [PubMed]
13. Sharma, K.; Arora, A.; Tripathi, S.K. Review of Supercapacitors: Materials and Devices. *J. Energy Storage* **2019**, *21*, 801–825. [CrossRef]
14. Wei, L.; Wu, M.; Yan, M.; Liu, S.; Cao, Q.; Wang, H. A Review on Electrothermal Modeling of Supercapacitors for Energy Storage Applications. *IEEE J. Emerg. Sel. Top. Power Electron.* **2019**, *7*, 1677–1690. [CrossRef]
15. Du, M.; Li, Q.; Zhao, Y.; Liu, C.-S.; Pang, H. A Review of Electrochemical Energy Storage Behaviors Based on Pristine Metal–Organic Frameworks and Their Composites. *Coord. Chem. Rev.* **2020**, *416*, 213341. [CrossRef]
16. Agudosi, E.S.; Abdullah, E.C.; Numan, A.; Mubarak, N.M.; Khalid, M.; Omar, N. A Review of the Graphene Synthesis Routes and Its Applications in Electrochemical Energy Storage. *Crit. Rev. Solid State Mater. Sci.* **2019**, *45*, 339–377. [CrossRef]
17. Alekseenko, S.V. Efficient production and use of energy: Novel energy rationing technologies in Russia. In *Sustainable Energy Technologies*; Hanjalic, K., van de Krol, R., Lekic, A., Eds.; Springer: Berlin/Heidelberg, Germany, 2008; pp. 51–74.
18. Higman, C.; Burgt, M. *Gasification*; Gulf Professional: Amsterdam, The Netherlands, 2003.
19. Young, G.C. Municipal solid waste to energy conversion processes: Economic, technical, and renewable comparisons. In *Plasma Arc Gasification of Municipal Solid Waste*; Circeo, L.J., Ed.; John Wiley &Sons, Inc.: Hoboken, NJ, USA; Georgia Tech Research Institute: Atlanta, GA, USA, 2010.
20. Landi, M.; Zivcak, M.; Sytar, O.; Brestic, M.; Allakhverdiev, S.I. Plasticity of photosynthetic processes and the accumulation of secondary metabolites in plants in response to monochromatic light environments: A review. *Biochim. Biophys. Acta (BBA) Bioenerg.* **2019**, *1861*, 148131. [CrossRef] [PubMed]
21. Grätzel, M. Recent Advances in Sensitized Mesoscopic Solar Cells. *Acc. Chem. Res.* **2009**, *42*, 1788–1798. [CrossRef] [PubMed]
22. Kiang, Y.H. Basic properties of fuels, biomass, refuse derived fuels, wastes, biosludge, and biocarbons. In *Fuel Property Estimation and Combustion Process Characterization*; Elsevier: Amsterdam, The Netherlands, 2018; pp. 41–65. [CrossRef]
23. Le Quéré, C.; Jackson, R.B.; Jones, M.W.; Smith, A.; Abernethy, S.; Andrew, R.M.; De-Gol, A.J.; Willis, D.R.; Shan, Y.; Canadell, J.G.; et al. Temporary reduction in daily global CO_2 emissions during the COVID-19 forced confinement. *Nat. Clim. Chang.* **2020**, *10*, 647–653. [CrossRef]
24. Barber, J.; Tran, P.D. From natural to artificial photosynthesis. *Photochem. Photobiol. Sci.* **2013**, *10*, 20120984. [CrossRef] [PubMed]
25. Allakhverdiev, S.I.; Casal, J.J.; Nagata, T. Photosynthesis from molecular perspectives: Towards future energy production. *Photochem. Photobiol. Sci.* **2009**, *8*, 137–138. [CrossRef] [PubMed]
26. Gaynanov, D.A.; Kashirina, E.S.; Khabirova, Y.F. On an effective ratio of conventional to renewable energy sources in the Russian electric-power industry. *Russ. Electr. Eng.* **2018**, *89*, 9–12. [CrossRef]
27. *Current Statistical Survey*; Federal State Statistics Service: Moscow, Russia, 2020.
28. SHARP. Official Site. Available online: https://global.sharp/solar/en/high-efficiency/ (accessed on 9 July 2021).
29. Heath, T.L. *The Thirteen Books of the Elements*, 2nd ed.; Dover Publications: Mineola, NY, USA, 1956; 464p.

30. Konnikov, E.; Konnikova, O.; Rodionov, D.; Yuldasheva, O. Analyzing Natural Digital Information in the Context of Market Research. *Information* **2021**, *12*, 387. [CrossRef]
31. Rodionov, D.; Rudskaia, I. Problems of infrastructural development of "industry 4.0" in Russia on sibur experience. In Proceedings of the 32nd International Business Information Management Association Conference, IBIMA 2018—Vision 2020: Sustainable Economic Development and Application of Innovation Management from Regional expansion to Global Growth, Seville, Spain, 15–16 November 2018; pp. 3534–3544.
32. Shabunina, T.V.; Shchelkina, S.P.; Rodionov, D.G. An innovative approach to the transformation of eco-economic space of a region based on the green economy principles. *Acad. Strateg. Manag. J.* **2017**, *16*, 176–185.

MDPI
St. Alban-Anlage 66
4052 Basel
Switzerland
Tel. +41 61 683 77 34
Fax +41 61 302 89 18
www.mdpi.com

Sustainability Editorial Office
E-mail: sustainability@mdpi.com
www.mdpi.com/journal/sustainability